The Adventurous Life of Tom "Iron Hand" Warren: Mountain Man

By

Terry Grosz

Print Edition
© Copyright 2017 Terry Grosz

Wolfpack Publishing
P.O. Box 620427
Las Vegas, NV 89162

All rights reserved. No part of this book may be reproduced by any means without the prior written consent of the publisher, other than brief quotes for reviews.

ISBN: 978-1-62918-774-7

Dedication

Years ago my younger son, Christopher, like his father, was able to marry the love of his life. Little did he realize that he, just like his father, had been given through the grace of God, an Angel! As a result of that union made in Heaven, Christopher and Lisa were graced with a short but wonderful marriage, a son, Gabriel, and a daughter, Laurel.

Christopher, a police officer, taken before his time at the age of 35, left his wife Lisa to carry on and to raise their two children. Since that day, my wife Donna and I have watched "Chris's Angel" raising two very beautiful children. Regardless of the long and arduous hours she puts in as a school principal and mother, Lisa always found time for her children. Time in exposing them to the wonders of our society, better understanding religion, enjoying the outdoors in camping, hunting and fishing, the teamwork found in sports, the values of education, and the art of living gracefully within our nations' history, humanity and heritage.

To her mother-in-law and father-in-law, the Angel Christopher saw in Lisa comes forth in a loving and understanding way that is exceptional! Lisa, being the loving and exceptional woman that she is, makes sure her children are shared in such a giving way that it is always love-based and extraordinary. In her own loving and caring way, Lisa is as close to Christopher's mother and father as is any caring, loving and much-loved biological daughter!

It is to this Renaissance Woman and Lady for all seasons, that I dedicate this book...

Table of Contents

Chapter 1: The Adventure Begins 1

Chapter 2: Medicine Lake Country And Unwanted Company!
.... 15

Chapter 3: Blackfeet and The Legend Of "Iron Hand"
Begins! 23

Chapter 4: Beaver Trapping And A Frontier "Doctor's
Remedy" 31

Chapter 5: Trouble, "Iron Hand" And Frontier Justice 43

Chapter 6: Fort Union And The Legend Of "Iron Hand"
Continues 55

Chapter 7: Unexpected Surprises and The Truce Holds 87

Chapter 8 – New Trapping Grounds, Deadly Gros Ventre
Surprises... 129

Chapter 9: The Return, The Gros Ventre and A Fortune Is
Made 193

Chapter 10: The Porcupine River, The "Bad Seed" And
Saving The Brothers "Dent" 277

Chapter 11: John Pierre And Sinopa's Revenge,
"Wambleeska"... 307

Chapter 12: A "Family" No More -- Old Potts, Big Foot And Crooked Hand Choose To Stay 329

Chapter 13: The Brothers York, The Arikara, White Eagle Comes of Age, Iron Hand, And St. Louis 339

Chapter 14: Keelboats, Reunited, "Buckskins" And Gabe's Rifle 367

Chapter 15: Flowers In The Graveyard And The 'Sixth Sense' 395

A Look at Josiah Pike by Terry Grosz 401

About the Author 402

The Adventurous Life of Tom
"Iron Hand" Warren:
Mountain Man

Chapter 1: The Adventure Begins

IN THE SPRING OF 1828, Kenneth McKenzie, Factor for John Jacob Astor, owner of the recently established "American Fur Company", posted an advertisement on the front page of the *Missouri Republican* newspaper in St. Louis indicating a need for men for work on the new frontier.

That article read, "Need 120 enterprising young men to ascend the Missouri River in keelboats and on horse to its source, there to be employed for a period of one, two or three years as hunters, trappers, woodcutters, blacksmiths, carpenters, and camp keepers for the American Fur Company. The criteria needed for employment is as follows: One needs to be masculine, adventurous, well-armed and able to work/trap for a period of from one to three years. Work materials, horses, mules, provisions and trapping equipment for company employees will be provided by the soon to be built Fort Union Trading Post, owned and operated by the American Fur Company. That trading post once constructed, will be located adjacent the headwaters of the Missouri River."

"Once Fort Union is completed, Company Trappers in the fall and winter months will ascend the Jefferson, Madison and Gallatin Rivers and their tributaries, for the purposes of trapping beaver, fox, river otter, muskrat, and other fine furs.

The Adventurous Life of Tom Warren

Come the following spring, Company Trappers will return to Fort Union to deposit their furs taken during the previous trapping season and resupply with provisions from company stores for the next fall and winter trapping seasons."

"All men wishing to apply for such opportunity must report to Factor Kenneth McKenzie at Third and Dowell Street at ten in the morning come Friday next. There, selections will be made of those meeting the above criteria, contracts drawn, signatures affixed and upriver travel arrangements to the Missouri headwaters made."

"Come the third day of March, this year of our Lord, three keelboats loaded with provisions and necessary equipment will leave Landing #4 and ascend the Missouri River by poling and cordelling (a French word describing the pulling of keelboats upstream with a long cable attached to the vessel's main mast by 20-30 men walking along the shore). The keelboats will be closely followed by the company's needed horse and mule herds being walked along the banks of the Missouri by new employees. That company of boats and men will proceed upriver until the headwaters of the Missouri are reached. Once reached, Fort Union will be constructed by the teams of men, so commerce can begin with the various Indian tribes in the area."

(Author's Note: Fort Union was not a military fort but one privately owned by New York businessman John Jacob Astor, for the commerce generated in trade with the Assiniboine, Crow, Cree, Ojibwa, Blackfoot, Hidatsa, and Lakota Tribes, in buffalo robes, furs and peltries in exchange for the white man's goods and alcohol. Of historic note, the Indians traded hides and furs that they considered were of common value, for white man's goods, which they considered highly desirable and valuable. In turn, the white man traded commonly valued goods from their culture, for that which he considered extremely valuable, namely furs, buffalo and bearskin robes and hides. Furs of which at the time were considered very valuable in European society because many of the native furbearing animals on that continent had been reduced or extirpated. American furs were also valuable because European cloth in that day and age was course, rough and colorless. On the other hand, furs were soft, colorful and warm, thus highly desired and valuable to European men and women of the day.)

Come 'sign-up' day, found an excited and noisy crowd of about 250 hopeful men milling around a table with one Kenneth McKenzie and a Company Clerk seated behind him conducting the business of enlistment. Both men were assessing each potential employee's abilities and values to John Jacob Astor's American Fur Company and in the process, explaining in detail what the business opportunity afforded, expressed the known associative dangers and hardships, and gathered up the signatures of those who were seriously interested, deemed of note and had passed muster.

As McKenzie personally interviewed each potential candidate standing in line for mental and physical fitness worthy of being employed by the American Fur Company, he could not help but noticing a giant of a man quietly waiting in line for his turn to be interviewed. That imposing figure was a very tall, heavily bearded and a massively framed individual, who stood at least six-and-a-half feet tall, or better than a foot taller than everyone else in the crowd! Intrigued over the physiognomy and dress of the giant of a man, McKenzie kept a casual examining eye cocked on the mysterious giant-sized individual until he finally stood ready at the front of the line for his turn to be interviewed.

It became quickly obvious to the sharp-eyed, experienced 'Man of the Mountains' McKenzie, that the person standing before him had been compelled, based on the coarseness of his dress and the sadness carried in his face, to be a subject of recent hard times. Without making it obvious and in keeping with the privacy of the men and the culture of the times, McKenzie quietly from the corner of his eyes, offhandedly examined the man standing before him. At first glance, the

The Adventurous Life of Tom Warren

giant of a man appeared to be extremely muscular, bronzed from long exposure to the sun, wore his hair shoulder length, and his general coarse makeup, other than his massive dark beard, made him almost appearing like that of a very large Indian.

It was then that McKenzie realized the giant of a man standing before him was silently examining him as much as he was being examined! Then McKenzie noticed the man's unusual eyes. They were deep, examining and piercing in manner from under his slouch hat. So much so, that they 'portrayed' to the experienced man of the frontier, McKenzie, unremitting vigilance, slightly masked by a deep sadness. On the whole, the giant stranger impressed McKenzie as taciturn, frontier-tested, yet one capable of extreme strength or tenderness, depending upon whatever moment in time was called for.

McKenzie's practiced eyes then quietly surveyed the man's array of needed frontier survival equipment. He noticed that the man possessed two pistols neatly tucked in his sash, a ten-inch-long cutting and gutting knife hanging from the belt on his right side, a much-used tomahawk as evidenced from its heavy wear neatly tucked under his belt on his left side, and his massive right hand carried a state-of-the-art, brand new, large bore Hawken rifle. McKenzie's practiced eyes also observed that the large bore Hawken, or "Rocky Mountain Rifle" as it was also called, weighing in at over ten pounds, yet was carried as easily and lightly as a 'willow twig' in the large man's massive and calloused hands!

When it was his turn interviewing with McKenzie, the man offered his physical services as a way to make it safely through Indian Country with the protection afforded by being included in a large and well-armed fur brigade. However, once there, the man advised he wished to remain on the frontier as a "Free Trapper" and not that of a company man. Then the giant of a man quietly advised that he was willing to work for 'found and wages' in order to earn his keep during the trip upriver and while helping to build the new fort. McKenzie, sensing the man's value as a laborer based on his size and possibly as a leader of men on the difficult and dangerous trip upstream based on his quiet manner and bearing, especially when it came to cordelling one of the heavy and awkward maneuvering keelboats upriver, agreed and happily signed him on as a laborer. But McKenzie also took the time and made a notation on the signing papers that the giant of a man wished to remain as a Free Trapper once the work at Fort Union had been completed and he contractually agreed. McKenzie also noticed that the man spoke in such a manner and his flourish of a signature upon the just signed papers indicated that a highly educated man was standing before him, which was an anomaly in that day and age.

With that bit of business completed, the huge man now known to McKenzie as Tom Warren, quietly faded off into the noisy and still excited crowd of other hopeful humanity hoping to be signed on, disappearing as silently and mysteriously as he had arrived. Little did McKenzie realize that his "Man Mountain Dean" just interviewed was a graduate of the U.S. Military Academy at West Point and an experienced Topographical Engineer, who was no longer in the military. Tom was also of independent means after the sale of his farm and a man who had just lost his wife and young child to a recent smallpox epidemic. Hence the look of great sadness spelled across the man's face and carried in his eyes.

As McKenzie would come to learn, Tom no longer longed for the life as an industrious colonist with family responsibilities but for that of a "Mountain Man". A Mountain Man leading a life where every footstep was beset with primitive enemies, angry beasts and evil men of his own kind, with many a moment pregnant with peril and life's challenges. To that end, Tom hopefully prayed that his great family loss of loved ones would now somehow be supplanted with the extreme dangers and challenges life in the far west were soon to offer him as that of a Free Trapper. At the end of that day,

The Adventurous Life of Tom Warren

McKenzie had signed up his quota of 120 men, and little did anyone truly realize that for one man, "The Adventurous Life of Tom "Iron Hand" Warren, Mountain Man", had just begun by affixing his signature to an innocent-appearing piece of paper...

Come the date set for departure, found Tom riding his favorite horse and leading his two fully loaded packhorses with what he figured he would need for the start of his new life as a Free Trapper on the new frontier. Upon his arrival at the pre-arranged 'jumping-off' site, he was surprised to discover that McKenzie had placed him in charge of leading the heavily packed horse and mule herd safely northward through dangerous Indian Country. In so doing, he was to slowly parallel the progress of the three keelboats moving upriver, under pole and cordelle. Toward that end, it was also Tom's responsibility as the leader of the horse and mule herd to move ahead at the end of each day, and pick a safe campsite for the tired keelboat men to land, eat supper and spend the night. That soon became the daily regimen for the next several months, as the fur brigade slowly made their way northwards towards the headwaters of the Missouri River and future home of the American Fur Company's Fort Union Trading Post.

Eventually, the headwater site was gained and the keelboats safely anchored on the western side of the river and their final camp made. Tom in turn, had his horse and mule wranglers swim the now unpacked herd of animals, whose packs had been transferred earlier to the keelboats, across the Missouri to join the keelboat boatmen camped near the site of the soon to be constructed Fort Union. The next day, the fur brigade surprisingly found themselves surrounded by hundreds of welcoming and curious Indians.

Indians of such character that the group soon discovered anything of value not tied down quickly disappeared! With that, McKenzie formed work parties to begin clearing the ground and building Fort Union. Additionally, he organized several other parties of men to 'round the clock' guard and make sure the provisions on the keelboats and the valuable herds of horses and mules did not 'walk off' in the hands of the milling crowds of curious Indians!

Thirty days later, the logs were cut and the walls went up around the huge fort, as work then began in earnest in constructing the house of the fort's Factor or *Bourgeois*. Following that piece of important construction, the places of residence for the Company Clerks along the inside of the fort's quadrangle were also constructed, as were the barracks for the *Engagés*. While one crew finished those buildings, another group of men constructed the storehouses across the square for storing the merchandise, provisions, furs (once they began flowing into the fort), and the other soon to follow peltries. Last and very important, up went the inside buildings designated to be used by the fort's blacksmiths. "Blacksmithing" being a frontier industry that was critical to the everyday functioning of the fort for making horseshoes and shoeing the mules and horses, repairing firearms, making door and window shutter hinges, barrel hoops, tools, nails, and other items of like importance used in furthering the overall industry of the trading post.

Then came the day when the closely guarded keelboats were finally offloaded and the goods and provisions brought within the fort's walls and placed into the storage houses for safer keeping. By now, over a hundred tepees were gathered around the outside of the walls of the fort and along the nearby river bottom, as their occupants waited for the fort to be opened for business and the trading and related commerce to begin.

Tom, now working alongside the rest of the common laborers, did so because of his word to McKenzie and that he realized he needed to 'top off' his supplies before he headed into the field for a season of fall and winter of trapping. By so doing, with the pay he was earning and the money he had from sale of his farm back in Missouri, he hoped to be able to purchase at the much-inflated frontier prices, his needed 'top-off' provisions like coffee, sugar, dried fruit, gunpowder, pigs

The Adventurous Life of Tom Warren

of lead, and the like. All things he had not brought with him because of the possibility of spoilage, theft or having to carry the excess weight of such items over those many hundreds of miles traveled.

In the interim, Tom had met and worked alongside several, whom he considered good men, who were also former Free Trappers from prior trapping expeditions into that area of the frontier. In so doing, they had decided to form their own small group of higher-classed Free Trappers for the safety in numbers that offered once they were out on their own in Indian Country.

One of the older men in their newly formed group was a much experienced trapper and prototypical "Mountain Man" who went by the name of "Old Potts". He had trapped with Vasquez in the first '07 fur trapping expedition on the upper Missouri at the old Fort Raymond. Another of Tom's newly formed party of friends and experienced Mountain Man was a man with only one eye who went by the moniker of "Big Foot Johnson". He had also been an experienced blacksmith from his former indentured life in civilized America, who knew firearms backwards and forwards as well as many of the techniques of ironworking. As such, "Big Foot" was constantly tinkering with the newly formed group of fur trappers' long guns and making replacement parts for their highly accurate .50 caliber Hawken rifles when he was not cutting and hauling logs for the fort. As for only having one eye, Big Foot had lost it in a saloon fight when his opponent had gotten him down and had gouged out his left eye with his thumb! Not to be out done, Big Foot got up off the floor at an opportune moment of the fight in a fury of pain and a load of adrenalin, and had killed his opponent by cutting out his liver with his long-bladed sheath knife! After that episode, Big Foot's reputation on the frontier was such that one did not want to mess with him or come at him from his blind side. The final member of Tom's little group of experienced Mountain Men was a man who went by the frontier name of "Crooked Hand

Harris" or "Crooked Hand" for short. The Indians had given Crooked Hand his frontier name, because in an earlier fight with the Blackfeet up on the Madison River, he had gotten excited when the killing had slid in too close to him for comfort and in hurriedly reloading his rifle at the time, a .40 caliber flintlock, he had inadvertently placed his hand over the muzzle and it had mistakenly discharged! In so doing, his off hand was badly damaged and remained crooked and deformed for the rest of his life. However, he had learned to adapt his broken hand in his reloading drills and could reload a rifle and shoot it with the best of them, especially his Hawken, which he named "Never Miss". He had named his rifle "Never Miss" because every time he lifted it to his shoulder, something always died...

In fact, both Crooked Hand and Tom were two of the best shooters at Fort Union with their rifles or single shot pistols, as time was to tell. A learned skill by Tom from his days with the U.S. Military, and for Crooked Hand, a skill learned under some of the most trying of times while defending his life as a trapper in Indian Country or providing food for himself from nature's vast storehouse of resources. Skills that were to come in useful time and time again once the men were out and about in Indian Country, trapping and surviving on their own...

Nearing the end of the most intensive portion of work in building the fort, Tom, based on his military background, figured it was time to make final plans to 'provision up' and head out onto the trapping grounds with his friends before the rest of the Company Trappers were turned loose and became the main competition for some of the best areas and the fort's many-times scarce supplies. Sitting down one evening with Old Potts, Big Foot and Crooked Hand, Tom laid out his hoped-for departure plans for the group from the fort and up onto the trapping grounds, solely based on his ex-military experience and the necessity of preparedness. He figured out loud among the group that since only Old Potts knew the territory from his previous trapping experiences with the 1807

The Adventurous Life of Tom Warren

expedition with Vasquez, that would be a built-in advantage over the rest of the Company Trappers. Plus, it would take them awhile to get located on the best beaver trapping grounds, get their winter camp constructed, set up their trapping territory and get to work. Additionally and without saying anything to the rest of his group, the quicker he could get into trapping on the dangerous frontier, maybe the faster he could forget about the loss of his wife and young son to smallpox...

All the men agreed and decided they would 'provision up' with what they figured the four of them would need for the year and with McKenzie's blessing, leave by week's end. The men figured with what money they had between them and the wages owed them by McKenzie, they would have enough for a 'stake' that would supply them for the coming year and into the following summer. Plus, they wanted to make sure they got what provisions they needed before the fort opened for business and then because of the trading factor, found that some supplies would soon be in short supply or unavailable until the annual keelboats arrived the following spring with new supplies. That concern and knowing the next supplies would not arrive at the fort until the spring during the period of high water when the keelboats could get over the many sandbars and safely upstream to the fort, the decision was made by the group to fill their supply needs while the getting was good.

The following day, Tom met with McKenzie who was currently overseeing the building of the fort's two blockhouses. "What can I do for you, Tom?" McKenzie asked upon seeing the quiet giant of a Free Trapper approaching.

"Mr. McKenzie, me, Crooked Hand Harris, Old Potts and Big Foot Johnson plan on leaving for the trapping grounds very soon. In fact, we plan on leaving once we 'provision up'. We would like to get on some decent beaver trapping grounds, get our winter quarters set up and get to trapping so we can start making some money before everyone else floods out into the field. The four of us would like to settle up with you and the company, get our provisions and head out before the first winter storms begin coming down from the north and making our travel difficult. Would that be alright with you if we did so?"

"Well, Tom, I knew I would lose you just as soon as the geese headed south and I guess now is the time, eh? I knew you wanted to be a Free Trapper but was hoping I could convince you to stay on as one of my main men and work for me and Mr. Astor. But I know the lure of the wilderness and once it begins 'calling', it is hard for anyone to resist. OK, let us meet tonight at the Clerk's house and I will see to it that we are square with each other in the wages department. You have been more than fair with me and the company, and I at least owe you that in return. Then if you would like, I will give you company credit for your labors in the form of the provisions you and your group will need. If that sits well with you, let us meet at sundown and we will get our business done to both of our satisfactions. However, you sure you will not stay on as one of my main men? I surely would hate to lose you or any of the hardworking and skilled men you mentioned that you have teamed up with for your proposed trapping expedition."

"No, we four have made up our minds. We came out here to see what lies over the next horizon and get in some trapping before the critters are all trapped out, make some money, see this neck of the woods throughout and that is that. We do however, thank you for letting us come north in the main brigade for the protections it offered and then employing us so we could make some money and be able to afford the provisions we needed," said Tom.

"OK, then. Have it your way. I will see you at sundown at the Clerk's house and we can settle up over what is owed the four of you men," said McKenzie with a smile.

With a handshake out of respect for McKenzie over what had just been proposed and agreed to, Tom strode back to his three fellow Free Trappers sawing away on blockhouse timbers with the good news.

The Adventurous Life of Tom Warren

That evening, Tom, Crooked Hand, Big Foot and Old Potts met with McKenzie, settled up and then headed for one of his warehouses holding a slew of provisions the men knew they would need for the coming year out on the trapping grounds and away from any vestiges of civilization.

Once in the warehouse, Tom, Crooked Hand and Big Foot began gathering up the needed provisions, as Old Potts read off a previously prepared list of what they would need and the amounts he figured they would need for the coming months. On and on droned Old Potts, as Tom, Crooked Hand and Big Foot gathered into a huge pile on the warehouse floor those things the men figured they would need until they could return to the fort the following summer and resupply.

As they did, Old Potts and one of the Company Clerks kept track of what the three other men were digging out from the vast stores at hand. Those provisions collected included two kegs of first quality gunpowder at a dollar a pound; 40 pounds of lead at one dollar per pound; 12 three-point blankets at nine dollars each; six butcher knives at seventy-five cents each; six sheet-iron kettles at two dollars and twenty-five cents per pound; four square axes at two dollars and fifty cents each; 40 beaver traps at nine dollars each; 50 pounds of sugar at one dollar per pound; 80 pounds of coffee at one dollar and twenty-five cents per pound; replacement horseshoes; five pounds of nails; six files; four kegs of rum; several spools of thread; iron rings; four extra bridles; 30 pounds of pipe tobacco; red ribbons for trade; three copper kettles; 12 iron buckles for strapping; a dozen fire steels; 30 pounds of dried apples; 20 pounds of dried raisins; two spools of brass wire for rifle stock repairs in case they were cracked or broken because of hard use; five pounds of washing soap; and on and on it went until the men figured they had everything they would need for the next eight to ten months living in the backcountry with no chance for resupply!

When everything was selected and accounted for, minus the costs of all those goods against the four trappers' wages

12

and the money they collectively possessed and contributed to the pot, they were twelve-and-a-half cents to the good when everything was said and done! So much said for the increased costs of goods in the far west, with company profits on all goods hauled up from the stores in St. Louis, running from 70 to 400% profit on each and every item!

The Adventurous Life of Tom Warren

Chapter 2: Medicine Lake Country And Unwanted Company!

BY NOON THE FOLLOWING DAY, with Old Potts in the lead since he had trapped in roughly the same country to which they were heading years earlier and with Tom bringing up the rear of their long and heavily packed string of horses, the four men planned on heading northwesterly up the Missouri River and riding along its north side. Little did they realize that when they left Fort Union, they created quite a stir in a Blackfoot encampment of many young men camped in the adjacent river bottoms. An encampment whose members had been trading their furs, bearskins and buffalo robes for the white man's goods, mainly liquor for the last two weeks! That river bottom encampment comprised a small component of the Medicine Lake Band of Blackfeet, an extremely aggressive group of Indians, who had a history of not getting along with a number of what they considered trespassing white men, especially after they had been drinking the white man's whiskey. However, in need of necessary supplies, especially powder, lead pigs, salt, and their extreme thirst for the white man's liquor, they were forced to 'pull in their horns' and trade with the much-hated Americans for those necessary provisions.

The Adventurous Life of Tom Warren

As Old Potts led, Tom and company left out from the Fort Union area trailing a fully loaded and valuable pack string of horses. As they did, a lone, young Blackfoot warrior quietly left the nearby Indian encampment from along the river... Several days of travel along the northern bank of the Missouri River brought the trappers to the confluence of the Big Muddy River. There they left the Missouri and turned north along the Big Muddy's eastern bank, heading for familiar beaver trapping grounds known to Old Potts from his previous trapping days working for an earlier trader, merchant, trapper and explorer from St. Louis, whose last name was Vasquez.

Several days later into their travels, Tom noticed that Old Potts had turned over the lead of his small pack string to Big Foot and was now casually riding down the remainder of the string of packed horses towards his location at the end of the group. Old Potts kind of wandered along as if inspecting the animals for any signs of injury or evidence of pack sores from the heavy packs they were carrying. Riding up alongside Tom and without showing any signs of alarm or discovery, Old Potts said, "Tom, we have picked up an Indian outrider following us. He is too far away for me to make out what tribe he is from, but we are deep in Blackfoot country and they can be killing son-of-a-bitches if they get all riled up. If I was a betting man, I would say he is a Blackfoot sent here to keep an eye on us, our direction of travel and our final destination for maybe a later ambush. After all, we represent quite a sizeable catch in horses and equipment, especially if they could catch us out in the open with our pants down. I think I may recognize him from his large size. I remember him because he is almost as big as you, Tom, and a muscular chap at that as well. I am not positive but I think he was back at the fort in that Blackfoot encampment down by the river. You know, the ones who were whooping and hollering it up every night after they got all liquored up after trading their furs for the fort's 'demon rum'."

Without looking in the direction of the Indian outrider trailing the men so as not to give away their knowledge that they were aware that he was 'dogging' them, Tom said, "I know. I have been watching him out of the corner of my eyes ever since we left the fort and you are right. He is sticking to our trail like the ugly on your mug. I also think you are right about who he is as well. He is the same Indian who kept paying such close attention to us as we packed up our horses back in our camp in preparation for leaving Fort Union. Pass the word along to the other two men in case they have not noticed our new 'friend' and in leading us, maybe keep us out from the brush of the river bottoms and out in the open as much as you can. That way, if he has any friends with bad intentions, we will not be so easy to ambush if that is what he and any of his friends have in store for us. But whatever it is, it cannot be good with only four of us and one hell of a long and valuable, heavily loaded string of packhorses and a mess of supplies moving out here in the middle of damn nowhere. Nothing like bringing flies to a jug of honey I would say," concluded Tom.

"I would agree. I will let the other two know that we have company if they don't already realize it, so they will be on their toes as well," said Old Potts, as he turned his horse around and slowly rode back to his own pack string and the head of the line like nothing out of the ordinary had just occurred.

Two days later, the men were at the confluence of the Medicine Lake outlet and turned east towards the familiar beaver waters Old Potts had personally trapped in '07. Shortly thereafter, in and among hundreds of small herds of quietly feeding and resting buffalo as far as the eye could see, the men made the western shore of Medicine Lake. Stopping near a mess of well-rubbed "Buffalo Rocks" left behind by the receding glaciers and in the shade of a large grove of aspens, the men reined up, lit down, stretched their tired and sore knees after so many long hours in the saddle, and made camp.

(Author's Note: Buffalo Rocks were so named by early pioneers, because over thousands of years of the buffalo rubbing off their itchy winter coats come summer against those rocks, the standing-upright glacier boulders had taken on a

The Adventurous Life of Tom Warren

smooth and well-burnished outer quality. They are still there to view in the Medicine Lake area of eastern Montana for those who have a keen eye and a knowledge of American history and what those rocks silently portray.)

Tom's practiced eyes noticed upon lighting down from his horse that their grove of aspens contained a small, clear running, spring-fed stream running through the center of the grove, which led down into the distant lake. Also of note, the dense grove of aspens provided some much-needed shade from the hot prairie sun and was so positioned that surrounding their grove of trees they had a hundred yards of open prairie running in all directions around their newly found sanctuary, giving them a clear, full field of fire if they were ever attacked by Indians.

Old Potts has chosen good ground for a defensive stand if our Indian outrider brings any of his friends for a not so casual visit, thought Tom with a grin, as he reached back into the memory banks of his prior military training. Then looking once again around his aspen grove surroundings, Tom realized white men had been there before. Not 20 yards away, Tom could see the old makings of a previous campsite near a small mountain of sheltering glacial boulders. Then he noticed Old Potts walking towards him with a big grin on his heavily whiskered face.

"What do you think, Tom? How is this for a great spot in which to make camp? In the summer we have the shade from the aspens, in the winter we have the shelter of a cave in that small mountain of boulders behind you, along with providing us a natural windbreak and in the case of trouble, have a clear field of fire all around us." Tom, ever the cautious one and being an ex-Topographical Engineer from late his stint in the Army, had already approvingly eyeballed the lay of the land surrounding their current location.

Then a happy Old Potts sounded off once again before Tom could answer the first question, with "Tom, go look in the cave in that jumble of boulders, up yonder. You will see my name spelled out on the west side of the rock wall inside the cave. That I did with a burnt piece of charcoal the first time me and my fellow trappers camped here in '07. How is that for having a good eye and nose for finding my old campsite?"

Tom just grinned over his friend's obvious enthusiasm and excitement over being back on familiar grounds once again. "Not too bad, Old Potts, but how come there is not any good buffalo steak roasting over an open fire when we got here?" asked Tom, as he chided his friend through his own heavily whiskered grin. Just then arriving near the cave site were Big Foot and Crooked Hand with their pack strings. Upon hearing Tom making a 'funny' at Old Potts's expense, all had a good laugh, generated by Tom's normally quiet and dry sense of humor.

"However, we still have our 'old friend' out there by that other pile of boulders on that distant hillside watching us. So, with that in mind and since this appears to be a good spot for our new home for the next few months, I say we get our tired horses unpacked, let them graze for a while, while it is still light and we can keep an eye on them. I would not want to come all this way, only to get lazy and have the local Indians make off with our stock and leave us a-foot in this country. In the meantime, I say we build up a corral here in the aspens next to our cave so we can make sure no damn Indian can easily run off with our horseflesh," said Tom quietly.

There was a murmur of agreement over Tom's cautionary words, as the men headed for their pack strings and marched them over to their new cave home in the rocks. There the animals were unpacked, double hobbled and turned out on the close at hand nutritious prairie grasses so they could put on the 'feed bag'. In the meantime, Old Potts and Tom stacked up their provisions into the far recesses of their spacious cave, dragged in a mess of firewood, made a better fire ring than the old one already on site from nearby rocks in their creek and began setting up the rest of their campsite. In the meantime, Big Foot and Crooked Hand built up a stout corral with cut

The Adventurous Life of Tom Warren

aspen timbers and rope, so their horse herd would be close at hand and semi-safe from being easily run off by any marauding Indians.

Casually looking up from his 'home-making' endeavors, Tom noticed their ever-watchful Indian who had been dogging their entire trip and had initially hidden himself in a mess of distant boulders, WAS NOW ABSENT FROM VIEW! Tom made sure everyone was aware of that fact and that everyone kept their 'smoke poles' near at hand and a cocked eye on their nearby feeding and valuable hobbled horse herd.

Finished with their earlier double horse hobbling and corral making duties, Big Foot and Crooked Hand rode out to the nearest small herd of buffalo and shot a nice fat cow. An hour later, the smell of roasting buffalo meat over an open fire tickled the men's primal senses. Soon, the only sounds coming from the trappers' new camp was that of the happy noises of men eating what they dearly loved and in amazing, semi-raw quantities!

For the next two days, the men posted up their camp, hauled in their winter's supply of firewood and made ready for the oncoming fall trapping season. Then for the safety offered in numbers, the men paired up, left camp and began exploring the various stream tributaries and marshes around Medicine Lake looking for the sign of beaver. There they found plenty, as well as a number of vast patches of willow growing nearby the lake in which to use for making hoops to stretch out the beaver hides in the subsequent shaping and drying process to come. Additionally, the men made good use of the near at hand buffalo herds when it came time to kill and 'make meat' in the form of sun and air-dried jerky for use during the long, northern latitudes' winter trapping season. Old Potts, being the better of the four men when it came to processing the buffalo meat into jerky, drew those duties, while Tom kept him supplied in good, non-resinous smoking wood used in the 'drying and jerking' process. Crooked Hand and Big Foot saw to it that Old Potts had all the best cuts of buffalo meat he could

20

process and that they could also cook and eat. Soon, the trappers had settled into a routine that satisfied each man's abilities.

On the side of their other duties, the men ventured forth numerous times into nearby draws, for the purpose of cutting and hauling even larger numbers of dry aspen logs back to their campsite for their winter firewood supply. Additionally, dry logs found around the lakeshore were also pulled back to camp by their horses and stacked for winter use as well. Soon the campsite took on a look of permanence and the men now felt that come the winter storms, they would be more than well-sheltered and ready for what challenges living out on the vast frontier would bring.

Then out came their beaver traps, which were duly smoked over an aspen fire to get rid of the man and metal smells, and hung from the trees for later use when the beaver were in their prime. With Crooked Hand standing guard because of his good set of eyes and shooting abilities, the other three men moved into several nearby draws and began cutting armloads of prairie hay and stashed several mounds of it into the deeper recesses of their cave for deep of winter use when their horse herd found getting enough to eat problematic because of drifting snows, howling winds and sub-freezing temperatures.

Arising one morning with a heavy frost in the air, Tom started a fire in their outside cooking firepit. Once the fire was more than blazing, he set their coffee pot on hanging irons and adjusted to a lower height over the fire's coals the earlier-placed roasting slabs of buffalo steaks for slower cooking. When those smells reached the sleeping area in the cave, the other three men awakened and stumbled forth with their tin cups in hand ready to fill with coffee and meet the day.

Sitting down on their recently dragged-in sitting logs around the fire, Old Potts said, "Boys, I think today is the day. What say we venture forth, set a few traps and see what the fur quality is like once we are looking at what we have trapped?"

The Adventurous Life of Tom Warren

Mumbles from the other three men with their mouths full of hot roasted, partially cooked buffalo meat greeted Old Potts's ears. "Then it is done. I say we start nearest our camp in the beaver waters and work our ways east as we trap out the closest of waters along the ponds and marshes by this side of the lake. Then we hit the river tributaries entering the lake and go from there. What does everyone have to say about that for a plan?" Once again mumbles greeted Old Potts's ears, as everyone kept filling their bellies with slabs of partially roasted buffalo meat and scalding hot, thick enough to float a mule's shoe, trapper's style of coffee.

With breakfast done, the rest of the day was spent setting out a number of beaver traps in likely looking places and then in scouting out the next day's areas to be trapped. Scouting the area by the men revealed the adjacent waters were loaded with many beaver and that year's production of beaver kits had been good as well. A buffalo hunt was held later in the day and once again, the wonderful smell of roasting meat soon filled the cooling night air back at their campsite. After supper that evening, a keg of rum was brought forth from their stores and the men topped off their evening with cups of the fiery liquid, as the fall winds gently rustled the many falling aspen leaves overhead. Then it was off to bed with visions of following up on their beaver trapping on everyone's minds, all fortified by the many cups of high proof, belly-warming rum imbibed earlier in the evening...

Chapter 3: Blackfeet and The Legend Of "Iron Hand" Begins!

AWAKENING RIGHT AT DAYLIGHT, Old Potts and Crooked Hand exited the shelter of the sleeping cave and headed in the pre-dawn darkness towards a thick stand of aspens near the horse corral to take care of a call of nature. Tom on the other hand, having already taken care of his call of nature and having risen earlier than the rest of the trappers, walked over to the woodpile, took up an ax and began chopping an armload of aspen wood for the morning's cooking fire. As Tom tended to the wood chopping and gathering duties, Big Foot, back at the firepit, tended to the coffee making duties.

Hefting up a large armful of wood as only a large man his size could carry, Tom turned and headed back towards the morning's fire in order to help in making the rest of their breakfast, since he had been designated the camp's cook and was their best biscuit maker going.

"ZZZIIPPPP—THUNK!" went a steel-tipped arrow into Tom's armload of wood that had been initially aimed and intended for his massive chest! Instead due to the poor light, that arrow had lodged deeply into his just recently shifted

The Adventurous Life of Tom Warren

upward load of firewood! It took just a split second for Tom to realize what had happened, then his close at hand field of view instantly filled with an enraged Indian rushing right at him with an upraised tomahawk! Instinctively, Tom heaved his huge armload of wood into the onrushing Indian's face, then quickly withdrew his ever present pistol from his sash and from just two feet away, shot his assailant in the top of his head with a load of buck and ball, as the unfortunate attacker was still bent over and trying to untangle himself from the huge mess of firewood!

By now, Tom was keenly aware of the heavy firing and fighting coming from all quarters of the aspen grove as the other trappers, having also been jumped simultaneously by the attacking Indians in a coordinated attack, were now in the fight of their lives! With empty pistol in hand, Tom sprinted for the mouth of their cave where the rest of his weapons were stored by his sleeping furs. However, his sprint for the cave was not to be! Another Indian materialized out from the pre-dawn darkness in front of Tom and with a vicious swipe of his tomahawk, sliced open Tom's right cheek to the bone! If Tom had not instinctively ducked when he saw the tomahawk being swung, he would have been killed on the spot!

Tom, in a rising internal fury he had never felt before, swung his empty pistol with such force in reaction to the tomahawk-swipe that had split open his entire right cheek, that he buried the whole barrel of the empty gun into his assailant's head, crushing his skull and exploding his brains all over his hands still holding the weapon! In so doing, Tom's face was also instantly splattered with the man's blood and flying chunks of brain as well! Plowing over his assailant's still falling body, Tom reached the cave's entrance, grabbed his remaining loaded pistol in one hand and his Hawken in the other, spun around and was INSTANTLY smashed into with the full force of another hard-charging Indian bodily slamming into him! Both men flew over backwards because of the Indian's impact velocity, spilling both of them violently onto

Terry Grosz

the cave's packed and now rock-hard dirt floor. Losing his Hawken rifle and pistol because of the surprising violent impact, Tom reached up, grabbed the Indian by his throat with his strong right hand and crushed the struggling man's windpipe and internal carotid artery in one fluid-crushing motion! As the man gurgled and in a spasm trembled away his last moments, Tom threw the dying man's body off to one side, sprang back onto his feet, quickly grabbed his Hawken rifle up off the floor of the cave and joined the ongoing desperate battle at their campsite. In so doing, he left his dropped pistol where it lay because of the intensity and confusion of the moment...

Storming out from the cave, Tom ran 'pell-mell' into another Indian coming his way at a dead run! In that collision, Tom dropped his rifle once again, grabbed his assailant and the two of them fell struggling to the ground. Tom in a still adrenalin-rising fury, bit down hard on the Indian's nose and part of his face and in the process, tore big chunks of the man's flesh off with his teeth! The man underneath Tom screamed out in pain in spite of his mouth now filling with blood from his gushing, ripped-off facial parts! Then the Indian's flying wide-open eyes showed extreme surprise, as Tom's ten-inch sheath knife was savagely ripped upward from the man's lower bowels clear into his heart and lung area! Throwing aside his dying opponent, Tom grabbed his Hawken up off the ground just in time to see Old Potts locked in a deadly knife swinging, one-sided combat with two Indians and their upraised tomahawks! Tom quickly threw his rifle up to his shoulder and snapped off a quick shot into one of Old Potts's assailants. Fortunately, it was a spinal shot and one of Old Potts's attackers dropped like a sack of rocks. Moments later, the other Indian, who had unwisely attacked Old Potts, had grabbed his last trapper... As for the Indian struggling on the ground with a broken spine, Old Potts matched that bullet wound with a knife wound running clear across that assailant's throat!

The Adventurous Life of Tom Warren

All at once another Indian filled Tom's view, running right at him with an upraised tomahawk! Tom, now in an absolute killing fury like he had never experienced before, ducked the Indian's tomahawk swipe, grabbed the heavyset Indian by the front of his throat with just his right hand, and LIFTED THE 200-POUND GURGLING MAN CLEAR OFF HIS FEET, MOMENTS LATER CHOKING HIM TO DEATH WITH A CRUSHED LARYNX AND RUPTURED INTERNAL CAROTID ARTERY, WITH JUST THE ADRENALIN-FUELED STRENGTH IN HIS RIGHT HAND!

As Tom dropped his still kicking and gurgling Indian at his feet with a crushed larynx, another Indian backed into him, screaming after having boiling coffee tossed into his fact by a frantic Big Foot! As the screaming Indian staggered backwards into Tom, Tom stepped off to one side, swung the barrel of his Hawken at the scalded man's head and felt the head bones crushing upon impact! In so doing, he also clearly heard above the sounds of battle, the strange squishing sounds a brain makes when it is violently turned to mush!

Then Tom noticed that Big Foot was now struggling under the weight of two Indians fighting from on top of him. Once again the bile of fury boiled up into Tom's throat, as he stepped over to the three struggling men on the ground, reached down and grabbed one of the Indians fighting with Big Foot. In an instant, that Indian had been jerked clean off his fight with Big Foot, lifted clear off his feet and once again, Tom's strong hands were wrapped around the throat of the now desperately struggling Indian. However, Tom's iron grip would not be loosened and that man, with feet kicking explosively in the air, violently died from a crushed larynx as well!

Witnessing that whole event was another young Indian man on his first raiding party standing just a few feet away. In so doing, that Indian looked on as Tom almost serenely but in a distinct killing mode, finished strangling his fellow Indian with just his right hand! That onlooking Indian, with a look of absolute terror, shock and disbelief on his face over seeing what Tom was doing to another of his Indian brothers with just his bare hand, broke away screaming and ran like the wind for his horse in order to escape the same violent death!

When that young Indian ran away from such a violent battle, he was able to carry that almost mystic memory back to his tribe, in which he presented a vision of Tom, a giant white man trapper, who was so strong that he could hold another man clear off the ground and crush the unfortunate's throat with just one hand! With that firsthand observation from one of their own and the story of such being told later around many Indians' campfires, the legend of a huge white trapper who killed even the strongest Indian with his bare hands drew sounds of amazement and almost disbelief from those hearing such words! However, from that moment on among the Blackfeet of the Northern Plains, the legend of a white trapper who killed with just his hands was now named "Iron Hand" and the legend was born! It was not long before the story of the almost mystical powers of strength and fierceness of that giant white man living in the cave by the spring with three other trappers, swept across the prairie grasses and throughout the numerous bands of Blackfeet, a nation of fierce warriors in and of themselves, with awe, respect and reverence...

Then except for the nervous shuffling of the horses in the corral over the sights and sounds of death all around them and the smell of fresh blood and burned black powder hanging heavy in the early fall morning air, all was once again eerily silent in the trappers' camp!

Looking all around for more danger with a killing look still flashing in his eyes, Tom saw only dead Indians lying scattered everywhere throughout their camp! Then he became aware of the blood still flowing freely down the front of his buckskin shirt from the tear in his right cheek from the earlier tomahawk swipe. Placing his hand alongside his split cheek, he became aware of the sticky hot feeling of a lot of blood and a row of his now painfully exposed teeth! Then his eyes once again swept the scene of the carnage from the surprise morning

The Adventurous Life of Tom Warren

attack by the Indians. Indians he was sure that had been led right to their camp by their earlier lone Indian rider who had dogged them during their entire journey from Fort Union to their Medicine Lake campsite.

Down by the horse corral, Tom saw Old Potts, all covered with his attacker's blood, rising up from his killing site where the Indians had attacked him as they tried to get at the trappers' horses. Then Tom's eyes flashed over to Big Foot. Hell, he was still in a killing fury and other than being put upon by a mess of savages and having to throw a pot of perfectly good, almost ready to drink hot coffee at one of his assailants, was busily scalping the last Indian who had attacked him! Then he took a quick look over to where he had last seen Crooked Hand in furious combat with two very large Indians to see how he was doing or if he had even survived. There he was, leaning against an aspen holding his hand over a bloody Indian's knife handle still sticking out from his thigh muscle! But other than that, all the trappers, aside from a bad case of fright and some superficial wounds, were still alive and kicking! However, the same could not be said for the small Blackfoot Indian raiding party.

"Did any of them red devils get away?" bellowed out Old Potts, as he furiously reloaded his pistol and rifle just in case there was further need.

"I saw one get away and eventually mount his horse and ride out across the prairie like the Devil was hot on his tail," said Tom with painful difficulty, as his sliced open face and heavy beard was beginning to get matted and hanging down under all the coagulated blood weight, pulling the wound down and open even more.

"Damn, Tom!" said Big Foot. "Get your ass up here and sit down so I can take a gander at that sliced open ugly face of yours. You too, Crooked Hand. You don't appear as stove up as is Tom, so I will take a look at him first to see what I can do."

Sitting down on a log next to the fire, Tom winced as Big Foot grabbed him by the lower jaw and turned his face sideways so he could get a better look at the damage on his face.

"Tom, I am gonna hafta take my knife to some of them lovely whiskers so I can get down to the skin and see about getting your ugly mug all sewed up," said Big Foot.

"Do what you gotta do," mumbled Tom, as his face began objecting to the rough handling it was getting from the hands of Big Foot.

Half-an-hour later, part of Tom's face had been roughly shaved clean with the knife used to scalp the two dead Indians, and Big Foot was now in the process of cleaning out the wound with some of their precious rum. Finished with the cleaning portion of the face fix-up detail, Big Foot then took a leather-mending needle and some heavy thread, soaked it into a cup of rum to 'slick it up some' and then sewed the two flaps of Tom's face back together! That needle and thread process however, took some doing. Tom kept flinching every time Big Foot jabbed his needle deeply into the damaged and tender facial flesh and began roughly pulling the loose flaps of skin together. But finally the deed was done and Tom could now sit back and listen to Crooked Hand howl as Big Foot took needle and thread to closing up the deep knife wound on his damaged thigh after pulling the Indian's knife from the leg with a hard jerk! But before he did, he once again reloaded the pistol and rifle lying at his side just in case other Indians came 'calling'...

"There. That is as good as I can mend you two fellas up so that will have to do. But I would stay off that leg, Crooked Hand, for a while and let me and Old Potts take care of the camp's chores. As for you Tom, I would try and not talk much and try to eat smaller chunks of buffalo come chow time," said Big Foot, as he sat back and quietly surveyed his attempts at 'frontier doctoring'...

The Adventurous Life of Tom Warren

With that bit of 'housework' out of the way, Old Potts and Big Foot stripped the Indian dead of any valuables, knives and guns they had brought to the fight and then with their horses, dragged their carcasses to a deep draw a quarter-mile from camp and left them there for the ever-present bears, coyotes, wolves, badgers, ravens, and magpies to feast upon. For the next two days, that deep draw was a popular place for all those four-legged meat eaters of the plains and their aerial friends... However, Big Foot, Crooked Hand and Old Potts did take notice of the fact that of the 14 Indians killed from the Blackfoot raiding party that morning, nine had been killed by Tom!

For the next week, Crooked Hand limped around and Tom was less conversational than his normally closed-mouthed self. However, the 'doctoring' by Big Foot was showing some signs of recovery, even though Crooked Hand tended to do a lot more limping around than normal for such a leg wound. However, sore face and leg aside, the men still had the serious business of beaver trapping lying ahead of them if they wished to survive on the frontier as trappers, so the preparation work continued unabated.

But a sharp eye was also now kept 'skinned' by the four men for any unwanted visitors just in case, sore face or swollen leg be damned...

Chapter 4: Beaver Trapping And A Frontier "Doctor's Remedy"

SITTING ON THEIR HORSES alongside the two pack animals they had brought to carry their traps and other gear, Tom, Big Foot and Old Potts surveyed the marshes lying below the rise upon which they now sat. However, that morning, they were minus Crooked Hand. His deep knife wound in the thigh received in the fight with the Blackfoot raiding party had taken a turn for the worse during the last week and was now festering yellow pus! The decision had been made collectively by the men for him to stay home that first day of beaver trapping, rest up and watch over the camp and the rest of their horses. Then when they returned, Big Foot would look once again at the puncture wound in the leg and see if there was anything he could do to alleviate the pain from the infection.

In the meantime, there was the pressing matter of their fall beaver trapping to attend to and time was of the essence. Leaving camp right at daylight, the men headed for the northeastern end of Medicine Lake. As they rode easterly, the men traveled through numerous herds of feeding buffalo and saw no sign of any Indians in the area. Arriving at the confluence of Medicine Lake's water source, the men paused

The Adventurous Life of Tom Warren

on a series of hills and once again surveyed their surrounding area for any signs of trouble. The previous week's Indian raid had brought forth the men's extreme caution and not wishing a repeat of such a surprise attack, they remained vigilant as to their surroundings. Seeing the way was clear of any signs of danger, the men relaxed and examined the country lying out below them. There below the series of hills upon which they sat stretched several miles of beaver ponds, their signature mounded stick-and-mud houses and water level-controlling dams.

Looking all around the riders' location once again in an extreme act of caution, Tom was the first to speak. "I suggest that Old Potts do the trapping and the two of us remained 'horsed' as lookouts in case we are discovered trapping by the Blackfeet. That way, Old Potts can safely get back to his horse and the two of us can provide covering fire if we are attacked. What say the two of you to that plan?"

"Sounds good to me," said Big Foot. "You are the best shooter among the three of us and I would feel better if you were the one of us who is standing guard."

Old Potts just nodded his head in agreement, as he nudged his horse and the two trailing pack animals down off the hill toward the first set of likely looking beaver ponds. Big Foot followed and Tom, before he left the rise, looked all around one more time for any signs of 'company', made sure a cap was on the nipple of his Hawken rifle and then followed.

Following behind Old Potts, the two designated 'guards' let him pick a likely looking site for his first beaver trap set. Old Potts, an experienced trapper from years past, ambled along with the horses until he was satisfied with a likely looking spot for that first set. There he stopped, stepped off his horse and handed the reins back to Big Foot. Examining the ground as he walked along the edge of the marsh, he soon knelt down and closely examined an area that he suspected was being used by the beaver in that pond as a travel way and slide.

As he did, Tom noticed that Old Potts had stopped near an obvious disturbed area near the edge of the pond, one showing signs of heavy and recent use by beaver, if their numerous tracks in the mud indicated anything to the serious trapper. Satisfied with what he was seeing, Old Potts rose, walked over to one of the packhorses carrying a number of beaver traps in a pannier and pulled one from the bunch. Turning, he walked out into the edge of the beaver pond near the end and off to one side of the slide. Taking his hatchet, he cut off a long and thick willow pole from along the bank, trimmed off its branches and laid it down in the water alongside his legs. Then Old Potts opened up the jaws of the trap, set the pan and carefully laid it down on the muddy bottom, under about four inches of water. Without a word and with his previous beaver trapping experience now in play, Old Potts carefully extended the length of the trap's chain out into the deeper water. Upon reaching the end of the trap's chain, he took the four-foot-long, freshly cut willow pole and using the back of his hatchet, drove one end deeply into the bottom of the beaver pond. Then taking out a piece of previously cut to length heavy twine, tied the ring on the end of the trap's chain to the bottom of the willow pole driven deeply into the bottom of the pond.

By now, never having trapped beaver before, Tom's attention was raptly watching the actions of Old Potts, instead of being on the lookout. So much so, that Big Foot had to embarrassingly remind Tom of his duty as a lookout and not as a beaver trapper observer...

Continuing, Old Potts walked out from the deeper water and back to where the trap lay under the water on the pond's bottom. There he carefully swirled the mud off the bottom into the water around the trap with his hand. After a few moments of swirling the water around the beaver trap, the waterborne silt settled down upon the trap in light layers, thereby hiding the trap from the eye of any animal to be lured into its jaws. Then walking over to the adjacent patch of willows, Old Potts took out his hatchet, cut off another limb and walked back to the now hidden trap. There he took the end of the pliable willow limb just cut, stuck one end diagonally into the bank next to the trap and extended the other end out over the pan of the trap in the pond. Old Potts then removed a bottle of beaver castoreum taken from the glands of a beaver that he had purchased from the stores back at Fort Union just before they had left, cut another smaller willow twig and daubed its end into the smelly liquid in the bottle. Removing the small willow twig from the bottle of liquid, Old Potts daubed the stinky mixture on the end of the twig onto the end of the willow limb sticking out over the pan of the trap. Once done, he put the small cork back into the end of the bottle holding the castoreum. Old Potts then carefully put the bottle back into his 'possibles bag' for use on the next trap set.

Walking out from the beaver pond, Old Potts could see Tom's eyes carefully watching his every move instead of alertly standing guard and had to smile. It was obvious that Tom being a novice beaver trapper was not going to miss one single action performed by Old Potts as he set the first beaver trap of their trapping season...

Old Potts, ever the teacher, smiled and said, "Tom, that is how one sets a beaver trap in these here waters. It is done that way because beaver are very territorial. If a beaver ever smells another beaver in its territory, it will swim to that smell and investigate. It will swim directly to the smell coming from the end of that stick I placed over the pan of the beaver trap. Once there, it will lift its nose up to the stick to smell the new beaver scent. In the process, it will set one or both of its feet down onto the bottom of the pond for balance and in so doing, hopefully place them onto the pan of the trap lying hidden under the light film of mud below. Once trapped, the beaver will panic and swim into deeper water pulling the trap and chain along with it. Of course, the length of the trap's chain will keep it from swimming away and once out into deeper water, the weight of the heavy trap and chain will eventually tire out the trying to escape beaver and cause it to eventually drown."

"There are so many beaver in this area that I would not be surprised on our return trip from the other end of where we are going to lay out our trap line, that some of these earlier set traps will have beaver in them. If that is the case, we will skin them on site in order to save our horses from having to carry such a heavy load of dead beaver carcasses back to our campsite. Or we may even take a few beaver back with us for our supper tonight. In fact, they are damn fine eating if one takes a notion to cook them up and try them," continued Old Potts, as he detected a 'learning moment' in his young beaver trapping protégé...

Continuing, Old Potts said as he mounted back up onto his horse, "Hauling those now skinned-out beaver *plus* (pronounced "plews") back to our camp and then our work really starts. We can put Crooked Hand if he be up to it, to gathering up some smaller willow limbs, flesh out our *plus* and then he can hoop them on those willows and tie them down with sinew so they can dry out right proper like. Then once they dry, we can fold them with the fur side in, pack them to about 60 to a bundle, and each of our packhorses can carry two of them bundles comfortably when we decide to return to Fort Union. There we can sell them, re-provision with them proceeds and then head back into God's country for more of the same. Asides, each of them packs of *plus* will bring anywhere from $300-600, and that buys a lot of rum drinking as well when we get back to Fort Union."

Once again, Tom found himself so interested in what Old Potts was saying and doing, that he forgot why he was there until Big Foot reminded him once again of his primary guard duties. Snapping back into the world at hand, Tom's eyes once again swept the countryside for any signs of danger. Seeing none but ashamed that he was failing at being a guardian of the defenseless Old Potts when he was setting out the traps, Tom never made that mistake ever again. That alertness was made even more important when he realized that Big Foot was damn good as a blacksmith and one who could fix their firearms at the drop of a hat. But he was one piss poor shooter when it came to the business end of his rifle at more than 50 yards, so 'why the necessary' of Tom being on the alert at all times because of being the better shooter... That was made even more 'necessary' if that first shot really had to count.

For the rest of that day, the three trappers set out the remaining 39 of their beaver traps in the numerous adjacent ponds and waterways. In fact, there was so much beaver sign in the area, that they only traveled another couple of miles until the remaining numbers of their traps had been set.

Then on the way back to camp as Old Potts had predicted, the men discovered that a number of the earlier set beaver traps already had animals in them! Once again, Tom got schooled in the fine art of how to quickly skin a beaver without wrecking the value of the pelt, while watching Big Foot in action. Then those already yielding traps were re-set and the men moved on.

But as the day progressed, Tom's well-developed 'sixth sense' got the better of him and in so doing, found him watching their back trail more and more frequently. However, nothing was seen to cause him any concern in the way of danger from an ambush, so he then began thinking maybe there was some kind of trouble back at camp. Maybe Crooked Hand had taken seriously ill or, even worse, maybe the Blackfeet had returned in even larger numbers. After all, one of their killing war party had escaped from the earlier battle and maybe he and others had returned...

By the time the men hove into view of their campsite, Tom found himself standing higher in his stirrups as if needing to see further into camp and see if his good friend Crooked Hand was up and moving about. To be frank, riding like that, Tom looked funny as hell. Looked funny as hell as all six-foot, seven-inches of his carcass standing that way all but dwarfed his horse... In fact, if there had been some ladies present, they

36

would have told Tom to get off his horse and let the horse ride alone for a change...

Finally, camp hove into view and there was Tom's friend, Crooked Hand, standing there by their horse corral waving his arm in recognition. But once in camp, things did not look or smell right! Crooked Hand's infection, after being stabbed in the thigh in their recent battle by a Blackfoot Indian with an obviously unclean knife, had gotten worse, infection-wise! His wound entry point had turned a deep ugly red and yellow pus was streaming out from the hole and running down his leg, especially every time he moved and his thigh muscle contracted! Crooked Hand's pain was so great, that he could hardly walk and it seemed that it even hurt for him to breathe.

Big Foot, after looking at the festering and running wound said, "Crooked Hand, I am afraid I am going to have to open up that wound, causing it to drain better and then have to cauterize it. To do that, I will need a red hot poker, which we don't have...or use the old tried and true gunpowder method to cauterize the wound."

"Do it with the gunpowder," said Crooked Hand firmly, who was familiar with the old frontier remedy of wound cauterization.

"OK," said Big Foot, "but first, I need to get you good and drunk so you don't feel so much pain when I treat the wound. However, if I don't treat the wound, you will surely lose your leg and most probably your life..."

"Let us get on with it," said Crooked Hand firmly, "the quicker it's done, the faster I can get back up on my feet and get to trapping."

Turning around, Tom saw Old Potts already coming from their cave with one of their jugs of rum. It was once again obvious that Old Potts, in his earlier days as a frontier trapper, knew what was coming next... Sitting down on a log placed around their fire ring for just such purposes, Old Potts took a drinking cup, filled it with rum and handed it to Crooked Hand.

The Adventurous Life of Tom Warren

Cup after cup soon followed until Crooked Hand fell off the sitting log some time later in a drunken stupor.

"Alright, let us get going before he comes back to this life," said Old Potts. Tom, unaware as to what was coming next, just stood off to one side and let his two experienced Mountain Man friends do what they had to do in order to save their friend=s life.

First, Big Foot and Old Potts removed Crooked Hand's buckskin pants. When they did, the smell and increased drainage of pus almost turned Tom's stomach. But he held firm as he and Old Potts laid out Crooked Hand on the ground. Then Big Foot removed his long-bladed cutting and gutting knife from the fire's coals and commenced opening up Crooked Hand's festering wound. When he did, his knife sizzled as it hit the stream of pus and bloody, puffy tissue of the inflamed thigh! Taking a deep breath, Big Foot thrust his knife downward until he felt it just tap against Crooked Hand's thigh bone! Then he twisted it around and around, cutting out a sizeable amount of the dead and dying putrid flesh surrounding the injury hole! When he did, Crooked Hand, still under the effects of the rum, groaned out in pain as his leg convulsed and began to spasm from the new damage being done to the healthy thigh muscle surrounding the dead and dying tissue!

However, Big Foot, having been there before on like wounds in his younger days with other trappers, kept cutting away the damaged and putrid flesh until he was content that he had removed the damaged tissue that was causing the festering! Then Old Potts handed Big Foot a powder horn normally used to refill their single shot pistols. Without missing a beat, Big Foot squeezed out the remaining blood and pus, poured the gaping wound hole in the thigh full of fresh gunpowder, just as Old Potts handed him a stick from the fire with a burning ember on its end.

"Stand back," cautioned Big Foot, just as he applied the burning end of the stick onto the mound of gunpowder filling the knife's wound channel. "FOOOFF" went the exposed gunpowder in a quick, upward burning explosion! With the application of the burning stick, the gunpowder exploded and burned its way upwards into the air, as well as clear down to Crooked Hand's thigh bone! When the flame erupted from the hole, Crooked Hand's body gave a shudder, which was followed by a loud groan from the man still in a drunken stupor, then silence. The only thing that remained of the 'deed now done' was the smell of burning flesh, steaming pus and the acrid smell of burned gunpowder!

With that, the cauterized wound was wrapped after being flushed out with rum and Crooked Hand was carefully removed from the fire ring area and placed under his sleeping furs in the cave. Then Big Foot placed his knife blade back into the fire's bed of coals for a few moments but not long enough to reduce its temper. It was then removed and dunked into their nearby creek, making a loud sizzling sound and emitting a small cloud of steam, which floated off into the now cooling night air... Now, Big Foot's knife was ready to cut chunks off his next buffalo steak when eating.

Without any further conversation, Big Foot put on their coffee pot to boil and Tom sliced several thick steaks off a hindquarter of a cow buffalo hanging from one of the trees in camp and supper was soon had by all. However, each man kept his pistols and rifle laid close at hand in case 'Blackfoot visitors decided to come uninvited to supper'...

The next morning, everyone was surprisingly awakened with Crooked Hand yelling for his morning coffee! Tom scrambled out from under his sleeping furs like he had been shot from out of a cannon, as did Old Potts and Big Foot! The men gathered around Crooked Hand to see how he was doing, only to be grumpily admonished for not having any coffee ready for him...

Big Foot rolled back Crooked Hand's sleeping furs and looked at his leg and the wound area. Damn if it did not look and smell better than it had the evening before. It was still

The Adventurous Life of Tom Warren

horribly swollen, reddish looking in color and ugly looking, but you could tell the 'frontier remedy' had more than done what it was supposed to do.

Several days later, Crooked Hand was up and limping about with the assist of a handmade crutch, but it was obvious he was on the mend and raring to go. However, the other three men kept him back at camp fleshing out the numerous fresh beaver and muskrat skins they were bringing back on a daily basis and hooping the same until his leg had healed even more.

Finally the day came when Crooked Hand tried to mount his horse so he could go trapping with his fellow trappers. However, he was unable to do so under his own steam due to his leg still being severely weakened from the knife wound-caused infection and the following 'gunpowder operation'. One week later, Crooked Hand tried once again to mount his horse and was still unable to do so. The hurt on his face was palpable at not being able to ride and seeing that on the face of his dear friend, Tom quietly and without fanfare, intervened.

"Try it again, Partner," said Tom as he stood by Crooked Hand's side. Once again Crooked Hand tried mounting his horse and that time, he was successful! Well, that and a strong boost from his good friend and the man with the "Iron Hand" giving him a strong assist upward. With that assist, the four men were once again united as trappers in the field, and now there were two very straight shooting men acting as guards when Old Potts ran the traps, instead of just Tom...

For the next two weeks until the first snows of winter began falling, the four men were reaping huge rewards in beaver, river otter and muskrat skins from their bountiful trapping area. Old Potts's idea to come clear up into the Medicine Lake area trapping had placed them away from the mess of the Fort Union's Company Trappers and of late, they had seen little sign of any roaming Blackfeet Indians. But for some reason, Tom's 'sixth sense' kept reminding him that all was not well and an ill wind was commencing to blow soon... An ill wind that he just could not put his finger on, but the sinister feeling remained ever present, just as sure as God had made green apples.

Crooked Hand's leg had healed nicely by the time the colder weather had arrived, leaving him with just a slight limp. A very large scar and a gaping indentation in the thigh muscle remained but for the most part, the pain was very much reduced and manageable. The health of the three remaining men had been good and other than being cold much of the time when wading about in the now icy cold waters as they set their traps, their world and life was good. That was especially apparent if one looked at the small mountain of prime furs stacked out of the weather in the deepest recesses of their cave. Furs that were waiting to be transported to and subsequently sold at Fort Union once summer arrived...

But that nagging 'sixth sense' still prevailed on a daily basis for Tom as he kept an extra sharp eye out for any sign of trouble, as the men continued their successful trapping operations. Then the 'sixth sense' made its appearance in real time and it was totally unexpected by everyone in the little trapping party of Old Potts, Big Foot, Crooked Hand and the big man with the "Iron Hand"...

The Adventurous Life of Tom Warren

Chapter 5: Trouble, "Iron Hand" And Frontier Justice

LEAVING THEIR CAMP EARLY ONE MORNING, the four trappers headed out to their usual trapping grounds in the Medicine Lake marshy areas to check their previously set traps. A new but gentle snow had been falling overnight and everything had been covered with six inches of 'white wilderness', making travel silent and easy as the trappers made their way to the northeastern end of their extensive trap line. Approaching their first beaver set, Old Potts noticed that the trap was empty and not sprung.

"Damn," said Old Potts. "I just knew that trap would hold a beaver this morning. There are so many beaver in this area, that trap should of had one by now."

Staying horsed, the four men rode to their next set and that trap did not hold a beaver either. Without a word, the men rode to their third set and once again, that trap was empty as well!

"Something is not right here," said Old Potts, as he stepped lightly from his saddle and walked over to the set. Kneeling down on the bank next to the beaver slide, Old Potts just examined with his fingers and eyes the now frozen ground for any keys it might reveal as to why the traps were running

The Adventurous Life of Tom Warren

empty. Keys that may or may not validate what dark thoughts he was now beginning to harbor in the recesses of his mind...

Rising and turning with a scowl spread clear across his heavily whiskered face, Old Potts said, "Boys, we have a trap robber in our midst and he is a white man! From the looks of the moccasin tracks in the now frozen mud, whoever it belongs to walks like a white man with a full footprint and not like an Indian does by walking on the outside of his foot! Whoever it is, that person came in after we had made our sets yesterday afternoon and cleaned out whatever catches we had 'slicker than cow slobbers'! Then that person, whoever he is, cleverly re-set our traps hoping not to get discovered by us for what he is doing. However, I need to see more of our other sets and look at what additional signs were left behind in order to know exactly what I am seeing and how many trap-robbing skunks are involved."

For the next hour, the four men backtracked checking their trap line, as Old Potts continued 'reading' the slight evidence of sign left behind by what was now discovered to be a number of very clever trap robbers. Trap robbers who, from what little sign had been left behind, were very experienced frontiersmen and beaver trappers as well. Finally Old Potts stopped on their fifteenth empty set, turned around facing the grim-faced men saying, "I can see from the sign they left, there are at least six of them trap robbers and they are all white men! They are riding shod horses and from those tracks that I can find not covered by this new snow, are trailing four pack animals as well. Since we did not cut any sign west of our camp and trap line, that means the men robbing our traps had to come in from the extreme northeastern side of the lake. That being the case, they are not Company Trappers from Fort Union but more than likely Hudson Bay men dropping down from their trapping grounds in the north, who have discovered and are robbing our traps. I suspect they have already trapped out their own beaver waters and are now looking for more promising grounds in our country. However, what they are doing in the way of trap robbing is a killing offense out here on the frontier to my way of thinking," growled Old Potts. "A killing offense because they are destroying our trapping livelihood and in essence, slowly killing us too!"

"Now, I think we need to catch these beaver-trap robbing thieves red-handed. The way I read the sign is that they are working our trap line in the dead of night. That way, the ground is good and frozen hard and they leave few tracks giving away what they are doing. They can do so because they now know where each of our traps are set and can ride right to them in the moonlight. The way I figure it, we have lost at least 20 good *plus* just this day alone! Once again to my way of thinking, robbing a man from what his traps are catching out here on the frontier is no better than stealing a man's horse and leaving him a-foot. A thief is a thief and needs killing, no matter where he comes from or who he is working for. I say we beat it out of here so we don't leave any more of our sign than we have to and come back after sundown. When we do, we can ride into that long draw down yonder, hide, and wait them out. From there, any riders robbing and riding our trap line will have to ride within 30 yards of where we would be hidden. When they do, we can deal with them as we see fit. What do you fellas say to that plan?" asked Old Potts.

"If we gonna be leaving so we are not discovered, I say we get to it," said Crooked Hand with an evil look scrawled clear across his weathered face. Without another word, the men left the area via the backside of a hillside so they would not leave any more sign of discovery along their trap line, which was currently strung out along the beaver-rich waters in the lake's marshes and creek tributaries in the bottoms.

That night and all through the following one, no one showed up to rob their previously set traps. So the men removed the beaver from their traps themselves like they normally would have done, re-set them so they could continue trapping in the beaver-rich area, and then returned to their camp to flesh and hoop out their catches.

The Adventurous Life of Tom Warren

"Do you think they saw us looking our empty traps over and have skedaddled?" asked Tom to the group of men sitting around eating breakfast the following morning, after not being able to catch their trap robbers red-handed from the night before.

"No," said Big Foot, as he wrestled down a rather large chunk of half-cooked buffalo meat. "I have a feeling that they are just letting us continue to trap in the area like nothing out of the ordinary has happened. Then they will hit us once again when they think our suspicions of trap-robbing are down. So we need to continue freezing our hind ends off each night until they return. And my big friend, rest assured, they will return. For sure they will return, especially when we are averaging almost a beaver to every trap that we have previously placed along the length of our trap line. Additionally, we will soon have trapped out this immediate area and they, whoever they are, will know that if they are beaver trappers like the rest of us. With that in mind, I would imagine they will soon tire of stealing from us once our catch begins dropping off," continued Big Foot.

That evening as the men prepared their horses for another long and cold night hidden in the wooded draw near their trap line, Tom found that his 'sixth sense' was bedeviling him once again. That time, Tom shared his unusual inner feelings with his partners. When he finished telling the men about those unusual feelings of foreboding as they continued saddling their horses, they all listened in respectful silence and said nothing.

Then Old Potts said, "Boys, make sure all your pistols are loaded with a double load of buck and ball. I also share some of Tom's concerns about tonight being the night of contact with our trap robbers and we sure had better be ready because there are more of them than there are of us. If that be the case, it is better to clear a man out from his saddle in the moonlight with a load of buck and ball from close range, than trying to hit him with a single ball from our rifles while riding on a moving horse." With those words, Old Potts began double loading and

46

charging both of his pistols by the light of their campfire. Moments later, the rest of the men followed suit...

Following that bit of deadly preparation, the four men left their camp riding single file, as they headed for the draw from which they hoped to ambush their trap robbers if they chose to appear that night and dispense some much-needed frontier justice. After half-an-hour of silently riding in the cold winter night lit up by three-quarters of a moon, the men arrived in their grove of trees and sat there quietly on their horses out of sight, wishing they were gathered around a warm fire instead of sitting in a cold saddle... As they did, each of the men more than likely had two thoughts on their minds. Would they fight bravely if their trap robbers bothered to show their cowardly faces that evening and put up a fight, and would this be 'their' last night...

Around two in the morning under a three-quarter moon sparkling brightly on the newly fallen snow, the men heard a horse whinny off in a distance! Their cold and stiffness of limbs was immediately forgotten over hearing that familiar sound, and even the horses they sat upon seemed to sense something special was about to happen, and were shivering in anticipation of the event to come as well.

Streaming across the bottom of the draw in which quietly sat the four aggrieved trappers silently moved a long caravan of men, mules and horses! Moving in a single file were ten 'trap-robbing' men, their riding stock and a fully loaded horse and mule pack string numbering 15 animals!

Then it dawned on the four trappers waiting in ambush. Their trap robbers were fully loaded because they were pulling out from a beaver-depleted area and moving into another trapping location holding more chances for trapping success. However, they were greedy and not going to leave the current area until they had picked clean all the traps set by Old Potts and his small beaver trapping crew!

But what to do now? flashed through Tom's mind. They were outnumbered by more than two to one and every one of

The Adventurous Life of Tom Warren

the men who had slowly filed by in front of the four hidden trappers in the bright moonlight, appeared to be heavily armed and ready to defend themselves at a moment's notice should the occasion arise!

Old Potts waited until the trap robbers' caravan stopped and removed a dead beaver from one of his nearby traps. Tying the dead beaver on the side of one of the packhorses along with a now stolen trap, that trap-robbing rider re-mounted his horse and the caravan continued on in the bright moonlit to the next of Old Potts's traps. When they robbed a beaver from that trap and stole the expensive trap as well, Old Potts had seen and had enough! *If it was not bad enough stealing a man's trappings, taking his traps was doing nothing but taking away his way of life on the frontier as well!* thought Old Potts through a gritted set of teeth and narrowly slotted eyes.

Waiting until the trap-robbing caravan began moving on once again towards the next trap holding a dead beaver, Old Potts quietly slipped in behind the caravan with his three compatriots close at hand. Together they rode along quietly until the trap robbers stopped to remove another furbearer from one of Old Potts's traps and all of their attention was focused on that beaver-removing event.

That was when Old Potts, with a left hand signal to the men riding quietly in the snow behind him to ride around him and be ready, began his move. Moments later after Old Potts's men had been motioned and moved around him to get into a better fighting location, he 'lit up the night and let her rip'!

"Hey! What the 'Sam Hill' are you doing robbing my traps!" bellowed Old Potts, who by then had moved quietly smack-dab into the middle of the mess of beaver thieves before they knew what was happening and unbeknownst to the rest of their kind.

Immediately, the men in the trap robbers' pack string 'blew up' in utter surprise at having a stranger so close at hand in their midst and catching them in their little beaver and trap-thieving act! Two of the men closest to Old Potts, immediately swung their rifles in his direction as if to shoot him from his saddle! That was their last act on earth as Old Potts blew both men from their saddles with his two double loaded horse pistols from just six feet away!

The man that Tom had quietly ridden up alongside in the long pack string, finally realizing the monster-sized man next to his horse was not one of his party, quickly whipped his rifle around as if to shoot Tom at close range. He did so, only to have his head smashed in from the stock of Tom's sharply thrust Hawken rifle, knocking that thief dead and under the hooves of his own horse! Another trap-robbing man sitting alongside the man Tom had just unhorsed by smashing in his head with the butt of his rifle, tried to knock Tom off his horse with just his hard-swinging rifle barrel! He did so because he was too close to shoot after his long-barreled rifle got tangled up with horse and man!

Reaching for the man swinging the barrel of his rifle trying to unhorse him and now aware of all the rifles and pistols firing up and down the line of trap robbers, Tom jerked him out from his saddle by the nape of his neck, just as the man was finally able to discharge his rifle in Tom's direction. The flame from the close-in shot from the trap robber's rifle burned one side of Tom's buckskin shirt all to hell! Fortunately, the bullet missed gut-shooting Tom by mere inches!

Feeling the fury rising up in him like he had felt when the Blackfoot raiding party had attacked their camp earlier in the year, Tom grabbed his man by his shoulder-length hair with his right hand, jerked him over to the side of Tom's horse and cleanly twist-snapped his neck, which made a sound like a muffled pistol shot! That man was quickly dropped under his nervous horse's hooves just as another problem quickly arose for Tom. Just then, another surprised and now fleeing trap robber tried to unhorse Tom by riding his speeding horse into the side of Tom's steed! That deadly move only got him cleaned clear out from his saddle with a blast from Tom's horse

The Adventurous Life of Tom Warren

pistol from just two feet away! That man was blown out from his saddle due to the closeness of the pistol's blast and fell down among the line of now nervously milling, fully packed mules and horses as well!

Then a smallish in size man tried dashing by Tom on foot on the opposite side of several nervously milling pack animals. As it turned out, he was the outlaw trap robber who had been cleaning out all the traps before Old Potts had 'blown up their little beaver thieving party'...

As the man ran by Tom, he fired a pistol in his direction from such close range that the flame from the weapon's discharge burned off most of one of Tom's eyebrows! That was the last earthly act committed by that individual as well, as Tom's tomahawk cut him down before he could run another step or take the time to unlimber the other pistol carried in his sash and finish Tom off...

Then silence reigned all up and down the line, other than the nervous shuffling and snorting of the trap robbers' stock train and the sound of Tom's rapidly beating heart. Once again in his fury, Tom had killed four men who deserved killing! One with the "Iron" in his hands by a snapped neck, one by Tom's rifle butt smashed into the side of a trap robber's head, one with a pistol shot at close range, and the last one with a well-thrown tomahawk!

"Is everyone alright?" yelled Old Potts, who had started off the trap robbers' 'Moonlight Prairie Dance'.

"I am," yelled back Big Foot. "I caught a pistol ball in my bad leg," groaned Crooked Hand, as his bad luck with his bad leg continued to hold out... "I am fine," said Tom, "just covered with blood and brain matter from the man I killed with my rifle butt," he continued.

"Tom, can you build us a fire so we can see what the damned hell we got ourselves into and I can look at Crooked Hand's leg?" yelled Big Foot.

50

"Sure can. Let me gather up some wood from the draw where we were hidden and I will be right back," said Tom as he started to head for the nearby draw.

"Tom, just build a fire back in the draw. We will meet you there with all these horses and pack animals once we get them all rounded up and settled down," yelled old Potts.

"OK," yelled Tom as he turned his horse around and headed for the draw. Soon, with his flint and steel, he had a blazing fire going so the men would know where to meet up.

Soon, Old Potts came into the light of the fire trailing four riding horses. He was followed by Big Foot leading another six riding horses. Lastly, Crooked Hand came, riding his horse and holding a bloody left leg. Tom helped him off the horse, removed a saddle blanket from one of the trap robbers' horses and made him as comfortable as possible on the ground next to the now blazing fire. Then after stoking up the fire even more, Tom went back out into the moonlight to help gather up some of the packhorses before they wandered off into the night, as Big Foot looked after Crooked Hand's bloody leg from being hit with a pistol ball fired from close range during the battle.

A half-hour later, all the trap robbers' livestock was secured in the draw, all the dead men's weapons were laid out on a horse blanket on the ground next to the fire, the dead were left where they had fallen, and the four men remained around the warming fire until the crack of dawn. That the four trappers did until with the arrival of daylight, so the men could see better as to what death and destruction they had wrought.

Back at the draw waiting for daylight, turned out that a pistol ball had blown clear through Crooked Hand's left or bad leg and had missed breaking any bones. Soon Big Foot had washed out the wound with cold water from his canteen and had tightly bound it with a wrap made from one of the dead trap-robbing men's shirts.

An hour later as the sun peeped over the horizon illuminating the scene of carnage, the men walked back to the location of the fight leading their horses in case they got jumped by any nearby Blackfeet overhearing all the earlier shooting and were coming to investigate. There Old Potts's wisdom of loading double charges of shot became evident in what carnage the men were viewing. In the first seconds of battle, Old Potts, Big Foot and Crooked Hand had killed six of the surprised trap robbers almost outright before they could fire any killing shots with their weapons! In so doing, Old Potts's crew had evened out the odds and made it a fair fight at the head of the line. Then Tom had killed the remaining four men who were trying to escape the ongoing firefight at the end of the pack animals' line. Rightfully so, Crooked Hand had killed the man who had shot him at such close range in the left leg. All in all, a good night's work according to Old Potts, and "Good riddance to a nest of suspected Hudson Bay, trap-robbing polecats!"

The rest of that morning was spent in stripping the dead of any valuables and weapons overlooked in the first go-around. Then Tom and Big Foot, with the aid of their horses, dragged off the ten men's carcasses over the ridge line and deposited them in a handy buffalo wallow. Three days later, the gray wolves, coyotes, grizzly bears not yet in hibernation, magpies, and ravens had pretty well, except for a mess of scattered large bones, scavenged and cleaned up the mess!

In stripping the dead, Old Potts later discovered papers carried by one of the men killed identifying him as a *Bourgeois* for the Hudson Bay Company! Then taking the entire pack string back to their campsite, the animals were unpacked, curried down, had their saddle blankets draped over the corral rails to dry out, hobbled and let out to feed adjacent the trappers' campsite. Come sundown, the horses and mules were watered and then herded into the now hastily enlarged corral to accommodate the additional 25 horses and mules taken from the Hudson Bay men in battle.

Then in amazement, the men began unpacking the packs and supplies carried by the now dead Hudson Bay men as they had been making ready to leave the area for good. Therein they discovered numerous *plus* of beaver, mink, fox, river otter, muskrat, bobcat, wolf and bear hides in the typically deerskin-wrapped, packed fur bundles! Additionally, other packs carried many hundreds of pounds of Indian trade items, rum, gunpowder, pigs of lead, blankets, *capotes*, and many other items the four trappers could use. Additionally, some packs included bags of salt, pepper, red pepper flakes, dried raisins, dried apples, and many other items that would turn out to be more than useful to the four Medicine Lake trappers as well!

With all that bounty laid out before them, the four trappers realized they had ambushed and settled a score with a major Hudson Bay pack train on their move to new trapping grounds. In so doing, they had not only protected their claim to their trapping grounds, but had wiped out a major trespassing competitor and had tripled the amount of furs the four men had so far accumulated to date! In short, in frontier wealth, Old Potts and his group of trappers were now very wealthy men in furs and saleable mules and horses once back at Fort Union! That is, if they could keep the Medicine Lake Band of Blackfeet Indians from stealing the trappers blind and killing all of them off in the process...

Then it was back to work as the men ran their old trap lines and then moved them further east when the beaver numbers were trapped out or severely reduced within their original trapping sites. Finally the dead of winter set in and the ice became too thick to easily trap through. When that event occurred, the men pulled their beaver traps and retreated back to their camp. There, they would re-outfit and repair their gear and with the full onset of winter, begin trapping for bobcat, lynx, coyotes and wolves, plus hunting buffalo for their meat, robes and hides until spring.

Then once again when spring had arrived and with 'ice out', trapping for beaver for a short period of time commenced again. Once those beaver went out of 'prime' in late spring,

The Adventurous Life of Tom Warren

trapping stopped and preparations were made for the return to Fort Union to re-provision their supplies and sell their accumulation of furs from the previous trapping season, as well as the extra Hudson Bay mules and horses... This they would do because they now had a treasure trove of furs and livestock from the Hudson Bay trap-robbing outlaws, and those combined with the four men's annual trappings, made them wealthy men and genuine members of the Free Trapper community in the eyes of those living and working out on the frontier...

Chapter 6: Fort Union And The Legend Of "Iron Hand" Continues

SEVERAL HOURS BEFORE DAYLIGHT on the day set aside for departure to Fort Union, Tom was up early and flying about preparing breakfast at his outside 'kitchen'. He had a blazing fire going, coffee boiling away making the strong brew known as 'trapper's coffee', and buffalo steaks staked out over the hot coals at one end of the firepit sizzling away. Thereon those coals also lay two Dutch ovens smelling heavenly of baking biscuits! Tom was more than eager to get back to Fort Union to see what a completed fort now looked like, see how much they could make off their furs and livestock sales, and once again be around a number of people in order to get caught up on the news from the outside world.

Then Tom paused in full stride, stood stone-cold still in the damp morning air and thought back to better times past. He had originally left civilization to forget about the loss of his wonderful wife Jeannie, and his first son Christopher, due to a smallpox epidemic. That was why he had joined McKenzie and Astor's fur brigade going up on the Missouri to establish a fort and trading post and to lose himself in the wilderness, trying to forget about the loss of his loved ones. In large part,

The Adventurous Life of Tom Warren

losing himself into the wilds of the frontier and its new adventures and dangers had caused the bitter loss memories of his family to almost fade away...almost. However, with beautiful soft summer days and their often times unusual cloud formations, stunning sunrises and sunsets, many of those memories of a wonderful previous home life, the associated love of a wonderful wife and the softness of a young son and a warm cabin, would still occasionally come flooding back...

Catching himself and shaking those thoughts of terrible family loss from his head, Tom resumed his duties, as he prepared to feed the rest of his fur-trapping partners. Partners who were now down at the new corral currying all the stock, saddling up their livestock and loading the pack animals with the many packs of furs gathered through their trapping efforts and those acquired after the deadly battle in the Medicine Lake marshes with the Hudson Bay crew of trap-robbing outlaws, discovered in the wrong place at the wrong time...

"Chow's ready!" yelled Tom loudly to his crew of fellow trappers working down by the corrals. Moments later, here came three trappers, all making noises like they were hungry as grizzly bears. Grabbing a spot on the sitting logs placed around their campfire and laying their rifles close at hand, the three men were all excitedly talking at once in anticipation of a great breakfast as only Tom could cook, and then the start on a much-awaited trip back to Fort Union. Tom, with a grin over the happiness he was now feeling through that of his friends, poured each man a steaming cup of trapper's coffee so thick that a mule shoe could stand upright in it! Those cups of 'bitter trapper's brew' were then followed with a bowl full of sugar so the men could make 'syrup' as Old Potts called it, once it was heavily spooned into their cups of coffee. Following that came a tin plate holding a couple of steaming hot Dutch oven biscuits and a buffalo steak, half-cooked as each of the men liked it cooked. Within moments, the only sounds surrounding the four men were that of a crackling fire and those made by hungry eaters as they shoveled into their mouths pieces of hot biscuits and steaming, blood-rare chunks of buffalo steak! After 20 minutes of wolfing down great gobs of 'rib-sticking' frontier victuals, the men slowed their eating and more of the excited talk regarding the upcoming trip to Fort Union resumed.

"Tom," said Crooked Hand, "we have loaded 30 packs of furs on the 15 horses and mules from the Hudson Bay outlaws' stock that we seized, and another 20 packs on our 10 pack animals! Damn, Man, we have a small fortune in furs if we can only get them back to Fort Union and them Blackfoot or Gros Ventre savages don't see and kill us in the meantime. That is not to mention the valuable 10 riding animals from the Hudson Bay men and our own riding stock."

"That is going to be a problem," said Old Potts slowly, as he put a damper on the happy ongoing conversation. "If that Medicine Lake Band of Blackfeet or their kissing cousins, the Gros Ventre, catch us trappers afoot out in the open, they will for sure be after us with a vengeance. Most especially the Blackfoot, since we killed such a passel of their kinfolk when they raided our camp early last year. If they spot us en route to Fort Union with such a big horse and mule herd, they will be on us like a mess of flies on a buffalo carcass laid out in the hot July sun for several days and that is a fact for sure!"

"What do you suggest?" asked the ever-cautious Big Foot. Then continuing as an afterthought, he said, "Maybe we best travel only at night or stick to the river and creek bottoms if we travel during the daylight hours?"

Tom, being the least experienced Mountain Man in such matters never having traveled under such circumstances before, just kept his mouth shut, listened and learned. But that was not to say he too was keenly interested on how they were going to travel and keep their 'topknots' and all of their valuable packs of furs and stocks of animals away from the Indians as well...

For the longest moments, the men's eyes had turned to the wisest and most experienced of their group, namely Old Potts.

The Adventurous Life of Tom Warren

Finally realizing the men, by their very eye movements towards him and now lack of discussion among themselves, had 'elected' him to address their now just thought of main issue of concern. That issue of concern being that of getting to Fort Union alive with such a very obvious valuable pack string, their hard-earned furs and their hair. Understanding the gravity of the moment at hand, Old Potts finally stirred once again.

"Well, Boys, we are deep in Blackfoot and Gros Ventre country and they can be murdering sons-of-a-bitches, as all of you know. I suggest when we get on the trail to Fort Union, we travel during daylight hours so we can cover ground the fastest ways possible and don't waste any time in getting there. That means up early, few fires, eat mostly jerky when we travel, keep our eyes 'slicked', and carry every loaded gun we have in case we get surprised. That we need to do, because if those 'war-hoops' catch us out in the open, they will mean to kill every one of us and take what valuables we have. That means we must be just as mean as them savages and kill every one of them just as quick and fast as we can if and when we have the chance. So, I propose that we take all the rifles and pistols we took from them dead Hudson Bay men, load them up and split them evenly among the four of us."

Pausing to take a sip of his coffee, Old Potts then continued saying, "Then, we take the first horse in our pack string assigned to each man and place them rifles and horse pistols alongside those animal's pack saddles so's we can get at them in a moment's notice once we 'run dry' with our personal arms that we each carry. Now, we collected up 12 good rifles and 20 pistols from those ten Hudson Bay men. I suggest that each man among us takes three of them rifles and five of them pistols and affixes them in handy positions on the first animal in each of our pack strings. That way if jumped by them savages, each man will have three rifles and five pistols to fire at any Indians that attacks us before he has to go to his own personal rifle and two pistols carried in hand or in our belts. That-a-way, if attacked, the four of us can fire a total of 32 shots at our attackers before any of us has to go to our own personal carried weapons. Then after that, may God almighty Himself help us... I would think with that kind of killing firepower close at hand, any Indians will think twice before they come at us for a second time."

"Remember," Old Potts said, "they know we normally only have one shot for each of our pistols and rifles. They will be expecting that and when we fire ourselves dry, here they will come at us in an organized charge. But with the extra Hudson Bay weapons which them murdering savages are not aware that we will be carrying, instead of only one shot and then having to reload, we can surprise the hell out of them and keep on shooting until we run dry with all our guns. They will not be expecting that amount of shooting and that will give us the edge on living till sunrise the next day! 'Specially when they first charge and we fire one shot and they know that we now have to reload our rifles. With that, they will keep-a-coming and be in close. When they do, we with our extra guns can pick off the closest Indians and that will surprise them and that should stop any further charges against us."

"Keep in mind, Indians don't like to lose people, 'specially their men folk. If they lose a few in their first charge and then come at us again only closer that second time, we will rain Indians onto the ground and like I said, they don't like losing their menfolk. Seeing us still shooting should dampen their spirits for a fight and back them off. Then maybe we can be on our way and safely arrive at Fort Union. However, the trick is to kill off a passel of them devils right at the first when they least expect it, so they realize it be best to leave us alone and move on," continued Old Potts.

"Remember, Indians is fighting all the time among themselves. Most tribes don't have even enough men to go around for every woman in their camps now, so they don't like losing any more men than they have to. So best we keep a sharp eye slicked and shoot straight if any attack comes our ways. I say that because if we do shoot straight and kill off a

The Adventurous Life of Tom Warren

passel of them Indians on their first charge, we stand a chance of keeping our hair and getting to Fort Union in one piece. Then if we do, we all can tip a cup or two of fine rum to our success of living just one more day on this damn scary most of the time frontier. But if this plan does not work and we get jumped by more than we can handle, make sure each of you keeps your two personal pistols loaded with buck and ball, so when they get in close to kill you, you can make sure you get a fair number of them afore they get an arrow or lance into you," suggested Old Potts.

For the longest moment the men mulled over Old Potts's suggestions and none of them having any better ideas, grunted their acceptance of his ideas as how to safely make it to Fort Union with all their 'topknots', packs of furs and valuable horses and mules...

With that bit of business out of the way, the four men returned to what they liked doing best, namely eating Tom's Dutch oven biscuits and the Indians' buffalo. That they did with much gusto that morning, because when out on the frontier, one never knew when or where the next meal was coming from. Come noontime, the men finally had all the livestock packed and were ready to leave for Fort Union. Those supplies that could be left in the cave were safely stored and left for their return since they had decided that they were going to return and trap the same area come fall. That decision had been made because there was still a goodly supply of beaver in the area and that spring's production of young of the year or 'kits', should be good for fall trapping successes as well. Additionally, their cave had been good for them as a safe and warm home base, the area was full of buffalo and so far, scarce on Blackfeet. Hence it was the four trappers' decision to return to the area and trap in it in the coming fall for at least one more trapping season.

With each man carefully leading a fully loaded pack string of ten horses and mules, the men set out heading down the Medicine Lake outlet toward the Big Muddy. However, as Old

Potts had suggested, each man had his first animal in his pack string as a heavily armed first horse in line. That meant at least three fully loaded and readily accessible rifles tied across the pack saddle, and five loaded, single shot pistols handily placed in buffalo skin leather holsters for quick retrieval as well. That gave every man at least 11 shots that could be fired before he completely 'ran dry'. That was providing he lived long enough, upon being attacked by a horde of Indians, to get to the weapons and get all his rounds off...

For the next several days, the trappers 'cold camped' at the end of each day (no campfires) and were horseback before daylight the next. They stayed to the creek bottoms and other such natural or rolling terrain types of cover as they could until they reached the Missouri. Once on the banks of the Missouri River, the four men with their long heavily loaded pack strings proceeded eastward on the north side of the river towards their final destination of distant Fort Union.

On the second night of their travels while proceeding easterly along the north bank of the Missouri River, the men noticed a faint light of a campfire down in an adjacent river bottom. Old Potts held the men up and sent Tom to sneak up to the mystery campfire and ascertain if it was that of another group of fur trappers on their way to Fort Union to trade in their furs or some Blackfoot or Gros Ventre Indians camped along the way, waiting to lure in and trap some unsuspecting trappers on their ways back to Fort Union.

Somewhat later, Tom sneaked up to the edges of the firelit camp and ascertained it was occupied by a small number of fur trappers. Quietly moving into the edge of the light from the campfire, Tom hailed the trappers and quickly identified himself so there would be no mistake as to who he was or fatal misunderstandings. The fur trappers around the campfire, alarmed at being surprised and hailed by a foreign voice and then the apparition of one damned huge individual suddenly appearing from out of the dense underbrush so close at hand,

The Adventurous Life of Tom Warren

sprang to their rifles and then covering him with leveled weapons at the ready, told Tom to advance and be recognized.

Tom did as he was instructed and once introduced to the camp's members, asked if he and his fellow trappers could join them. Tom was instructed to bring his fellow trappers in and camp with them for the added protection in numbers that offered the entire group of trappers. Returning to his fellow trappers somewhat later, Tom advised that the campfire was indeed that of a fellow band of American Fur Company Trappers and that their group had been invited to join them.

Twenty minutes later, Tom and his band of trappers joined the eight American Fur Company men and their Indian women around the campfire. After introductions were made all around, Old Potts took his band and they made camp near the other trappers, so all were not crowded together and their livestock intermingled. All their horses were then unpacked, hobbled and turned out to feed, and then the new arrivals prepared their campsite as well. Soon they had their first fire in a number of days and some fresh buffalo killed from the day before, merrily roasting on green willow stakes over the edge of their welcome fire.

It was then that an unexpected explosion of events played out between the two camps of trappers. Tom was busily currying down their mules before they were hobbled and let out to pasture so they would not 'sore up' under the weight of their heavy packs, when he heard a spate of yelling and a woman crying out in terror and pain! Looking toward the adjacent fur trappers' camp, Tom observed a huge man built like a small grizzly bear, taking a large piece of limb wood and savagely beating a young and smaller in stature Indian woman lying on the ground, who was trying to cover up her head as the blows savagely rained down upon her!

Tom saw 'red' in what was happening, dropped his curry comb and ran to the scene of the larger in size man beating the tiny and helpless woman. "Hey! Hey!" Tom yelled. "Don't you beat that woman!" he yelled as he continued running to her aid.

Surprised at the intervention of such a massively sized stranger just met earlier when he had quietly arrived into their camp, the huge man stopped momentarily beating the young Indian woman. Then the huge man recovered from his surprise at being interrupted over the beating he was administering to the woman, turned and faced Tom as he burst through the brush.

"Mind your own business, Stranger. This is my woman! I paid for her fair and square from another trapper who caught her out berry picking alongside a creek and stole her away from her people. She just 'spilt' a mess of stew on me and now by damn, she is going to get a damn good beating for being so clumsy. So back off, Stranger, unless you want some of the same! She is mine to do with as I please and that I am aiming to do right now," replied the monster-sized trapper with the limb still upraised in his hand.

With those words, the burly man once again slammed the stick he was holding down upon the Indian woman lying crying on the ground! That was when his upraised right arm holding the stick to render another blow was grabbed with a vise-like grip that was so strong, it forced the man to drop the stick of limb wood from a now quickly going lifeless hand! However, the grabbed man then quickly went for his sheath knife with his other hand, only to find his neck, like his other hand, in another vise-like grip that completely cut off all of his wind and the blood flowing to his brain!

"Hey, you son-of-a-bitch! Let go of my friend and put him down right now or you will die where you are standing!" yelled another fur trapper from the big man's camp, as he rushed at Tom with an upraised pistol in hand.

Tom dropped the Indian woman-beating huge man who had now passed out in the chokehold, turned in a fury, grabbed that man's hand waving the pistol in his face with his left hand and then grabbed him by the whole face with his right hand, 'crunching' the hell out of it with his "Iron Handed" facial grip until the surprised pistol-waving man screamed to be let go!

By now, both fur trappers' camps were in an uproar over the violent events unfolding between Tom and the two unfortunate, mean-assed fur trappers who had crossed him! The trappers from the original camp were rushing to aid what they suspected was a savage attack on two of their own, and Old Potts, Big Foot and Crooked Hand were simultaneously rushing armed to Tom's defense. The three trappers did not realize why they were rushing to Tom's defense because he had proven that he could more than take care of himself, but they just knew it would be for good reason if he was somehow involved because of his normal behavior of even handedness and reserved nature.

Tom then let the pistol-waving man go, whose face he had been holding in his crushing grip. In fact, his grip had been so crushing, that in just the few moments he had the man's face in his right hand in such a crushing iron grip, THAT IT HAD BLACKENED BOTH OF THE MAN=S EYES AND HAD BLOODIED HIS NOSE!

Then the larger in size trapper who had been beating the tiny defenseless Indian woman came to after being choked out, with a bull buffalo in rut-like bellow! Rising to his feet, he went for his sheath knife once again and drawing his knife, made a menacing move towards Tom. As he did, he found his knife hand grabbed in another vise-like grip that was so strong, that Tom broke three of the man's fingers on that hand!

"OWWWHEE!" screamed the man with the crushed fingers, as his knife dropped from the broken portion of his hand! By now with the two camps in an uproar and threats and counter threats flying back and forth, Old Potts finally gained control over the situation.

However when he did, Tom coming down from his killing high, slowly reached down and with one hand placed under the Indian woman's shoulders, lifted up her tiny frame and held her next to his body until she could stand on her own free accord. Then Tom's memory banks flooded open and he realized it had been a long time since he had held a woman... Especially a woman as lithe, supple and trim as the one he had just lifted up off the ground. A woman who had the same build and weight of his much-beloved wife that he had lost to smallpox in the civilized world back in Missouri...

"Hey, you big son-of-a-bitch! Let that woman go or I will shoot you dead myself!" yelled the grizzly bear-sized man who had been beating the Indian woman earlier and was now holding his broken fingers. A man Tom would soon come to learn was named "John Pierre". A man who then quickly grabbed his rifle up off the ground and aimed it at Tom with his one remaining, unbroken good hand...

It was then that Old Potts slipped in behind the woman-beating bully of a man and touched the cutting edge of his always razor sharp cutting and gutting knife to the front of the man=s throat and quietly whispered into his ear, "Hey, your miserable life for his, Stranger. Your call..." It was amazing just how quickly the bully dropped that rifle to the ground at his feet...

Soon, cooler heads prevailed and calm more or less slowly returned to the two camps. However, Tom did not intend to return the badly beaten woman to John Pierre! Instead, Tom walked up to the face of the 300-pound bully saying, "You, Sir, don't deserve to live. No man should be allowed to beat a woman like you would a cur dog. She will stay in our camp until we get to Fort Union. Then we will let Factor McKenzie decide what is to be done with her. Since the man you purchased her from stole her away from her people, McKenzie will decide what to do with her. This issue will be handled that way because if her people ever discover that she was taken against her will by a stupid white man, they will run hog wild over the land, killing every trapper they can find in revenge. If that happens, I don't cotton to being in this country when that kind of a killing rampage occurs because of some stupid act you furthered when you bought her out of bondage. So,

The Adventurous Life of Tom Warren

McKenzie will decide what is to be done with her. But I can tell you this, Sir. Touch her again in my presence and you will never again set another beaver trap or see a sunrise!"

Little did Tom realize that the Indian woman was intently watching and listening to him speaking to her captor and understanding those words of warning being spoken to her previous captor and beater, one John Pierre. Little did Tom realize that those words of warning being directed at the trapper who had previously beaten the woman would someday be rewarded in the future when his own life was in jeopardy as well.

With that, Tom reached down once again, picked up the badly beaten woman and cradling her in his massive arms, took her back to his camp as the fury in his body slowly began subsiding over the trapper's indecent beating of a woman... That evening, the badly beaten woman remained in Old Potts's camp and no one came over from the other trappers' camp in an attempt to bargain for the return of the woman. Especially after seeing the cold and disarming look in Tom's eyes, when he quietly, with iron-edged wording, laid the law down to everyone in the other trappers' camp on how the woman's fate would be subsequently decided...

The next morning, John Pierre came over to Old Potts's camp and in a whining tone of voice asked for what he called his woman back. After responding to John Pierre in a few short and sharp words, Old Potts sent him packing... With that and a lot of evil looking-back at Old Potts's crew of trappers, John Pierre and his crew of men hurriedly broke camp and it was plain that Tom had just made some serious enemies. As Tom and the others packed their livestock for the trip once again on the way to Fort Union, Old Potts thought to himself that he and his mates had best keep a sharp eye peeled out for Tom's back. That was especially as it related to John Pierre and his murderous looking crew of trappers. He quietly figured that with enemies like that, he, Big Foot and Crooked Hand would have to watch them like a cat does a mouse in the future. If not, Tom could fall prey to their obvious now murderous intent over their embarrassing loss of the valued Indian woman and their loss of 'face' to the smaller group of Old Potts's trappers...

For the next several days, Old Potts and company continued their measured travel along the Missouri River en route to Fort Union. They took their time in order not to stress the pack train because of the heavy loads of furs they were carrying, plus they were still in Indian and grizzly bear country, so caution was the word of the day. Finally come the day, Fort Union loomed into view. Stopping a short distance away on an overlook, the four men just sat on their horses and looked on in wonder at the drastic changes that had been made to the fort and trading post and how things now looked, compared to what it had looked like when they had left the summer before.

Tom could hardly believe what he was now looking at! When his group of fur trappers had left the fort, it was just four palisaded log walls with a few skimpy log buildings inside. Now its four walls were heralded by the presence of two impressive-looking, roofed-over blockhouses on opposite corners of the fort's walls that were two stories high! The fort now also showcased a huge set of heavy plank wooden front doors and even from a short distance away, shooting loopholes were readily discernible located along the fort's walls and on the sides of the two blockhouses. Men could be seen positioned standing guard along the inner walls, as they watched over the fort's outside activities and the coming and going of various Indian traders and fur trappers.

There were now dozens of tepees scattered about on the fort's grounds, as well as numerous quietly feeding horse herds scattered about belonging to the Indians and the American Fur Company. Off to one side of the fort's outside walls was a large garden being currently tended by several men from the fort, and there was a constant line of Indians and fur trappers streaming into and out from the fort's massive front entrance portal. The whole area was a constant beehive of human

The Adventurous Life of Tom Warren

activity both comings and goings, as well as moving around inside the cavernous fort's spacious interior.

Shortly thereafter, into that swirl of activity rode Old Potts and his crew of very successful fur trappers, as evidenced by their long strings of riding stock and heavily packed mules. Approaching the front gate, their presence was announced by a gate guard and within moments, they were heartily greeted by one Kenneth McKenzie, the Factor for Fort Union.

Standing in front of the fort's gate, the look on McKenzie's face showed nothing but amazement! In part, it had to be his last remembrance as Old Potts and his crew had left the fort with a string of only 18 riding and pack animals. Now those same trappers were returning with several strings numbering around 40 very valuable riding and pack animals! Additionally, as Old Potts and his crew approached the fort's outer walls, numerous Indians admiring the frontier wealth in livestock and packs of furs, took notice and approached for a closer look-see as well.

"Welcome, you men! Free Trappers are always welcome at Fort Union," said a very happy McKenzie over what he was seeing in the way of the return of old friends, and a wealth of furs and always scarce on the frontier, horses. It was then that McKenzie realized that all the animals he was now seeing loaded with packs of furs, belonged to just the four trappers bringing them in!

"How the hell did you men gather unto to you all those horses, mules and packs of furs since you left the fort only a year ago with so little in the way of a pack string compared to this?" asked an obviously still amazed over what he was seeing, Kenneth McKenzie.

"It is a long story," said Old Potts, "best told over several cups of your good rum, if you still have any of that left in your fort's warehouses."

"Well, there is only one way to find out. Let me get together several of my American Fur Company Clerks so they can take a gander at what you have brought me, and then we

68

can talk," said an obviously very happy Factor, as he looked at all the furred wealth the men were bringing into his trading post to sell or trade for provisions.

Moving their pack strings into the fort's interior, Tom noticed they were 'greeted' by a small cannon placed in the courtyard, where it could be counted upon to control anyone entering the front gate. Looking around still in amazement over the changes made since his last presence, Tom observed all manner of buildings now located along the fort's interior walls. Then he noticed that center to the square was a huge fur press to be used in taking loose furs and compacting them into tightly bound traveling packs weighing about one 100 pounds each. That way, when the furs were shipped by boat downriver to St. Louis, they could be more easily and compactly stowed aboard the keelboat's limited cargo spaces.

As their horses and mules were unpacked and the fort's Clerks began the huge chore of grading and counting all the assortment of furs Old Potts and his crew of Free Trappers had brought in, McKenzie noticed Tom tenderly removing a young Indian maiden from the saddle of one of the extra riding horses and gently standing her onto the ground like a gentleman would do a lady back in the civilized world. With that move, it became readily apparent to McKenzie that the Indian woman was someone important to Tom and should be treated like a lady. McKenzie also noticed that the woman appeared to be in total awe of the size and newness of the white man's fort and especially of Tom, and his gentle manner and protective treatment of her...

As Tom, Big Foot, Crooked Hand and Old Potts closely watched the Clerks grading and counting the furs from each of their 50 packs, the young Indian woman continued standing in the center of the fort's inner courtyard, looking all around at her new surroundings in awe and wonder. McKenzie could tell that she was in total awe over what she was observing obviously for the first time in her young life. That in mind, he walked over to her and because of her dress which identified

The Adventurous Life of Tom Warren

her as to her tribe, addressed her in the tongue of the Blackfoot, asking her if she would like to sit in the shade of the porch and have some water to drink. At first, the young Indian woman just looked at the obviously important white man speaking to her in the tongue of the Blackfeet, then smiling and avoiding looking directly into McKenzie's eyes, demurred and did not respond to McKenzie's question.

McKenzie, familiar with the Blackfoot culture and the shyness associated with its women around strangers, especially white men or men in power, called Tom over and asked him to let the young Indian woman know it was alright to go and sit on the porch in the shade and have a drink of water while the trappers' furs were being graded and counted.

Tom just laughed and then said, "McKenzie, you can talk to her in English. She speaks English as good as you and me, having learned it from her captors. She was captured last year by a rogue trapper, who repeatedly raped and then later sold her to another trapper, namely one of yours, named John Pierre. He too and some of his friends have abused her and as such, she is understandably shy around anyone white. Especially after all they did to her from stealing her away from her people when she was out berry picking, to all those dark things I spoke about. Her Blackfoot name is "Sinopa". That means "Fox Kitten" in her native tongue and I have discovered that her demeanor matches her name."

"I rescued her from being savagely beaten by that John Pierre fellow I told you about, and she has been stuck to me ever since like glue the Indians make from the hooves of a buffalo. I am not sure what the four of us are going to do with her and that is why she is here with us today. I forcefully took her away from John Pierre several nights ago, while he was in the process of beating her for spilling some buffalo stew on him during supper. I stopped the beating and then told John that if he ever touched her again, I would kill him and I meant it! He came to us later the following day and wanted her back because he said he had paid the other trapper fair and square for her. I told him I would leave that decision as to what would happen to her up to you once we got to Fort Union. I did that because you being the Factor and your word being essentially the law in the territory, I figured you would do for her what was for the best," said Tom quietly.

"I also did so because if that first trapper, whoever he is, forcefully took her away from her people and they found out that was what had been done by a white man, I figured there might be a bloodbath for all us trappers clear across the Northern Plains by irate, white men-hating Indians. So I leave what happens to her up to your good graces and wisdom. But if the surrounding tribes you trade with on a daily basis discover what has happened to her by some evil white trappers, that may cost all of us our lives, and you and Mr. Astor the loss of a damn good fur and trade business here at Fort Union," continued Tom, with an 'I am glad this decision is up to you' twinkle in his dark eyes.

"So, Mr. McKenzie, it is up to your good judgement as to what to do with her. But I will tell you this, if you decide to give her back to John Pierre, I will be forced to kill him and bodily take all our furs down to St. Louis in the future to sell, instead of selling them to you here at the fort," said Tom, with more than just a touch of finality in the tone and tenor of his voice and a decided lack of 'twinkle' now showing in his eyes upon uttering those words just spoken...

"Well, Tom," said McKenzie, "that does not leave me much latitude as to my decision now, does it? You know, when I first saw you that day so long ago when I was signing up help for Mr. Astor's American Fur Company to come upriver and build this trading post, I saw in you something that I did not see in any of the other rough frontier types who were also signing up for the adventure of a lifetime. I don't mean to pry into your personal life but on that day so long ago, I saw a lot of sadness in your eyes, along with a certain touch of steel and yet a tenderness that matched all of what I figured factored into your other qualities. That was one of the reasons as to why I

The Adventurous Life of Tom Warren

hired you, even though you wanted to be released as a Free Trapper once we arrived at the Missouri's headwaters. Well, that and you were so damn big, I figured you could kill a grizzly with a switch and make it on your own anyway no matter whatever came your way."

"All of those personal judgments I made about you that day so long ago came flooding back to me today, when I saw just how tenderly you removed that Indian woman from her horse and set her down so gently onto the grounds inside the fort. I will tell you right now, I have no plans on turning Sinopa over to that brute John Pierre, just so he and his friends can do with her as they please. I am not that kind of a man. Ever since I signed John Pierre on as a Company Trapper, he has been nothing but pure trouble and lazy to the letter of the law on any and all work around here at the fort. Oh, don't get me wrong regarding the man. He is a damn good trapper but he has a touch of snake running throughout his soul and if I were you, I would be very careful in everything you did around him in the future. You crossed him when you took away 'his woman' Sinopa, and he and his 'snakes' den' of fellow trappers will kill you by either shooting you in the back or a knife to the guts when you are not looking and least expect it. With that personal assessment of what you are facing from him and his kind, that you can rest assured. Knowing him as I do, he means to have her back. So I would stay clear of him or at least watch your back all hours of the day as long as he and his kind are in country or here at the fort. Those words of caution go for the rest of your party as well," coldly advised McKenzie, "because 'what is good for the goose is also good for the gander' to their twisted and evil way of thinking."

"Tom, Sinopa is yours to watch over and keep as you see fit. I have made that decision because I know you will treat her right and care for her if you decide to keep her as yours. But mind yourself, full well knowing that John means to get her back, 'come hell or high water'. Now, I will let John and his cohorts know of my decision and also let them know that if they cause any trouble over this issue, on the fort's grounds now or hereafter, I will have them removed without any chance of re-supplying their provisions at this fort, much less sell their furs. That, plus they will never be allowed on these grounds ever again for any reason as long as I am the fort's Factor. But that will not stop him from cold tracking you down and killing you and your kind if he gets just half a chance..." quietly advised McKenzie.

"Now, after my Clerks get through counting and grading your furs, I would be honored if you and your Free Trapper friends would have supper with me this evening. And when you do, I am dying to hear how the four of you came into possession of all those mules, riding horses and extra packs of furs over and above what a successful trapper can acquire through normal trapping in a season. Especially since I saw several marks on those packs that clearly indicated they came from the Hudson Bay Fur Company. Knowing the likes of that bunch of scoundrels, I am sure they did not just up and give them to you," said McKenzie with a suspicious and yet telling grin...

"Oh, by the way. Please bring Sinopa, or Fox Kitten as you call her, with you and your fellows for supper tonight. I aim to show her that not all white men are evil. And by the way, I will ask her once again, in English this time, if she would like to sit on my porch in the shade and in so doing, I will also let her know that you said it would be alright..."

"Oh, one other thing. Keep Sinopa close to you at all times. As I said, knowing Pierre as I do, he aims to get her back. He and his cohorts stole her once before and they will do so again if the opportunity arises," said McKenzie over his shoulder, as he walked over to the Indian woman with thoughts of inviting her to sit on his porch in the shade while the business of fur counting and grading at hand was occurring...

That evening, Old Potts, Big Foot, Crooked Hand, Tom and Sinopa were honored guests at Factor McKenzie's supper table. There they were served buffalo steak, fresh potatoes

The Adventurous Life of Tom Warren

fried with onions from the fort's garden with real butter and fresh homemade bread, courtesy of McKenzie's personal Chinese baker. That supper was then topped off with a fresh blackberry pie baked in a wood-fired oven and all the first class rum the men could hold. As McKenzie figured, based on her earlier demeanor in the courtyard of the fort, Sinopa was as 'quiet as a mouse pissing on a ball of cotton' during the whole meal. Additionally, she seemed amazed over what she was seeing and the things she ate, especially surprised with the use of all the fine china and silverware. Lastly, it was readily apparent that she really loved the freshly baked homemade bread and blackberry pie...

After supper, several cups of first class rum and some fine cigars that came all the way from New Orleans, Old Potts shared the Hudson Bay Company story with McKenzie. Throughout the telling, McKenzie never said a word, just listened intently with a more than serious look on his face. When Old Potts had finished, McKenzie said straight away, "Served those bastards right!" Then he indicated that just a week earlier, a Hudson Bay Company representative had come to Fort Union with a story about a number of their missing Hudson Bay Company fur trappers, along with their best *Bourgeois* from the Wolf River trading post. According to the Hudson Bay Company representative, the group of trappers and their *Bourgeois* were long overdue their arrival at their new trapping grounds on the upper end of the Porcupine River... McKenzie told the Hudson Bay Company man to try talking to some of the surrounding bands of Indians around the fort to see if they had any information on the disappearance of his men. Following that, McKenzie rather ingloriously sent him on his way. McKenzie indicated he had done so because his company represented serious and direct competition with the American Fur Company for the rapidly dwindling stocks of available furs. Then if outside competition was not enough of a problem, McKenzie had come to learn that a quarter of his Company Trappers that he sent forth the year before had disappeared and

74

were never heard from again! In that, McKenzie suspected accidents, horse wrecks, problems with the Indians, bad weather and grizzly bear attacks as the main culprits involved in the disappearance of so many of his trappers in just their first year out on the frontier.

(Author's Note: Historically speaking, around a quarter of all trappers who ventured forth on the western frontier annually from 1807 until around 1845 as fur trappers, were never heard from ever again because of fatal horse wrecks, being killed by Indians, killed by wild animals, accidents, freezing to death, starving to death, drownings, and the like!)

After supper that evening, McKenzie suggested the men sleep in one of the empty Clerks' cabins inside the walls of the fort that evening, as well as Sinopa doing the same, knowing that would preclude any trouble being caused by Pierre and his men, who were sleeping just outside the fort's walls after they had traded in their furs several days earlier.

The next morning, the men initially settled up with McKenzie over their haul of furs, robes and hides. Then taking Sinopa along with them, they visited one of McKenzie's warehouses and spent the day procuring most of the provisions they figured they needed for the coming trapping season. When they did, Tom saw to it that Sinopa was treated to a bolt of red and blue cloth, several packages of vermillion, and all the blue and red beads and metal cookware that she wanted. Once again, her eyes grew as big as dinner plates once she saw all the wealth of the white man's goods that were contained in the fort's warehouses! That evening, the Free Trappers (the highest class of frontiersman other than the Factor or *Bourgeois*—another name for the one in charge of the trading post) and Sinopa were treated to supper as honored guests once again with McKenzie at his home within the walls of the fort.

(Author's Note: Supper with the Factor or *Bourgeois* was an honor that was only accorded to the highest ranking visitors, company leadership, and some very successful Free Trappers in the tradition of the more established fur trading posts.)

The Adventurous Life of Tom Warren

It was during that meal that McKenzie announced the coming arrival of a great Blackfoot Indian chief named Gray Wolf or in the name of the Blackfoot, "Chief Mingan". When McKenzie made the announcement in a rather offhanded fashion, Sinopa bolted upright in her chair like she had just been shot and her countenance froze. Seeing the immediate change in her demeanor upon hearing that coming arrival announcement of the fierce Blackfoot Chief Mingan from the Medicine Lake Band, McKenzie asked her if everything was alright. The look in her eyes showed pure amazement over McKenzie's announcement, then she caught herself and went almost blank as she silently sat there during the rest of the meal... As it was, Sinopa did not finish her supper and when she and the trappers went to bed that evening, she was as quiet as an Indian burial site... The trappers and McKenzie thought that behavior strange, but not totally understanding of her kind, they collectively backed off and let her have her space.

The next morning, Old Potts, Big Foot, Crooked Hand and Tom finally settled up with McKenzie on what he still owed them for selling him their 50 packs of assorted furs, hides, buffalo, and bear robes. As it turned out, the American Fur Company owed the men just a bit over $40,000! To be sure, that amount of money was a fortune in those days and an amount that was not readily held in the fort's safe. Realizing that, the men agreed to being issued a note of credit from the American Fur Company that would be good upon demand in St. Louis for that amount, whenever they decided to return to what they called "The Civilized World". With that, the four trappers realized they were now very wealthy men if they chose to return to the civilized portion of the United States. Plus, that figure did not include what they were yet to receive when they sold off their extra mule and horse herd to McKenzie for use by his Company Trappers as well! That evening, the men sat down over a jug of first class rum, while Sinopa sat quietly and watched the four men, still in wide-eyed

76

wonder over the events swirling around her, and in anticipation of those moments in time yet to come...

It was during that evening's session over those cups of rum that the men finally decided they would go back to their original trapping grounds in the Medicine Lake area as planned earlier and continue trapping there for at least another year. Then more rum was had by all except Sinopa, who just quietly sat on her bed and watched what was going on around her, still in 'wide-eyed' wonder over what she had been observing in the white man's 'civilized' world.

Then her darkest memories flashed back over her and in so doing, she was also quietly thankful she was not being sexually mauled by John Pierre or some of his drunken fur trappers that evening... Then the full shame of what had happened to her, from her original capture by a trapper to the beating she had recently taken at the hands of John Pierre, flooded over her once again. With that, as if to hide that shame, she crawled into her bed, pulled the covers over her head and suffered in silence, full well knowing what was yet to come on the morrow...

Around noon the next day, there was great excitement and commotion in and around the grounds of Fort Union. The great Blackfoot Chief Mingan or Gray Wolf, from the Medicine Lake Band was coming into the fort to do his annual trading of furs and robes for the white man's goods that his people coveted. Goods such as guns, powder, salt, vermillion, colored glass beads, metal cookware and most of all, the fiery liquid that made many of them 'crazy in the head', namely rum and especially whiskey! Soon the men inside the fort could hear the numerous excited noises made among the fort's surrounding and almost resident Indian population. Then into view from those watching in the fort came a dust cloud of loosely organized, streaming masses of horses, barking dogs and Indian humanity, all of which were representative of Gray Wolf's Medicine Lake Band of proud Blackfeet.

The Adventurous Life of Tom Warren

Standing up on the walkways located along the inner walls overseeing the fort's grounds below, Old Potts's crew of Free Trappers watched in awe at the mass of Indian humanity slowly moving their way. Bypassing the fort, Gray Wolf's band of Blackfeet moved into the bottoms along the Missouri and noisily made their camp. Surprisingly, Sinopa did not accompany the men but stayed in her bedroom when the men went out to observe the oncoming spectacle of Indians, huge horse herds and barking dogs as they moved around the fort into the Missouri River bottoms.

Then the inside of the fort erupted into all kinds of activity. Numerous men scurried about setting up a huge fire in the fort's center firepit, as another number of men dressed for roasting over an open fire a young, in preparation for the great event, freshly killed cow buffalo. Other company men hurriedly set up long tables in the fort's courtyard for the meal to soon be served to the important chief and numerous sub-chiefs from the Blackfoot Indian encampment. As McKenzie's preparation for a great feast was unfolding around them in the fort's courtyard in honor of the great Blackfoot Chief, Old Potts and his fellow trappers sat on McKenzie's front porch drinking rum and watched the activities swirling around them with a keen interest as they waiting for the great evening meal to get underway. Soon great smells of roasting buffalo meat and beans cooking in huge cast-iron pots began filling the air, as McKenzie's three Chinese cooks scurried around the outdoor firepit with their helpers, making in addition to many other things, small mountains of the great smelling sourdough Dutch oven biscuits in the coals of the firepit.

About three hours later, numerous excited noises could be heard once again outside the fort's front gates and soon in streamed dozens of magnificently dressed and obviously very proud, Blackfeet Indian men. Many were obviously dressed in their finest dress, all decked out with double train eagle feather headdresses, war shields, gaily decorated coup sticks and rifles. At the head of that stream of primitive humanity proudly strode Gray Wolf. He stood at least six feet in height and was a physical specimen to behold. He was strikingly handsome, deeply bronzed, extremely muscular, wearing red-dyed leggings showing that he walked "The Red Road" (the road of goodness), blue-beaded moccasins and a magnificent double train, spotted eagle headdress that almost dragged on the ground! There were no two ways about it, Gray Wolf was the chief of the Medicine Lake Band of the mighty Blackfoot Nation!

As the Indians streamed into the courtyard, out came McKenzie dressed in his finest suit of black clothing and wearing a beaver skin top hat obviously meant to impress. Once all were gathered in the courtyard in silence, McKenzie made a long speech in the Blackfoot language welcoming Chief Mingan, his sub-chiefs and other important warriors. Then Chief Mingan gave a short speech about the brotherhood of the Blackfeet Nation and the great American trader, McKenzie. It was then that Tom noticed the main event of the day, at least to the Indians, was now being set into motion. Six company men brought out three wooden kegs of whiskey, set them up on their stands and began dispensing the fiery liquid from wooden taps into eagerly held out tin cups for the now very happy Indians.

(Author's Note: Such whiskey-fueled events were soon to be prohibited by the U.S. Government in all frontier forts and trading posts because of the adverse effects it had on the Indians. Additionally, that prohibition was mandated because the whiskey and its adverse effects on the Indians allowed them to be more easily cheated in their subsequent trades of furs and robes by many of the unscrupulous white traders scattered throughout those numerous trading posts on the frontier.)

With cups full of the much anticipated liquid, McKenzie and Chief Mingan sat down at the head of the long table and soon were followed by the ranking Company Clerks, *Engagés* and a number of Free Trappers of note, including Free

The Adventurous Life of Tom Warren

Trappers Old Potts, Crooked Hand, Big Foot and Tom. Soon much happiness could be heard emanating from the interior of the fort, as the hungry men were served huge bowls of baked beans loaded with chunks of bacon and onions, fresh biscuits and slabs of still cooking, steaming hot buffalo meat. Then even more happiness was heard among the celebrants, as the whiskey began taking effect as did the good food being served by McKenzie's company men.

All of a sudden, Tom looked up from his place at the table and noticed Sinopa standing on the steps of the Clerk's house where they had been sleeping for the last two nights. She was beautifully dressed in all of her finery, including some of the vermillion and colored beads that Tom had given her days earlier when the men were gathering up their supplies for the coming trapping season. As she stood there in plain view with her long black hair shining in the sunlight, the happy noises of the men feasting in celebration of the arrival of Chief Mingan's band of Blackfeet, upon seeing beautiful Sinopa standing quietly in view, began to quiet down until almost total silence reigned in the fort's courtyard over her surprise appearance...

It was then that Tom saw Chief Mingan slowly rise from his seat at the head of the table and then after a long pause looking at the young Indian woman, gracefully stride across the courtyard over to where Sinopa stood on the lowest step of the porch's stairs. For the longest time Sinopa and Chief Mingan just stared at each other. Then the chief with a shout of great joy, scooped up the tiny Sinopa in his arms and began whirling her around and around in obviously great joy!

Not knowing what the hell was happening Tom rose from his seat at the table as if to see if Sinopa needed his help. However, he was quickly stopped with the calming hand of Old Potts being gently laid on top of his. "Tom," said Old Potts, "she is alright. I think they know each other. Maybe she was his wife before the trapper caught her berry picking and took her away. But don't you make a move until we see what the hell is going on, or you could create a bloodbath right here in this courtyard if your actions are taken unkindly by all of Gray Wolf's warriors."

Slowly sitting back down, Tom kept his eyes on Sinopa to see if she was in any kind of danger in the arms of Chief Mingan, because if she was... It was then that Tom and his fellow Free Trappers, usually not surprised by about much of anything out on the great frontier, got the surprise of their lives!

All of a sudden, the chief put Sinopa down and holding her hand, eagerly led her over to where Tom and his fellow trappers were quietly sitting trying to figure out what the hell was happening. As they approached, Tom could see that Sinopa had been crying but was now wearing a huge smile and looking right at him with a big smile on her face. Walking up to Tom, Sinopa reached out, took one of his massive hands in hers and said in perfectly good English, "Tom, this is my brother! He has been looking for me for over a year with a broken heart after the evil trapper caught me berry picking and then stole me away. He figured I was with the "Cloud People" when he could not find me. Now he has found me and his heart is glad. He does not speak very good English but has asked me to take him to the man who saved and then has returned me safely to him and "The People". I told him about how you saved me from John Pierre and he wants to thank you for making his heart whole once again."

With that, Sinopa introduced Tom to her brother in the Blackfoot language, as everyone sat or stood there in amazement over the surprising turn of events unfolding in front of them! Then Chief Mingan spoke to Tom in the Blackfoot language as his steely dark eyes stared directly into Tom's. For the longest time he spoke forcefully to Tom, then stopped and looked over to his sister to translate what he had just said, so Tom would understand what had just been said to him.

"Tom, my brother, the chief of the powerful Medicine Lake Band of the Blackfoot Nation, says he has been made very happy today because a white man trapper has found his sister and safely returned her to him and The People. He says

The Adventurous Life of Tom Warren

his heart is very happy and wishes to let the white trapper named Tom, know that he is forever allowed to trap and hunt in Chief Mingan's land without the fear of warring with any of the Blackfoot People. My brother also says he forgives "Iron Hand" for taking the lives of some of his unwise young braves who drank the "Fire Water", then tried to kill the white trappers at their cave home over a year ago, so they could rob them and steal their horses to sell here at the fort for more whiskey. He says they were wrong to do so and found out that the man from Medicine Lake, who survived the raid against the trappers, saw you killing without effort and great strength, while only using your hands! He reported back to the tribe, that the giant of the 'white man trapper' who killed with just his hands should now be called "Iron Hand" among The People! My Brother also says the reason he did not come after you and your trapper friends after you took the lives of some of his young and unwise braves, was because The Great Spirit warned him and his People through his Medicine Man, that Iron Hand has many magical powers in the strength of his hands and also in his heart. The Great Spirit also told his Medicine Man that Iron Hand is destined to do great things someday for Chief Mingan when it comes to his family! That is why Chief Mingan did not come looking for you and your trappers after your kind killed some of his braves. So now it has come to pass as The Great Spirit has said, when Iron Hand returned my sister from her darkness back to her brother. It has also been said by The Great Spirit that Iron Hand is to be protected by the great Blackfeet Nation or bad things will come to The People! Chief Mingan now considers you his "New Brother" and wishes to seal that bond of life and friendship between the two of you warriors with the mixing of blood."

When Sinopa had finished with the translation, she stopped and looked back up at her still stern-looking brother, letting him know with her eyes that she had finished with the translation. With that, Chief Mingan drew his sheath knife to the gasps of surprise of most everyone in the crowd of diners who saw his move fearing death was close at hand, then surprisingly drew it across the palm of his hand and then stood there looking at Tom as Gray Wolf's hand dripped blood onto the dirt of the courtyard at his feet.

"Tom, you have got to do the same thing," said Old Potts, who spoke and understood what had been said and now needed to be done. Without a word, Tom then slowly drew his knife from its sheath, drew the blade across his open palm and as his blood oozed, he and the powerful chief clasped their bloody hands together in friendship!

Then Chief Mingan looking down at his little sister spoke once again in the Blackfoot tongue. When he finished, Sinopa turned once again to Tom saying, "My Brother wants you to know that as long as he shall live, the two of you are brothers and he shall never lift the hand of death against you or your family of trappers. However, he says those who did to me what they did, shall die many deaths before their bodies return to Mother Earth and become food for the little people, the ants. He also says that as long as the aspen leaves move in the summer winds and fall to the ground when the cold and snow comes, you shall be forever known among The People as "Iron Hand"!" When Sinopa finished translating, she once again looked back up at her brother, Chief Mingan, with a look indicating that his powerful words had been heard by Iron Hand. Chief Mingan then reached out and placed both hands upon Tom's shoulders and drew him into a hug between the two giants of men as a lasting sign of friendship.

With that move, the rest of the Indians sitting or standing around the supper table broke out into a chorus of approval, as did the other trappers sitting there as well. The whole time, McKenzie looked upon the event with concern at first and then relief, when he realized what was historically and culturally occurring. It was then that the supper and party really got going as the whiskey flowed and great amounts of food were consumed by the Indians and fur trappers alike, as the celebration honoring the arrival of the powerful Chief Mingan

The Adventurous Life of Tom Warren

and his people uproariously continued long into the evening hours.

Come nightfall when the whiskey was gone and the food had been consumed, the group broke up and noisily went their ways. However, Chief Mingan had some final words for Tom and asked his sister to translate for him once again. After he had his say in the language of the Blackfoot, Sinopa once again translated into English saying, "My Brother has requested that Iron Hand and his friends return to the Medicine Lake area together so he can protect his new brother and his fellow trappers from other bad Indians in their travels. He also says that he will send out tribal runners far and wide, bringing The Great Spirit's words and Chief Mingan's promise that Iron Hand and his people are friends and are to be respected, welcomed and watched over in whatever they chose to do in the land of the Blackfoot."

Looking over at Old Potts and getting an affirmative nod, Iron Hand told Sinopa that he and his fellow trappers would be honored to travel with Chief Mingan back to their trapping grounds in the Medicine Lake area. When Sinopa translated Tom's 'words' back to Chief Mingan, he got a big smile on his face and once again hugged his 'new brother'.

Then Sinopa also hugged Tom and told him that she must now go with her brother back to her people and would also do so when the decision was made to leave Fort Union and return to their ancestral wintering grounds. Tom smiled and then reached down and gave little Sinopa a big ole grizzly bear hug as only Tom could give, much to the enjoyment of both Chief Mingan and Sinopa, and the rest of the remaining Indians and trappers who had gathered for the celebration. Then the chief and his sister disappeared into the darkness, as they headed for the Indians' camp along the Missouri River bottoms...

For the longest time, Iron Hand and his fellow trappers stood there quietly in wonder over the day's events. Finally, they retired to their sleeping quarters and drank many cups of rum in a quiet celebration of their own. A quiet celebration in

Terry Grosz

honor of the 'naming' of one of their own by the Indians and the comforting knowledge that they would now be safely trapping and hunting in the Medicine Lake area under the protection of Chief Mingan and the rest of the Blackfoot Nation...

The Adventurous Life of Tom Warren

Terry Grosz

Chapter 7: Unexpected Surprises and The Truce Holds

FOR THE NEXT TWO WEEKS, peace reigned at the fort with Iron Hand and his partners. McKenzie, true to his word, met with John Pierre and the rest of his American Fur Company Trapper cohorts. When he did, he laid down the law on the settlement issue involving Sinopa, warning that if any trouble developed over his decision, their entire group would be unwelcome around the fort as long as he was the Factor. Those warnings and the Sinopa decision, as McKenzie had suspected, did not sit well with John Pierre, and the man made his unhappiness known regarding the issue over his 'lost and stolen woman'. When he did and the word got back to McKenzie, the Factor had enough of the trapper's acrimony and boiled over! John Pierre and his men were ordered to re-provision and return immediately to their trapping grounds before any further trouble developed between Iron Hand, Chief Mingan, McKenzie, and John Pierre's large group of fellow trappers and dedicated followers. Two days later, John Pierre and his group of trappers disappeared back into the vastness of the frontier as they had been ordered by their 'boss' and

87

American Fur Company Factor at Fort Union, Kenneth McKenzie...

After that still-upset and warring group had left, Old Potts's crew got down to making their final departure preparations as well. They sold most of the Hudson Bay Company string of horses and mules to other Free Trappers arriving daily at the fort in need of livestock replacements. Those not sold to Free Trappers in need were then sold to McKenzie for use by his Company Trappers who had lost, had stolen by the Indians or had crippled up their horses or mules during the last trapping season. However, when they sold off their excess stock animals, they kept the best animals back for themselves and a few for a surprise present to be made at a later date. Then they had all their remaining livestock re-shod by the fort's blacksmiths in preparation for the many long and hard months lying ahead of the trappers and their animals. Additionally, with some of the money from the sale of their horses and mules to other trappers in need, the men purchased extra horseshoes and nails in case any of their stock threw their shoes while in the outback and needed replacing in the field. Time was also spent by all of the men casting up a 'small mountain' of bullets for their 'lead-eating' Hawken .50 caliber rifles, and additional bullet-making mold blocks were purchased to replace those lost or misplaced for those weapons as well. Then two of the newest models of pistols were purchased for each member of their party, along with the appropriate bullet molds, ignition caps, pigs of lead and new powder horns. Additionally, another 20 of the newest beaver traps, along with 10, larger in size, toothed wolf traps were purchased to go with those older traps left back in their cave in the Medicine Lake country when the men had originally left for Fort Union to trade in their first season's furs. Finally, extra leather strapping, buckles and new pack saddles to replace those that were worn out were purchased from the fort's vast stores for the group's new pack string. Lastly, another trip was made to the fort's warehouse to stock up on extra kegs of first class rum and gather up the last of the needed supplies, especially spices, to top off their long list of needed provisions.

Come the announced day of departure, Chief Mingan's people were ready to leave and the trappers were up early and had eaten a huge final breakfast with their friend McKenzie. That they did in typical trapper style, because they were not sure of the quality or quantity of their next meal on the road. After that, they made their farewells until they met again with their friends and finally streamed out from the walls of the fort with their pack strings and hooked up with Chief Mingan's 'streams of horseflesh and humanity'. Once all groups were together, they planned on heading northwesterly along the Missouri on their trek back to the Indians' historical camping and buffalo hunting grounds.

As the trappers' small group met up with Chief Mingan's people, Iron Hand spotted Sinopa riding in front of the horde of happy to be going home, Medicine Lake Band of Blackfeet. Seeing that, he then spurred his horse and leading a string of four horses captured from the Hudson Bay Company Trappers, rode up to the head of the line of Indians and bid his 'new brother', Chief Mingan, "Good Morning". Then without another word and a smile on his face, he spurred his horse over to little Sinopa riding by her brother, who was now wearing a happy smile over seeing her favorite trapper riding her way leading four spirited riding horses. Riding alongside Sinopa, Iron Hand bid her "Good Morning" and then in a surprise move, handed Sinopa the lead rope to the string of four horses that he had been leading!

"These are for you, Sinopa. They are a gift from us four trappers so you can have a start to your very own horse herd." Sinopa, in utter surprise over receiving such a valuable gift, just smiled at her 'savior from a life of hell' and with that and a tip of his slouch hat, Iron Hand rode back to his group of trappers and took back one of the lead ropes from Big Foot who had been leading two strings of pack animals. Big Foot just grinned a big ole dumb grin at his giant friend, who was

The Adventurous Life of Tom Warren

equally embarrassed over his public gesture of respect, and maybe something more...for little Sinopa. Without a word being spoken between Iron Hand and Big Foot since none were necessary over what had just occurred, they just rode on together in silence...

However, the unique gesture by a giant of a white man trapper, now respectfully named Iron Hand by Chief Mingan's band of Medicine Lake Blackfoot Indians, was not lost on the great chief... Looking over at his very happy little sister, Chief Mingan smiled a knowing smile as well. He now had his much-loved sister back with his people and it was obvious to anyone who looked at her, she was now once again very happy, especially in the presence of her favorite white man trapper...

Once the group of travelers reached the confluence of the Big Muddy River several days later, they turned north. Several more days of travel and two buffalo hunts later to supply food for the large numbers of travelers, the group was at the confluence of the outlet river from Medicine Lake. There the group of Indians and fur trappers split their routes of travel, with the band of Indians heading to their ancestral camping site to the southwest of Medicine Lake. As for the trappers, they continued heading due east towards Medicine Lake and their lakeside grove of aspens and a moraine of rocks sporting a cave called 'home' for the duration of the coming trapping season.

But before the two groups separated, Sinopa and Chief Mingan paused to bid the four men from another culture, "Good Bye". Once again, Chief Mingan bid his 'new brother' Iron Hand a warm "Good Bye", as he did the other three trappers. However, Sinopa rode her horse over to the one Iron Hand was sitting upon, reached across and gave the gentle giant a warm and surprising public hug. She then once again thanked Iron Hand for rescuing her from a fate worse than death. Then almost as if an afterthought, Sinopa stretched over from her saddle one more time and gave Iron Hand another very warm and long hug. None of the other three trappers got any such personal hugs however...

Quietly sitting there resting their horses, the trappers watched their new friends moving off to the southwest in a 'noisy' cloud of dust, barking dogs and happy people. "Best we keep moving, Boys," said Old Potts, ever the 'mother hen' and anxious to be on his way as 'Man of the Mountain'. "We have a fer piece to travel and if we want to get there afore dark, we best keep moving," he continued. With those words of wisdom, the four trappers and their string of heavily loaded horses turned and headed for their familiar aspen grove and home for as long as they wished to remain in that part of the beaver country as Free Trappers.

Right at dusk, the four trappers approached their familiar grove of aspens leading up to their cave-home. As they did, their horses in concert began acting all crazy-like and became somewhat unmanageable. However, a set of spurs to their flanks and good bridle control soon regained some semblance of control over the entire pack string. However, the horses were still acting up all nervous-like and not happy with the approach to their old home site and the corral located nearby.

As the men finally entered the lower portion of their aspen grove, they were met with a very strong and rancid smell hanging heavy in the warm summer air. "Damn," said Iron Hand, "did an old bull buffalo wander into our grove of trees and up and die? Because if an old buffalo did, he sure is stinking up the place."

"Sure smells like it," said Crooked Hand, as he once again was trying to regain control of his very skittish and now 'crow-hopping' horse as they approached the entrance to their home site.

"Damn," said Old Potts, as he tried keeping his normally very mild-mannered riding horse under control. "Any of you boys see that damn old stinking dead thing anywhere? If you do, someone needs to throw a loop around its head and drag it far away from here so we can get back to living in a place that is fit for us humans," bellowed Old Potts.

The Adventurous Life of Tom Warren

"It can only be a dead and rotten buffalo," said Crooked Hand, as he reined in his now almost totally out of control riding horse and pack string just in front of their cave's entrance.

"URRRGH!" ROARED A HUGE GRIZZLY BEAR, AS IT CHARGED OUT FROM THE CAVE IT HAD BEEN LIVING IN EVER SINCE THE TRAPPERS HAD LEFT FOR FORT UNION EARLIER IN THE SUMMER! In a second, the great bear had swarmed all over Crooked Hand's horse and him all in the same motion, mauling, biting and snapping its savage teeth, jaws and six-inch claws every which way onto anything that was soft and moving! When the bear had exploded in its surprised and angry charge from the mouth of the cave, every terrified, madly bucking horse and desperately trying to remain in their saddles, riders, were scattered like loose grouse feathers in a howling prairie wind!

Old Potts was bucked clean off his now 'sun-fishing' horse in an instant, and was promptly deposited back-first onto a pile of boulders left behind by the retreating glaciers! When he was tossed, he landed hard on the rockpile with a savage and body-rendering sounding "THUMP"!

Iron Hand's horse 'exploded' under him with such force, that he was thrown into a nearby aspen and his flying weight and velocity snapped a six-inch-thick aspen tree right off at its base!

As for Big Foot and his last position in the string of trailing horses, he was simply bucked off, fell underneath their flashing hooves and was stomped among the stampeding-the-hell-out-of-there, pack-animals' flying feet!

Recovering from his collision with the aspen, Iron Hand fortunately had been unhorsed with his Hawken rifle in hand. Realizing the danger at hand and quickly staggering back into the fight for survival, Iron Hand charged into the savage fury of a much-surprised and violently angry boar grizzly bear! As he did, his actions were accompanied by the terrified screams of pain, as Crooked Hand was being savagely mauled along with the terrifying sounds his horse was making, as it was being torn asunder into smaller pieces while still living!

Running right up to the grizzly as it was savagely occupied with mauling Crooked Hand and his horse, Iron Hand jammed the end of his rifle barrel into the front shoulder of the bear and touched it off! BOOOM! went the heavy rifle and in an instant after it had been fired, the bear with its paw swiped Iron Hand across his left shoulder so hard, that he flew over backwards and landed in the rocks in their nearby creek with a hard "CRUNCH"! Iron Hand had hit so hard, that for a moment he could not get his wind and all he saw were bright flashes of light running in streams across the backs of his eyes! But he could still hear Crooked Hand screaming in terrible pain and his horse making the sounds of abject terror, as it was being torn to pieces by the now, out of its mind in pain, wounded grizzly bear, as a result of Iron Hand's close at hand, heavy caliber rifle's blast into the shoulder of the bear!

Crawling out from the creek with his left shoulder not wanting to work because of the bear's savage swipe with his massive paw, Iron Hand ran back to the ten-foot-tall grizzly bear now standing on its hind legs and roaring out in pain! Bloody-white foam was now spewing from the bear's open and gaping mouth as a result of his rifle-shot damaged lungs and his eyes appeared to Iron Hand to be almost devil-red in color!

Once again, running on the internal fury he had felt when first attacked by the Blackfoot once he got his iron-strong hands around the Indian's necks, Iron Hand ran right up to the maddened standing beast, thrust his horse pistol into its gaping open mouth hoping it would still fire after being immersed in the creek and pulled the trigger. "POOOFF" went the pistol as it misfired and was instantly knocked from Iron Hand's right hand by the bear's paw! By now, Iron Hand's left shoulder was coming back to life after the powerful swipe from the bear it had taken moments earlier and removing his second pistol from his belt, thrust the barrel of that pistol right into the bear's mouth as he had done the first time with the pistol that had just misfired. Iron Hand felt the bear's jaws quickly snap shut on the pistol's barrel just as he pulled the trigger on his last loaded weapon. BOOOM! went his recently purchased, heavy caliber pistol and with that explosion, the bear instantly dropped like a stone, DIRECTLY ON TOP OF IRON HAND, who had now just fallen backwards flat onto his back!

Smashed flat under the bear's 800 pounds of dead weight, Iron Hand found himself almost unable to move and finding it difficult to even breathe. Especially when breathing in the great stench from the bear's open and bloody foam-spewing mouth just inches from his face! Gagging under the bear's faceful of soured hot breath, bloody foam and now involuntary vomit being emitted up from its foul-smelling stomach, Iron Hand tried desperately to crawl out from under the animal's great weight...

Finally, with a face and now mouth full of the dying bear's involuntary vomiting of its acidic vile stomach contents from a meal of a long dead buffalo, Iron Hand managed to drag himself out from under the great weight of the bear and its increasing involuntary spasms of death. Now vomiting himself from the terror of the close encounter and covered with the bear's own thick oatmeal-like in consistency vomit, Iron Hand vomited the remains from his breakfast all over the front of himself! Still in the emotion of the moment, Iron Hand drew his ten-inch sheath knife and prepared to continue the fight if necessary, vomit, snot and blood-covered or not.

It was then that Iron Hand saw Big Foot, dragging one of his legs behind him, bravely staggering up on the bear with a pistol in hand. Dragging himself right up to the bear's massive still-thrashing head, Big Foot discharged the pistol directly into the animal's cranium from just two feet away! With the blast of that heavy horse pistol ringing throughout the aspen grove and the air filling with the acrid smell of black powder smoke, the bear gave one final involuntary shudder and then lay still forever...

Iron Hand then staggered over with his knife in hand to the huge mound of what was once a monster grizzly bear to make sure all danger from the animal was past. Satisfied the terror of the moment was over, Iron Hand turned and made a survey of the damage done to his fellow trappers. In his case, he had a sprained left shoulder from being swiped by the bear's huge and powerful paw and other than being covered with the bear's and his own stinking vomit, he was still functional and would live to fight another day.

As for Big Foot, a quick look showed that he had a badly sprained knee after being bucked off his horse and landing awkwardly on his leg. That and a number of large black and blue bruises covered most of his body, as a result of the stomping he had taken after falling under his horse's hooves and most of those of the scared all to hell pack string he had been leading.

About then, Old Potts came limping into camp, all bent over with an obvious damaged and sprained back when he had been bucked off his horse onto a pile of boulders. He too was black and blue all across his back and the back of his legs, from the hard impact he had suffered when he had landed in a rockpile after being bucked off. But other than being pissed off over the 'hurrah' the damn bear had caused, in time he would heal as well without any apparent lasting damage.

As for Crooked Hand, his plight was a 'horse of another color'. He had ridden right up to their cave's entrance in order to make offloading all his heavily loaded pack string of horses easier. When he did, the noises of his arrival had suddenly awakened a soundly sleeping grizzly in his day bed in the cave. Coming out from a deep sleep and being surprised as he was, the bear had reacted accordingly as surprised grizzlies are prone to do, namely blowing up and charging what he figured was the source of his danger! In so doing, the bear had charged right into Crooked Hand, his horse and part of the pack string he had been leading. When he did, the bear had swarmed all over Crooked Hand's horse, it being the closest and broadside

The Adventurous Life of Tom Warren

to the charge and then had bitten down on the back of the horse's neck. That set the horse to bucking frantically, in the process tossing Crooked Hand right over the saddle on top of the head of the grizzly as it was biting down on his horse's neck. When that happened, the bear, figuring it was under attack from above when the unfortunate trapper had landed on its head, reacted accordingly and bit down on the nearest thing human, which turned out to be Crooked Hand's previously badly damaged left leg from the year-earlier knife fight with the Blackfeet! Then adding insult to injury, the bear after biting down on Crooked Hand's left leg, had slung him off into the brush with a mighty head swing!

The assessment of the trapper's plight over, and satisfied that everyone would live to trap another day, Iron Hand began tending to the wounded, starting with 'bad luck' Crooked Hand. Under Big Foot's tutelage and knowledge of frontier injury remedies, Iron Hand set to work. Removing Crooked Hand=s bloody and torn buckskins, Iron Hand saw a large gash on the previously damaged, knife-wounded thigh from their earlier fight with the Blackfeet Indians. "Well, Crooked Hand, it appears your bad luck is still holding when it comes to tearing your bad leg all to hell," said Iron Hand with a smile in his heart, realizing that his good friend was not badly hurt.

"Iron Hand, find the packhorse that is carrying all of our jugs of rum, grab one from the pack and bring it here," said Big Foot, who was sitting on a log by their old firepit nursing his sprained knee. Iron Hand finally located the pack string carrying the rum among many other items, and brought them into their horse corral for safe keeping. Removing a jug of rum from its pack, Iron Hand took a swig to calm down his still emotion-filled high and then walked back to their firepit in front of a still bear-stinking cave. There as instructed by Big Foot, Iron Hand opened up Crooked Hand's thigh wound with his fingers, gouged out the now coagulated lumps of blood, daubed the wound dry with a rag and then after a warning, poured some of the 100-proof rum into the open bite wound!

"YEOOOW!" yelled Crooked Hand, as he almost stood up off his sitting log from pain in his open wound in his leg caused by the burning treatment of the high proof rum. "This is for your own darn good, Crooked Hand," said Iron Hand with a grin. "Next time you get into a scrape, stick your good leg in the way instead of this damned old left one that is 'gowed' all to hell," he continued with a 'grin' in the tone and tenor of his sympathetic voice.

Then it was time to check out Big Foot's injury, which turned out to just be a badly sprained knee, along with one hell of a mess of bruises. Iron Hand, realizing there was nothing he could do for his friend's bumps and bruises, just took a tin cup from out of their bear-stinking cave, filled it to the rim with strong run and dispensed that as a treatment for Big Foot's sore knee and ugly black and blue body-looks, which he was all too happy to accept...

Then it was Old Potts's turn in the 'frontier doctor's office', as Iron Hand carefully seated him upon another sitting log around their old firepit. Removing Old Potts's buckskin shirt, Iron Hand was greeted with a back that was entirely blue-black in color from all the contusions suffered by the old man when he landed on his back onto a mess of unforgiving boulders after being bucked off his horse when the grizzly charged out from the cave's entrance... Old Potts also got a tin cup filled to the brim of the high proof rum to help him withstand the pain in his back from 'meeting' a mess of boulders at a high rate of speed!

Not to be outdone, Iron Hand helped himself to a full cup of rum as well, before he built a fire for the injured men to sit around, as he began cleaning out all the rank and stinking dead animal refuse and dung the bear had accumulated in their cave during the summer's absence by the trappers. Then in the light of the fire, Iron Hand brought in the packhorses and unloaded all their heavy packs. Following that, bad shoulder and all, he distributed the valuable supplies into various parts of their cave for safer keeping. Lastly, all the animals were watered and put into the corrals without being allowed to graze that evening because of the lateness of the hour. That was except for Iron Hand's horse. He was brought back into camp and used to drag off Crooked Hand's dead horse which Iron Hand had cut into halves to make for the dragging off by his horse easier. Then it came the dead bear's turn. There Iron Hand removed the best cuts of meat from a very fat animal, staked them over an open fire and while those choice cuts of bear meat cooked away, Iron Hand, with a great deal of difficulty from his skittish horse because of the bear's smell and its fear of the animal being towed behind it, dragged off the remains and left it with the parts of the dead horse for the scavengers to enjoy.

Iron Hand then laid out each man's sleeping furs where they had normally slept in their earlier days of occupation in the cave before the bear had taken over. When the tired and sore men went to bed that evening, Iron Hand burned up the pile of offal he had collected that the grizzly bear had accumulated in their cave. It was only then that Iron Hand sat down around the camp's fire, poured himself another full cup of rum and quietly enjoyed the sounds of night and the friendly crackling sounds of the fire. He did so while off in the distance, he could hear all the 'eating disagreements' taking place by a pack of wolves over by the grizzly bear and dead horse dump site out on the prairie a short distance away from their camp...

The following morning, Iron Hand was up early and had their familiar trapper's coffee boiling away over an open fire, as the remaining bear steaks from the evening before were merrily fat-sizzling away at the edge of the fire's coals. Moments later, his three stove-up friends stumbled and hobbled out from their cave home, sat down on the sitting logs by the fire and had Iron Hand hand each of them a steaming cup of his brand of strongly brewed coffee. Each sore and stove-up man almost simultaneously took a sip of Iron Hand's style of coffee and then got big grins on their whiskered faces. Big grins because they all had discovered the coffee fixed for their breakfast that morning had been laced with a generous amount of rum to help patch up what ailed each man...

After breakfast, Iron Hand, at Big Foot's direction, once again cleaned out Crooked Hand's thigh injury over lots of howling, then heavily laced it with a generous amount of rum and finally wrapped it up once again. Then Iron Hand hobbled all their livestock and turned them out onto the grassy plains next to their camp so they could put on the 'feed bag'. Finally, it was back to camp, all the while keeping an eye on the trappers' horse herd, while he rearranged the packs and their gear in the cave for safer keeping and daily use as they saw fit. Iron Hand also reloaded each man's pistols and rifles in case trouble came a-brewing and the men needed to protect themselves or their valuable horse herd from those who had a case of 'light fingers'...

Come the early afternoon, Iron Hand brought in their livestock, placing them into their corral after they had a chance to tank up on the clear and cold spring water running through their aspen grove. Following that, after checking the men who had all gone back to sleep in their cave to help in the healing-up process, Iron Hand saddled up his horse and a packhorse and left camp. One shot from his .50 caliber Hawken and two hours later, Iron Hand returned back to camp with a packhorse carrying a heavy load of fresh buffalo meat. Unloading the hindquarters, hump ribs and backstraps, Iron Hand hung some of the meat in a close at hand aspen and set aside the rest for their meals to come.

As it turned out, it was a good thing that he had killed and brought back to camp such a huge amount of camp meat. When the men emerged from the cave in the afternoon on the second day after their unfortunate run in with the sour-mannered grizzly, they were as hungry as the mean-assed critter they had killed the day before! With boiling coffee, staked buffalo meat and Dutch oven biscuits by the score, the men's appetites were finally satiated.

The Adventurous Life of Tom Warren

For the next two weeks, that became Iron Hand's daily routine. That was, he found himself tending to the livestock, hauling in drinking water, cooking, killing the occasional buffalo, bringing that meat back to camp, and making coffee and biscuits for his fellow trappers, as they slowly healed up and became more and more mobile with each passing day. At the end of that two-week period, Iron Hand's sore left shoulder was a distant memory, Old Potts was up and moving around almost like he used to, Big Foot's bad knee was healing up well, and Crooked Hand's deep laceration in his thigh was responding well to the daily cleaning and applications of high proof rum used as a disinfectant. Additionally, Iron Hand had been breaking in one of their packhorses to ride as Crooked Hand's replacement horse for the one the grizzly had taken out in the surprise attack the day of their arrival back at their aspen grove camp.

Things were looking up and it was almost time to begin checking the beaver trapping areas to ascertain what kind of annual production the animals had managed, the overall health of the populations and making a long trip easterly to Lake River, to ascertain the potential trapping waters along the main feeder source for Medicine Lake.

"Today we get off our dead hind ends and hit the beaver trail!" bellowed out Old Potts, as he emerged from his sleeping furs in the cave early one morning. Then he hobbled over to the ever-boiling pot of coffee hanging over the coals of the campfire, poured himself a steaming cup and then thankfully sat down upon a nearby sitting log to rest his weak and still recovering lower back. Then spying several previously cooked buffalo steaks lying on a flat rock near the firepit keeping them warm on a 'fire-rock', speared one with his sheath knife and began making a big steak into a little one...

Iron Hand, with a grin over Old Potts's healing-up behavior, staked several more buffalo steaks on green willow limbs over the fire and within moments, they were sputtering out rivulets of liquefied fat as they began cooking. Moments later, Big Foot and Crooked Hand emerged from within the cave's confines, as the smells of fresh coffee and cooking meat graced the morning's air, their nostrils, and 'made a promise' to their stomachs of better things to come.

"Damn, if that don't smell like a little bit of Heaven," said Crooked Hand, as he poured his tin cup full of the strong as an 'angry mule's kick' trapper's coffee.

"Yeah, like you will ever know what Heaven looks or smells like," said his close friend Big Foot, with his always infectious, heavily whiskered grin.

Iron Hand turned the staked buffalo slab of meat cooking merrily away and just had to grin once again. With all the good-natured bantering going on around their firepit that morning, Iron Hand came to realize that his partners were finally and happily on the downhill side of the mend from all of their 'grizzly bear-induced' injuries.

Forty minutes later, the men had finished with their morning meal and Iron Hand had two packhorses fully loaded with 40 beaver traps in their panniers. That way, if any beaver had been trapped by late in the day, their fresh skins could be hauled back to camp in the now empty panniers, fleshed out and hooped for drying. Besides, today would be his day to shine. Old Potts had talked with Iron Hand the evening before and had instructed him once again in the fine art of beaver trapping.

With that latest lesson under his belt and what he had learned the year previously watching the old man when it came to setting and tending his traps, Iron Hand figured it was his time now to become the group's main trapper. Especially in light of Old Potts's still bad back, Crooked Hand's bad leg and Big Foot's seemingly forever tender knee. Physical weaknesses that Iron Hand had decided did not lend itself well to one carrying around such limitations, all the while walking around in the sticky muddy bottoms and soon to be icy waters of beaver ponds while tending the traps.

The Adventurous Life of Tom Warren

Leaving camp, the trappers headed for their original spot where the first trap had been set the year before and for good luck, would be the same area chosen for the first trap set of their second beaver trapping season. Reining up his horse by the lake's marshes some time later, Iron Hand stepped heavily from the saddle due to his large frame and walked back to the first horse in their pack string. Removing a beaver trap from the pannier and under the 'close eye' and supervision of Old Potts who remained in his saddle, Iron Hand made the first beaver set of their second trapping season. When finished, the look he got from Old Potts was one of confirmation of a job well done.

For the next six hours, Iron Hand, with a few additional suggestions thrown in by Old Potts along the way, set their remaining 39 beaver traps in the marshes, along beaver dams and next to the evidence of heavily used slides. As he did, even though promised that a truce now existed between the trappers and Chief Mingan in their trapping area, Old Potts, Big Foot and Crooked Hand kept a sharp watch out for any signs of danger. After all, they still were close to the dreaded and fierce, trapper-killing Gros Ventre Indians' home range, as well as the hated Hudson Bay fur trapping country located just to their north and west. All four trappers had learned early on that the way of the west, if one wanted to live long, was to keep a sharp eye peeled, a quick ear tuned and a loaded rifle always at the ready.

Finished with their trap setting at the end of their trap line, the men scouted out some new areas in which to trap beaver when they had trapped them out in their present location. A short time later, Crooked Hand had killed a cow buffalo and as the men butchered the freshly killed animal, they all feasted on their personal choices of raw meat. Finished with butchering and just taking the choicest cuts of meat, the men then headed back to their previously set trap line to see if they had any critters in their traps.

In so doing, the men discovered 17 recently trapped dead of the very territorial beaver that had already been caught in their 'castoreum-scented' trap sets. As Iron Hand removed the beaver and reset his traps, Big Foot, who was their best and fastest skinner, saw to it the carcasses were skinned on the spot and their fresh skins or *plus* (pronounced "plews") dropped into a pannier carried by a packhorse for later fleshing out and hooping back at camp. Finished, the men, still keeping a sharp eye peeled for any hostile Indians, headed for their camp and the very necessary fur-processing duties that were to come.

Upon their return to camp, the rest of their stock still in the corral were double hobbled and turned out to feed, while the men fleshed out their recent catches, hooped the same with small willow branches and set them out around their cabin and up on their roof to dry. As they did, Old Potts set out the coffee pot to boil and soon had their supper staked out around the campfire with some of the meat from the cow buffalo the men had killed and butchered while out trapping beaver during the morning. Finished with fleshing and hooping the beaver skins, the men retired to their supper and then brought in their horse herd, putting them back into their corral for safer keeping. Then the trappers adjourned to their sitting logs around their campfire, smoked their pipes and relaxed from their day's long and hard labors.

Thus began their routine labors for the month and the next and the next, as the stacks of beaver *plus*, muskrat and river otter skins grew in the musty-smelling safety of the back of their cave. The bottom line, the four successful and already wealthy Free Trappers from their last season were becoming even richer and were trapping and hunting in an area that was now safe for them just as long as the truce held between the four men and the mighty Blackfoot Chief, Mingan.

One morning as the men prepared to leave and run their trap line, Iron Hand and the others noticed the late fall weather clouds looming to the northwest were such that they promised foul weather before the day was done. Bracing for the same, the men packed along their heavier winter *capotes* on the packhorses and set out to make their daily run on the trap line. That was when their day somewhat later turned as foul as the coming weather had been so predicted that morning...

Iron Hand, now the official trapper of the bunch, seeing his trap was missing as he bailed off his horse, waded out to see if he could find the dead beaver drowned in the deeper water at the end of that trap's chain. THAT WAS, IF THERE WAS STILL A BEAVER TRAP IN THE AREA! Iron Hand quickly noticed that the trap was missing and there was no dead beaver still attached that had drowned in the deeper water! After much looking around for the expensive beaver trap ($9 each back at Fort Union!), Iron Hand concluded that the entire set, chain and all, was mysteriously missing...

Stepping off his horse after Iron Hand had made known the situation, Old Potts, suspicious as always, knelt down and closely examined the ground around the beaver slide where the trap had been previously set. There were tracks there alright, but they were not those from a beaver but moccasin footprints from that of a white man! Standing back up, Old Potts's keen eye discovered that they were now not alone in beaver country. Off to one side of where the trap had been set were the tracks of six horses! Walking over and kneeling down, Old Potts discovered the horse tracks were from shod horses! More than likely horses being ridden by the trap and beaver-robbing, white men thieves. Looking eastward along their trap line, Old Potts took a bearing on the mystery horse tracks' direction of travel and said, "Mount up, Boys. We have company who have discovered our trap line, stolen not only our beaver but our expensive and hard to replace trap as well. Let's ride because the way I figure it, this is not going to be the first trap or dead beaver of ours that they have stolen this day."

Hoping against hope, the men rode up to the next four of their beaver sets, only to discover they too were missing as well! Now with that deadly information hanging over them, 'there was going to be war in camp'! Old Potts said, "Boys, our trap robbers have not only discovered our line of traps but have taken every one of them they have discovered as well. Let us see if we can set a course and head them off before they steal us blind. We need to get rolling, because we have a winter storm approaching that will hit us in all of its fury by this afternoon. If we want to catch these guys, we need to hurry afore the snow comes and makes tracking these bastards all but impossible."

Removing their heavier winter clothing from the panniers and staking out their two pack animals in a nearby grove of trees out of sight so they could make better time, the four men then rode 'hellbent for leather' along the suspect trap thieves' shod horse track-line before the afternoon snows made tracking the robbers almost impossible and catching them doubtful. Around four in the afternoon, their 'weather' luck that had held until then, ran out. The snows predicted earlier that morning now hit the four men and the countryside with a typical northern prairie winter's vengeance. Soon those snows were heavily swirling around the men, who were now walking their horses instead of riding them, so they could better follow the quickly disappearing tracks of their 'trap robbers'. However, the trap-robbing thieves had not figured pursuit was so close at hand and were still walking their horses and now heavily loaded pack animals in a northeasterly direction. That was their second mistake, their first being the stealing of the traps and dead beaver belonging to Old Potts and his crew... Once again to the men's collective thinking, stealing a beaver or two was bad enough. But stealing a man's livelihood out on the frontier, namely his traps or horses, well, to their collective way of thinking that was a killing offense...

Come dusk, the four trappers were still hot on the fast-vanishing trail of their trap line thieves, now reasoned out to being some men from the hated Hudson Bay Fur Company because of the direction in which they were heading! But Old Potts still had a trick or two up his sleeve. He knew of an old but well-used isolated grove of dense coniferous timber out on the prairie that was routinely used by white men and Indian alike. A grove of trees that were used by Indians and fur trappers alike for the protection it offered from the elements during the sudden occasion of bad weather events out on the Northern Plains. He had remembered the same from his earlier trapping experiences in the area during his first expedition into the same area trapping beaver with Vasquez. Like a long-tailed weasel hot on the trail of a cottontail rabbit, Old Potts was heading directly for that area in the heavy timber, now that the dark of the night and swirling snows made direct tracking of their trap robbers all but impossible.

Several hours later found the four trappers resting their horses, as they looked into the area of dark timber Old Potts figured his thieves might hole up within during the bad weather. Sure as buffalo made some of the best eating going, a faint flicker of light from a small campfire was observed in the darkness of the heavy timber, now that the heavy and swirling snows had somewhat abated.

Dismounting and staking their horses in the deep timber so as they approached the suspect camp of beaver and trap thieves, their horses would not scent those horses of the 'now hunted' and give out a betraying warning whinny. With that, the four trappers clutching their rifles began their stalk in the sound dampening layer of fresh snow towards the suspect trap thieves' campsite.

Thirty minutes of stalking, found the four men standing silently just outside the light of the suspect trap-robbing thieves' campfire. Nearby, were staked six still unsuspecting horses and around the campfire were observed four men beginning to cook their supper. Two of the men were white fur trappers speaking with heavy French-Canadian accents and according to Old Potts's whispers, the other two men were more than likely their partners, two dreaded Gros Ventre Indians! Gros Ventre Indians who were known across the Northern Plains for their savagery when it came to mistreating any captured trappers or any white man in general.

Off to one side of the thieves' staked horses was a pile of beaver traps that Iron Hand figured were those of their group's recently stolen traps. Nearby sat two panniers stuffed clear full of fresh beaver skins, which Iron Hand suspected were the day's trappings removed from their traps that had been stolen from them as well. Carefully moving in even closer, Old Potts's group could finally make out what the men were saying. When they did, everything began clearing up as to what had occurred, as they listened in on what the four suspect outlaw trappers had to say about their day stealing someone else's trapped beaver and traps in the Medicine Lake area.

As their conversations confirmed, the four suspect men around the campfire were trappers from the much-hated Hudson Bay Fur Company. They were Company Trappers who had been running their new trap line down from their normal trapping territory located further to the north, southward toward Medicine Lake. After setting their traps in their previous northern trapping location, the four men had scouted further south toward Medicine Lake looking for new trappings, once they had trapped out their old trapping grounds. In so doing, they had discovered the end of Iron Hand's set of beaver traps upon locating a floating and dead beaver carcass on the end of a trap's chain. It was then that the four outlaw trappers realized they had stumbled across another trapper's trap line other than one of those from the Hudson Bay Fur Company. It then was just a simple matter of continuing into another trapper's area, locating each of his sets by following Old Potts's group's horses' tracks, the floating dead beaver carcasses, removing the animal, and stealing the traps to frustrate and remove that trapper and his competition from the area by wrecking his livelihood.

Upon hearing those incriminating words coming from the four suspects around their campfire, Old Potts saw red! Motioning around his three men with his right hand held low so his movement would not attract attention, he quietly set out a battle plan to take back what was rightfully their property.

The Adventurous Life of Tom Warren

Then following his plan, set the four Hudson Bay men afoot in Indian country and see how they liked having their livelihood challenged. With the plan made, Old Potts held his ground as his three cohorts disappeared quietly into the snowy darkness of the coniferous forest to carry out their part of the battle plan.

After waiting a few long moments for his three fellow trappers to get into their positions, Old Potts made his grand entry. Quietly moving into the firelight, he figured he would get the drop on the four men, take their gear and then send them packing in the dead of night, so they could warn the other Hudson Bay men they would eventually run into to steer clear of the beaver trapping areas in the Medicine Lake area.

"Evening," said Old Potts, as he moved deeper into the direct light of his trap-thieving suspects' campfire! For a second, none of the four men moved as they stared in disbelief at the unknown and unanticipated fur trapper standing there just yards away with his rifle in hand and held at the ready! Then the trap-stealing outlaws' camp 'exploded' in four scrambling men making moves for their weaponry! A huge Gros Ventre Indian stood up quickly and in one fluid motion, drew and threw his tomahawk in a split second directly at Old Potts. Ducking the man's high throw, Old Potts promptly shot him in the face with his rifle! The explosive spew of blood and brains from the Indian's head after being hit with a .50 caliber lead ball from Old Potts's Hawken sent the other trap-stealing outlaws into extreme motion. One of the French-Canadian fur trappers took off running right at Old Potts! Being an experienced Mountain Man himself, he realized that Old Potts had fired his one and only rifle shot. Now that his rifle was empty, to his way of thinking, now was the time to physically attack an unarmed man! In his hand the Frenchman held a long sheath knife that he had been using cutting up some fresh buffalo meat! But now he figured its keen blade was needed to take care of Old Potts...

However, before Old Potts could react and draw one of his pistols in self-defense, the side of the charging Frenchman's face exploded into a million bits of bone, blood and brain tissue, as Crooked Hand drilled him dead center in his left ear with a shot from his Hawken! That charging Frenchman's head and falling body splattered into the white snow, turning it into a crimson red splotch!

Simultaneously, the remaining Gros Ventre Indian back at the fire ducked his head and took off running into the darkened forest cover before anyone could blink! However, that man ran directly into Iron Hand and in so doing, instantly and viciously swung his tomahawk, knocking the rifle from his hands before he could raise it and shoot! That got the Gros Ventre a broken neck snapped by two very strong hands of 'iron' before the hard-charging Indian could use his tomahawk once again against the giant fur trapper!

The remaining Frenchman jumped to his feet as the black powder smoke filled the air around the campfire, drew one of his pistols, pointed it towards Old Potts and had his face exploded into flying snot, blood and bone fragments by Big Foot's close at hand shot with his Hawken! In fact, Big Foot was so close to the scrambling-for-his-life Frenchman, that when he shot him, his heavy slug tore clear through the man's face, exited the far side of his head and killed one of the Hudson Bay men's pack animals looking on at the fast and furious action as it stood quietly in a makeshift rope and stick corral!

Later that night, Big Foot and Iron Hand backtracked themselves and brought their previously staked-out riding horses into the Hudson Bay camp for safer keeping. They then dragged off the dead men's bodies, along with the four fresh Indian scalps surprisingly discovered on the two Gros Ventre Indians, a distance from camp and left them for the critters. Later that evening, the four Medicine Lake trappers feasted on the buffalo steak supper that had been made previously for the deceased Hudson Bay men, and then they slept in their sleeping furs until dawn the next morning. However, not much sleep was had by the four men because a pack of wolves,

The Adventurous Life of Tom Warren

smelling all the dead men's blood, had noisily feasted and fought over the newly discovered 'morsels' throughout most of the night. That plus Old Potts's group had a hard time sleeping in the dead men's sleeping furs. Smelling badly was a common thing when living out on the frontier. However, the smell being emitted from the dead men's sleeping furs reeked of 'polecat', making sleep problematic for the four victorious 'Old Potts's trappers'.

The next morning, Iron Hand and Old Potts butchered out the horse that Big Foot had killed inadvertently when his rifle ball had blasted clear through the head of the last Frenchman trying to kill Old Potts and into one of the Hudson Bay men's horses staked nearby. Removing a mess of steaks from the horse's hindquarter, they were staked out over the fire and later consumed for breakfast. Then the men went through the Hudson Bay men's belongings, took what they could use, gathered up their riding and packhorses, burned the remaining camp's gear, and left the area. However, as the men searched through the dead men's belongings, the discovery of the four fresh Indian scalps discovered on the bodies of the Gros Ventre kept running through their minds... Little did any of the men realize that mystery of the fresh scalps would soon be resolved, almost resulting in mistaken bloodshed between the four trappers and some of their nearby Medicine Lake Blackfoot Indians.

When the four trappers left the once deadly scene of the previous evening's battle, they were leading an additional five horses, carrying four extra rifles and five pistols, not to mention 21 beaver skins the Hudson Bay men had removed from Old Potts's trap line from the day before. On their way back into their previous trapping grounds, the men stopped and retrieved their two packhorses they had left staked out in a grove of aspens so they could travel lighter and faster after their trap robbers. Then back in their old beaver trapping area, Iron Hand once again set all of their traps in likely looking

110

places and when they arrived at the location for the last trap-set, the four trappers got an unexpected surprise!

Rounding a low ridge line, the four men were unexpectedly and quickly surrounded by 20 fiercely painted and mounted, hostile-appearing Blackfoot warriors! Stopping and raising his arm in the universal sign of peace, the trappers were still mobbed by the warriors with a lot of aggressive horse-bumping, curdling whoops and threatening gestures! Old Potts, who fluently spoke the Blackfoot language, was finally able to get everyone calmed down, identify his group of trappers as friends to Chief Mingan and in a subsequent discussion with the war party's leader, discovered what the issue was regarding the Indians' openly aggressive behavior and reason for the mistaken identity.

It seemed that the day before, four of their tribe's younger men had been out buffalo hunting nearby and had killed several of the beasts. Then from all indications of sign left behind at the kill site, four men on shod horses had ambushed and killed the young Indian buffalo hunters! When the young men did not return home that evening, several family members went looking for the hunters and had discovered their scalped and mutilated bodies, just as four unknown and suspect 'trappers' were sighted riding off to the northeast. The Indians' main camp was alerted to the killing and a war party assembled and sent looking for the four trappers suspected of killing the young Indian men on their first buffalo hunt.

As it turned out, the war party had subsequently a day later run into Old Potts and his group of trappers returning from the Hudson Bay men's battle, thinking they were the killers of the four young Indian buffalo hunters. By the time it was all sorted out, Old Potts had figured out what had really happened as well. He suspected that the four Hudson Bay trappers were the culprits who had killed the young Indian buffalo hunters. That was because that group of trappers had been in that same area the day before stealing all of Old Potts's and his men's dead beaver and traps! The one and same men who had been

The Adventurous Life of Tom Warren

cooking fresh buffalo steaks when they were surprised and killed by Old Potts and his group of aggrieved trappers, after they had been tracked down to their campsite in the heavy timber the night before.

Talking fast to avoid any further confusion and possible deadly misidentification consequences, Old Potts explained to the leader of the war party what he suspected had happened. He then identified Iron Hand, Brother to Chief Mingan, to the group for further validation of the truthfulness of what he had just said. Immediately, the young leader of the war party, one who had not been at the 'arrival celebration' at Fort Union earlier in the year, did however recognize the significance of Iron Hand's name and his new 'relationship' to Chief Mingan. Instantly the tone and tenor of the war party changed to one of respect and admiration for the giant of a man quietly sitting on his horse in their presence.

Then Old Potts advised the young war party leader that the four evil Hudson Bay trappers had been killed by Iron Hand's people the night before and their bodies left for the wolves! With those words, the Indian leader's eyes quickly revealed a new respect for the trappers in his midst who had done such a great deed. Old Potts, realizing the value the Blackfoot placed on possessing good horseflesh, looked over his shoulder and said, "Big Foot, bring those horses we took from the Hudson Bay men forward."

When Old Potts spoke those words, all the eyes of those in the war party, now swung onto Big Foot as he pushed his horse through the tight throng of men, trailing five horses to Old Potts's side.

"These horses all belonged to the evil trappers who took the lives of your four young men yesterday. Since we killed those evil trappers, they no longer have any use for good horses. Therefore, we are giving them to you to bring back to Chief Mingan as a gesture of our good will and for distribution to his people as he sees fit. Additionally, we are also giving to you the four rifles and five pistols the bad trappers were

112

carrying and probably used to kill your young men. Again, since dead men have no use for good firearms, we are giving them to our Indian brothers and Iron Hand's Brother, Chief Mingan, for use and distribution as he also sees fit. Lastly, we are returning the scalps taken by the Gros Ventre Hudson Bay men from your young hunters who were killed, so they can be returned to their bodies so they will not have to forever walk in the "Happy Hunting Grounds" looking for the rest of their 'beings' before they can rest."

With those words, all the anger and fight went out from the group of Indians assembled and the trappers could now see renewed respect for not only them but their relationship with their powerful chief as well.

Then the strangest damn thing happened, which took all four of the trappers by surprise. The young leader of the war party, a man in his early twenties and one who was very well-built, named "Spotted Eagle", cousin to Chief Mingan, said to Old Potts, "Is that big man who sits on his horse like a tree, the one my tribe and chief holds in such high esteem for the big power he holds in his hands?"

Old Potts just grinned, remembering what Iron Hand had done to the Blackfoot attackers of their camp with his bare hands. He then said, "Yes, he is the one who can kill with just his bare hands if he is unjustly attacked." When he spoke those words, he could see the normally quiet Iron Hand sitting on his horse, now appearing to be very uncomfortable since he had now become the unwanted center of attention.

"Show me," said the young Indian leader pointedly and in so doing, obviously trying to impress the other young warriors in his war party who were following him...

"How do you mean?" asked Old Potts, now concerned over the buildup he had given Iron Hand, and with what could possibly happen when an Indian tried to show off and was embarrassed in the process, especially in front of his friends.

"I will show," said the young and very muscular young warrior, as he kicked his animal in the flanks over to that of the

The Adventurous Life of Tom Warren

side of Iron Hand's horse. Sitting there on his horse, looking Iron Hand directly into his eyes as if trying to intimidate him, he thrust out his right hand. When he did, the young Indian leader kept his hand out as if expecting something from Iron Hand.

Then Old Potts finally realized what the young Indian warrior was trying to do with his gesture. He wanted to test Iron Hand's strength and in so doing, impress the young warriors following him, especially if he could hold his own with a man who had become a legend among his band. "Careful there, Iron Hand. Don't make him look too bad or we may have our own little war right here and now if you piss him off and embarrass him all to hell," warned Old Potts with a cautionary tone to his voice.

"What do you want me to do?" asked Iron Hand, with a now concerned look on his face as well.

"He wants to shake your hand like all white men do when greeting one another. Just do it but be aware, he may squeeze the hell out of your hand showing off to all the other young warriors looking on, to show how strong he is," said a now very concerned Old Potts.

Iron Hand did as he was told and was immediately surprised over the power the young Indian leader had in his hand, as he squeezed down, HARD! "Now what do I do?" asked Iron Hand still holding the young Indian leader's hand, as he looked over to Old Potts for some guidance on what to do next in a ticklish situation not of his making.

"Squeeze his hand so he knows he has been bested by a better man than he, but for God's sake, don't injure him," said Old Potts with a bit of worry now creeping into his voice.

With those words of direction, Iron Hand produced a big ole bearded grin over at the young Indian man trying to impress his followers, and then CLAMPED DOWN HARD ON THE MAN'S HAND, AS HE KEPT GRINNING A DEVIOUS GRIN AT THE MAN TRYING IRON HAND 'ON FOR SIZE'!

What happened next surprised even Old Potts and the rest of the onlooking assembled group of trappers and Indians. Iron Hand clamped down so hard on the young Indian warrior's hand that he literally exploded off his horse and onto the ground, all the while loudly yelling over the pain Iron Hand was inflicting on the young man's hand!

With that surprising reaction from the young man in obvious extreme pain, Iron Hand quickly let go but kept on grinning at the warrior like nothing out of the ordinary had just happened. With that, the young Indian's hand hurt so much, that he was hopping around and limping in pain as if it was his foot that hurt instead of his right hand and did so, all in the same motion of movement...

When that happened, there was a loud "OOOOOHHH" emitted from all the rest of the now amazed and impressed war party over what they had just witnessed happening to their very strong in and of himself, leader. Finally, the young warrior chief got control over his hand's pain, leaped back onto his horse and tried looking like everything was alright. However, it did not take the assembled groups of Indians and trappers long to realize that the young Indian warrior leader was holding his horse's reins with his off hand and not the one that Iron Hand had just about crushed...

Then as if to dampen the young warrior's embarrassment over what had just happened, Old Potts handed the Hudson Bay men's horse string lead rope to the young leader as their weapons were distributed to the rest of the men still looking on in amazement over what had just happened to their young leader. A young leader who in times past had physically bested every one of them who were following him that morning!

With the basic issue of pursuit of the guilty parties over since they had already been killed, the two groups separated and went their ways in friendship. As they did, it was obvious that the young leader of the war party was still holding his horse's reins with his off hand.

The Adventurous Life of Tom Warren

Later back at their camp, the four trappers double hobbled all of their horse string and let them out onto the prairie to graze. Since they had carried a portion of the horse meat from the Hudson Bay camp back to their home site after a long day, a fire was quickly constructed and soon coffee was boiling and the horse meat was cooking away. After supper and having brought their horses back to their corral for safer keeping, the men spent a number of hours into the night fleshing out beaver by the light of their fire and hooping those skins retrieved from the Hudson Bay men before retiring.

For the next month until just before freeze up, the men trapped beaver from daylight until dark regardless of what the weather brought to the Northern Plains. Upon freeze up, in came the beaver traps and out went the wolf traps placed around several freshly killed buffalo carcasses being used as bait. Soon a substantial number of thick and beautifully furred wolf skins were being processed and later stored in their cave along with the rest of the men's furs. By now, deep winter on the Northern Plains had set in with a vengeance, and for many days the men were now confined to their camp because of the extremely cold and blizzard-like conditions howling across the Northern Plains.

One morning, Iron Hand went forth into their grove of trees to procure some firewood for their cooking fire so he could make coffee and breakfast. Walking back to their camp, his 'sixth sense' kicked in, letting him know something was 'blowing' in the wind. With those inner feelings 'stoking his fire', Iron Hand brought forth his Hawken from the sleeping area in the cave and laid it next to where he was fixing breakfast by the campfire, just in case. Looking upward, Iron Hand could see that the day was going to be extremely cold and the sky was a deep blue, as one only finds on the great Northern Plains in the quiet and cold of the dead of winter. However, the sun was 'smiling' weakly and it would be another good day to go out and check their wolf traps. As Iron Hand moved their coffee pot off the hanging irons from directly over the fire and set it off to one side to cool on the designated 'cooling rock', his 'sixth sense' inside him told him to 'look up'.

When he did, Iron Hand spotted a lone Indian dressed in his heavy winter clothing, sitting on his horse not 30 yards from their camp, quietly watching him! Glancing over making sure exactly where his rifle was leaning against a sitting log, Iron Hand looked back up and when he did, THE INDIAN THAT HAD BEEN QUIETLY SITTING THERE ON HIS HORSE WATCHING HIM, HAD DISAPPEARED! Standing up and looking all around sensing danger was when Iron Hand realized the Indian was now working his horse slowly through their grove of aspens and coming his way. Walking over to the sitting log next to the fire, Iron Hand sat down alongside his Hawken and just watched his mystery Indian visitor, as he continued slowly walking his horse through their leafless aspen grove towards their campfire like a man on a mission.

Finally, the mystery Indian walked his horse right into their camp and then just sat there in the saddle looking intently down at Iron Hand. Fortunately for the moment in time, Old Potts had been teaching Iron Hand the ways, culture and language of the Blackfoot Nation for the last six months, so that he now felt comfortable conversing in their language. Without alerting his fellow trappers still sleeping in their cave, Iron Hand welcomed the Indian to the camp of the trappers in the Blackfoot language and told him to light down and warm up by the fire. For the longest time, the Indian remained seated in his saddle like he didn't understand what had just been said and continued looking down at Iron Hand in the strangest of ways.

Then Iron Hand suddenly recognized the young Indian man sitting on the horse. He was the one whose hand he had almost crushed in a handshaking event to see who was the toughest of the lot among them, back earlier in the fall when his group of trappers had been intercepted by a number of 'on

The Adventurous Life of Tom Warren

the hunt' Blackfoot warriors. Warriors who were hunting a number of unknown fur trappers in the area, who had recently killed four young Indian men from their tribe on their first buffalo hunt nearby. Warriors who were too late in their mission of revenge because Old Potts and his band of trappers had already tracked down those killers, who were also the ones stealing their beaver traps and had killed the lot...

By now, Iron Hand could hear the coffee still slow boiling merrily away in their old pot as it sat next to him on a flat rock, just removed from the fire irons and now in the process of cooling. Standing up and moving away from his rifle as a gesture of friendship and good will, Iron Hand took a rag, removed the pot from the cooling rock and poured a steaming cup full of the 'Devil's brew' and held it out towards the young Indian man. Iron Hand held out that hot cup of coffee to a man in appreciation for one who had to be damned-near frozen after making what had to have been a long ride on a horse in the dead of winter over to the fur trappers' camp.

With that gesture of friendship, the Indian lightly leapt from his horse, walked over and thankfully took the cup of hot coffee, smelled it and then took a careful sip of the still hot brew. Iron Hand then poured himself a cup and sat down on the sitting log by the fire, next to his rifle, and beckoned for the young Indian man to do the same. Without any hesitation, the young man sat down by the fire, obviously cold from his long trip a-horse from somewhere distant, and then just looked long and hard into the fire without talking and obviously deep into his thoughts on what he had to say to the giant fur trapper.

Remembering what Old Potts had taught him about the Blackfoot culture, Iron Hand just quietly waited for the young man to speak his piece, as he quietly took a sip from his own coffee cup. Long moments later, it soon became evident as to why the visit from the young Blackfoot warrior into the camp of the white men fur trappers had been made.

Turning and looking intently right at Iron Hand, the young Blackfoot warrior, one who now introduced himself formally

118

as "Spotted Eagle", began speaking (Author's Note: A 'spotted eagle' is an immature golden eagle--one whose tail feathers are highly prized by most Native Americans, tail feathers that many times illegally sell for up to $40 apiece). Once he began, Iron Hand came to quickly realize that the young man had rehearsed what he needed to say in front of the white man and was not to be interrupted while he spoke on so serious a matter. So, sitting back and looking directly into the young man's eyes, as he was in turn looking into Iron Hand's, he quietly listened to the serious sounding young man. In the sing-song language of the Blackfoot, Spotted Eagle said, "My Father had once been a great warrior and my mother had been one of the most beautiful women in my village. Both had since been taken by the white man's disease that puts red spots all over the body and makes the body seem like it is burning up (smallpox). When they died, I had to raise myself from a young age and was now older and wise as the horned owl. Now, I am a successful warrior and highly respected by the old people in my village as well as by my uncle, the great Chief, Mingan."

"But I have fallen in love with a young woman from my village and talked to my uncle about my love for this person and my wish to marry her. However, my uncle told me that maybe the young woman I love had eyes for another man, namely a white fur trapper of renown and greatness. My uncle told me that I needed to talk to the Great White Trapper, the one who my people have named "Iron Hand", and see if he wanted the same woman as did I. He asked me to do so, because you have already given the one I love four very valuable horses as a gift, like a man in love would do. If so, my uncle said Iron Hand needed to be the one to marry the woman since he had saved her life, had given her such a valuable gift and in so doing, had spoken for her in the way of the Blackfeet. With Chief Mingan's words of wisdom in my ears and heart, he told me where I could find you so we could talk. So that is why I am here. I wish to take Sinopa as my

The Adventurous Life of Tom Warren

wife, the great chief's younger sister. But my uncle says I must check with you first to see if you wish to be the one to take her as your wife. If you do wish to have her as your wife, let me know so that I may leave with a heavy heart," continued Spotted Eagle, whose dark eyes never left those of Iron Hand's! With those much-practiced words out and 'into the wind', Spotted Eagle turned away and once again, looked deeply into the fire and waited for Iron Hand's response... Then as if he could not stand the suspense of waiting any longer for Iron Hand's response, turned and looked directly back into Iron Hand's eyes for his answer of greatest importance.

Iron Hand just smiled over the words spoken by the young warrior and then said, "Spotted Eagle, I love little Sinopa but just like a much-loved little sister. I want her to find and marry a great warrior of her own choosing and from her own people, for the happiness that will bring to her. I gave her those fine horses as a sign of my respect for the great heart and bravery she showed, after the evil white trappers had taken her from her people against her will. True, I saved her from those evil men but only for her to find the love of her life and live many moons with the mighty warrior of her own choosing."

With those words, Iron Hand saw a look spread across Spotted Eagle's face as if a great weight had been lifted from his very heart and soul. With that, he jumped up and said, "Iron Hand, like my uncle has said, you are a great warrior! I want you to know that I will love Sinopa as no other man ever will and will care for her like she is your much-loved little sister. I will always remember your words and like my uncle, will be a 'Brother' to you for as long as The Great Spirit allows me to be!" With that, he vigorously shook Iron Hand's hand up and down like a pump handle on a well, then quickly vaulted onto the back of his almost now spooked horse and stormed out across the prairie back towards what had to be his tribal winter encampment, like an evil spirit was on his tail...

120

Iron Hand, with a shake of his head and a smile over the happy events that had just occurred, picked up Spotted Eagle's spilled coffee cup and cleaned it off. Then he became aware that his friends were all standing at the mouth of the cave and had quietly overheard everything that had been said. Almost embarrassed over the recent turn of events, Iron Hand asked, "Any of you 'boneheads' up for some damn fine coffee this morning as only I can make it?"

For the following deep cold winter months, the trappers ran their trap lines for the ever elusive wolf, hunted buffalo for meat, made jerky for the trail soon to be traveled back to Fort Union, repaired their equipment, and tended to their horses. Come the spring, it was back to trapping beaver once the ice went out and until the critters went out of prime. Then it was back to hunting buffalo for the meat and making jerky for the long early summer trip back to Fort Union to sell their furs and re-provision their supplies for their third year as Free Trappers on the western frontier. However, a decision had been made by the men not to return to the Medicine Lake area to trap beaver for a third season. They had trapped the area so intensively, that few beaver remained and it would not pay to continue trapping in that area any longer.

No final decision had been made as to which area to trap in the coming season but Old Potts was 'itching' to see what kind of trapping country lay to the west of the Big Muddy and in the end, the rest of his crew were so inclined to think 'westward' as well. However, there was a problem in moving their trapping endeavors further to the west. Therein to the west lay the hunting grounds of the dreaded Gros Ventre Indians, a tribe not known for its love of the American fur trappers. They, however, did get along with the Hudson Bay folks because that company freely supplied the Gros Ventre with rifles, whiskey, powder and pigs of lead for making bullets, in exchange for their trade in furs, hides and robes...

Whereas, years earlier at Fort Raymond with St. Louis businessman Vasquez, he had refused to supply rifles, powder

The Adventurous Life of Tom Warren

and lead to many of the Blackfoot and Gros Ventre bands of Indians. All the other Indians like the Sioux that traded at Fort Raymond were able to purchase firearms but not members from those two other tribes. As it turned out, Vasquez did not trust many of the Blackfeet or Gros Ventre because of their historical killing ways. However, when he traded and supplied all the other local Indian tribes with weapons and not the Blackfeet or Gros Ventre, that gave them the edge, and the tribal balances of power shifted to the Sioux, Crow and the like tribes because they now had the most advanced weaponry such as firearms. Hence, after that treatment, many of the bands of Blackfeet and most of the Gros Ventre considered the white man the enemy and treated them accordingly.

However, Old Potts had heard from other Free Trappers that if one could 'keep his hair' and steer clear of the Gros Ventre, the beaver trapping to the west of the Big Muddy on the Poplar River was unbelievable! He had heard tell of numerous trappers who had sent their fur down the Missouri to St. Louis, telling tales of numerous 'blanket-sized *plus*' coming from beaver, some of which weighed at least 100 pounds, live weight! They also told tales of catches many times that accounted for a beaver for every trap that was set! As it turned out, the more Old Potts told of the beaver waters to the west, the more excited the men became to giving such magically sounding waters a try. So much so, that come the time to head for Fort Union and take care of business that summer, the four trappers were figuring pretty hard, that 'moving west was their next course of empire'...

With the arrival of the last of the month of April, when the heaviest spring snows had finally stopped sweeping the northern prairies, found Old Potts and his fellow trappers en route to Fort Union. The men traversed their old familiar trail of moving westerly along the Medicine Lake outflow, then south along the eastern side of the Big Muddy until they reached the mighty Missouri. Then they traveled easterly

122

along the northern bank of the Missouri until they reached their destination of Fort Union.

Along the way toward Fort Union, the men met other groups of trappers heading in the same direction. In so doing, they bunched up for the mutual protection such groups of mountain men offered, and finally and safely ventured onto the grounds of Fort Union.

Once on site, their camp was set up along the Missouri River bottoms, and soon the men were inside the fort's walls deep into watching the Clerks counting and grading their furs, hides and robes. While there, Old Potts met up with some other Free Trappers from his days with Vasquez, who were selling their furs as well and gleaned additional information regarding the beaver trapping grounds to the west of the Big Muddy. That information gleaned from some of Old Potts's fur trapping buddies relative to the 'trappings' west of the Big Muddy on the Poplar River, did nothing but encourage his desire to see more of the west... But he also discovered that about a quarter of the company's fur trappers and Free Trappers, on an annual basis, who were trapping in that area, always seemed to disappear just about the time they were heading to Fort Union to trade in their furs and robes. The word among the surviving trappers familiar with that Gros Ventre-'infested' area was that those who disappeared, many times did so at the hands of the savage Indians hellbent on robbing the trappers of their hard-won gains and taking their scalps. Those few surviving trappers suspicioned that those stolen furs taken from other less fortunate trappers, would then be traded to the much-hated Hudson Bay Company, who did not question the Indians as to where those furs had come from. In fact, the word from friendly Indians was that the Hudson Bay Company encouraged theft of furs from any American Fur Company Trappers, because they figured it would reduce the competition for the rapidly disappearing in many places from being over harvested, valuable fur species...

The Adventurous Life of Tom Warren

When Old Potts later gathered together with Crooked Hand, Big Foot and Iron Hand over the information he had recently gleaned from those mountain men who had trapped in the heart of the Gros Ventre Indian Country and had survived, they listened in silence as if weighing their odds of trapping success along with survival. Then Big Foot and Crooked Hand more or less advised Old Potts that they could not live forever, and that it sure would be nice to see and trap some new country, regardless of who 'owned' the real estate! Then Iron Hand, who had remained silent while he weighed Old Potts's information said, "Well, when in Indian country trapping beaver and hunting buffalo, one has to take his chances. As for the Hudson Bay people, well, we have a surefire remedy on how to take care of those so affiliated who wish to treat us wrongly... So count me in, if for no other reason then to be there and be the one to patch up Crooked Hand's bad leg when he busts it up again and he will, or to help Big Foot in his 'doctoring'. Besides, I now make the best coffee and biscuits among the four of us and as such, I sure as hell would not like to see the three of you go hungry or without your morning coffee! Because if you do, I figures all of you and a mean-assed grizzly bear end up being a lot alike..."

"Then it is agreed," said Old Potts with a grin of anticipation settling across his whiskered old face. "However, since where we want to go is a far piece away from our trading post, I say we pick up three extra riding and pack animals in case some of what we have are stolen by the Indians or we cripple them up as a matter of course. I would also say we load heavy on our weapons cache as well. I think we need to maybe pick up an extra rifle and pistol apiece, just in case. That way, we can lay a mess of extra 'smoke-poles' on our first trailing animal in our pack strings like we did when we brought that huge pack string into Fort Union the first time. Then if we are outnumbered and attacked, we would have a number of extra shots to stand off any attackers. In so doing, since they would be figuring on us to only have one shot apiece, we could fool the hell out of them and kill that many more when they closed in for the kill. Remember, in any kind of battle, Indians don't like losing any of their menfolk. That also means we best load up on powder, caps and lead for our 'lead-eating' Hawkens if we are going further west into hostile Indian country. In fact, Iron Hand, that will be your responsibility to see that we have enough firepower to hold off an army of them damn killing Gros Ventre if the good Lord causes that to happen when we are out on the trail. No two ways about it, the best medicine against a mess of them damn savages is best served with a dose of hot lead in the right places of their carcasses..."

"Big Foot, with your background as a blacksmith, I would suggest that you see to it that we have all the things 'iron' we might need once we get way out there where we are a-going. That means extra horseshoes, horseshoe nails, files, fire making equipment, some extra traps in case some of ours are lost or stolen, a mold block or two for our rifles and pistols in case we lose one, some extra cast-iron cookware and several new axes because our old ones are getting a bit worn and hickory-handle thin. Make sure the axes you buy are the best quality, because where we will be a-going, we will not have a cave to crawl into like back at Medicine Lake. We will be a-building a cabin or lean-to of sorts to keep the weather off us, and that means we will be cutting a mess of logs in order to build a cabin to accommodate all of us, our furs and provisions. So be sure our axes are good ones, and maybe a saw or two and a couple of files would be a wise purchase as well," continued Old Potts.

"Crooked Hand, you are our horse expert. How about you pick up three more riding and packhorses from the fort's horse herd. Make sure one of them new riding horses is big enough to pack Iron Hand because he tends to gets grouchy if he has to walk all the time on those damn big feet of his. When you make your selections, make sure all of them are 'shoed' up proper like as well. With those extra pack animals, maybe we can load them down with those giant-sized beaver all my old

The Adventurous Life of Tom Warren

buddies are talking about, eh? Also, make sure we have extra leather for repairing our riding saddles, pack saddles and the like," said Old Potts.

Continuing, he said, "The four of us have plenty of money from this year's trappings but if you need more, hit up McKenzie and take some of the money the American Fur Company is holding for us from last year's trappings and use that to settle us up on the company's ledgers. Hell, where we are a-going, we probably will not live long enough to use up all that money from last year no-how's.@

"Me, I will see to our rum supply and double it first off in case Crooked Hand needs more leg treatment for infections, a goodly number of beeswax candles cause they are the best for holding the flame, a whole lot more salt than we had last time, more black pepper, another bag of hot pepper flakes for our venison stews, dried fruit and even more flour than we had last time 'cause I like them biscuits that Iron Hand makes. I will also see to it that we have more sacks of beans, rice, coffee, sugar, tobacco, and some more of them bottles of castoreum, since we seem to be always breaking them afore the trapping season ends. I will also see that all of our old Hudson Bay heavy blankets are replaced as well as all of our *capotes*." With those instructions off his chest, Old Potts seemed to run down on what he had to say to the men. Then as an afterthought he said, "Can any of you think of what we might ought to get in addition to what I have suggested?"

"We could use a new coffee pot 'cause the old one is getting to look a little ragged and it is just a matter of time before it is 'holed' on the bottom from all the open fire it sits over. I would also suggest get me a third six-quart Dutch oven for biscuit and pie makings. I would hate to break my old ones but if I do, we will need a replacement. I would also suggest we get some more leather for making bridles and the like as a reminder. Buffalo hide just does not do it up right and proper like it needs to be for a number of our needs," said Iron Hand.

With those words, the men fell silent and then Old Potts said, "Well, each of you has his marching orders. I would suggest we get to it because for the distance we need to travel to get where we are a-going never having been there afore and then have to build a cabin to live in and store our furs and provisions, we will need to leave the fort earlier this summer. Asides, we need to make sure we get what we need from the fort's stores afore they run out and then we end up having to make do without," said Old Potts.

For the next two days, the four men went about their ways fulfilling the orders Old Potts had suggested, or they figured they would need on such a long and dangerous journey they were planning. Finally, the men had acquired about what they figured they would need for the adventure lying ahead for each of them. For the next two days, the men visited with other fellow trappers and old friends who had survived, the members of the fort, their friend McKenzie, and made the rounds among the many visiting Indian camps, looking for much-decorated buckskin shirts, beaded moccasins, beaded knife sheaths, and fringed shirts that many of the Free Trapper breed favored when it came to showing off their status as one of the free and some of the wealthiest trappers on the frontier.

Finally, they spent their last night in the shadow of the fort, safe from all the many ways of finding death out on the frontier. Iron Hand prepared their final supper under the protection of the fort's walls and found himself liking being able to cook without looking over his shoulder every time he made any kind of a 'cooking move' around his campfire. Supper that evening was staked buffalo steaks, his favorite brand of sugared biscuits, fried spuds with onions supplied by the fort's huge garden, and an apple pie hot from one of his Dutch ovens. Then the men lounged around their campfire smoking their pipes and drinking more than several cups of rum, as they discussed what was lying before them on their journey out from the safety of the fort and into the unknown for the coming trapping season. Then it was under their

The Adventurous Life of Tom Warren

sleeping furs and the hordes of mosquitoes supplied by the Missouri River bottomlands...

Chapter 8 – New Trapping Grounds, Deadly Gros Ventre Surprises

HAVING SOLD OFF THEIR FURS, hides and robes and re-provisioned with the goods needed for yet another year on the frontier, Old Potts and crew prepared to leave Fort Union en route to their new and yet to be discovered trapping grounds further west in the potentially deadly ancestral country of the Gros Ventre. Breakfasting with McKenzie on their last day in the fort, the men feasted on roasted buffalo hump ribs, freshly fried potatoes mixed with onions from the fort's garden, sourdough biscuits with homemade wild plum jam, and all the coffee loaded with sugar that they could tolerate. Or as Old Potts called their favorite 'brand' of trapper's coffee, loaded with mounds of sugar like that being served at McKenzie's breakfast table, as "syrup".

Then lining up their stock strings of 16 animals plus those ridden by each man, out the front gate of Fort Union they streamed heading for a land of unknowns like the 'lordly' Free Trappers without a care that they represented. However, as per Old Potts's instructions, the first animal in each man's pack string, now in addition to its packs of provisions, was hooked

The Adventurous Life of Tom Warren

an extra, fully loaded rifle and two additional pistols fastened onto the pack saddle for quick and easy retrieval.

That way, with what each man was personally carrying in the way of firepower and what was hooked onto their first animals in the pack string, each man was capable of being able to fire six shots before they 'ran dry' if they were given the opportunity! Additionally, each of the men's personally carried pistols were loaded with deadly buck and ball for close in fighting if things went 'downhill' in a fight and got dicey...

Understanding that they would be a valuable target for any Indians on the loose who were 'horse hungry', especially with their valuable animal string and what they were carrying in the way of provisions, the men began their everyday ritual of remaining alert to their surroundings, just as soon as they left the front gates of the fort or any other form of safety. For the next several days, they slowly trailed their pack animals as they headed westerly along the northern side of the Missouri River. Upon reaching the Big Muddy River, the men forded at a set of low riffles and then continued westerly once again into the unknown prairies and hunting grounds of the dreaded Gros Ventre. For the next several days, the men passed herd after herd of peacefully feeding or resting buffalo, as they crossed the rolling hills of the verdant prairie en route to the Poplar River country and its fabled, much-rumored beaver trapping grounds.

Finally arriving at the confluence of the Poplar River where it entered the Missouri River, the men rested and grazed their double hobbled livestock under their watchful eyes on the lookout for any sign of horse-stealing Indians. Numerous times during their travels to date, the men had crossed trails of numerous unshod horses moving to and fro throughout the prairie, but as of yet had not seen hide nor hair of any of their riders. However, the men knew the Indians were in country and close at hand, based on the freshness of their horses' droppings, campfire signs and dead and partially butchered buffalo carcasses left out on the prairie after the Indians had made a kill and left what they did not need.

By the third morning with their horses now well rested, the men crossed the Poplar River at a shallow ford and proceeded northwesterly up the Poplar until they came to a string of wooded hills on the western side of the river. There they paused out of sight of any hostile watching eyes out on the open prairie in a swale, rested their livestock and began looking for a secluded place nearby in the timbered hills in which to build a cabin. This they did because all along the Poplar River and its adjacent marshy areas, they had observed numerous signs of beaver. Beaver ponds abounded, signature mounded houses of the animals were everywhere, and mud and stick dams were numerously interspersed throughout the entire low-watered areas. Additionally, there were even heavier beaver signs all along the entire Poplar River that the men had observed as they slowly traveled along upon entering the country they planned on spending a season trapping. All in all, the country abounded with beaver sign as other Free Trappers had so informed Old Potts in his previous conversations with them recently back at Fort Union. Conversations from a number of Free Trappers who had trapped the area but decided to leave all those good trappings behind because of the heavy Gros Ventre Indian presence at what seemed to be every turn in the trail. With all those excellent beaver signs, the men had decided the placement of their new home would be so dictated near those areas to be trapped but off the main beaten trail of that historically traveled by the numerous local Indian bands moving around ahead of the trappers' line of travel.

After several days of looking and finally discovering a flat spot in the wooded terrain adjacent a small but vigorously flowing spring, Old Potts informed the group that to his way of thinking, that was the best out-of-the-way and secluded spot they had seen for the building of their new cabin. Sitting there resting on his horse, Iron Hand took a 'gander' at his new home site to be as well. They had a good flat piece of ground upon

The Adventurous Life of Tom Warren

which for their cabin to sit, there was an abundance of firewood nearby, they were not out in the open where just any passing Indian would notice their presence, the area was open enough so it could be fairly easily defended from the shooting holes soon to be built into the cabin's walls, and a good flow of water and a meadow were nearby for the horses. From all of what Iron Hand could observe, Old Potts's past experience as a trapper had just been brought to bear in the selection of their new home site.

However, now that they were there in their new home site location for the trapping season, the hard work was now looming on their horizon. The men unpacked all their livestock and using the packs of goods and their saddles, made a small 'walled' fort around the base of two pine trees for their general protection in case of a surprise attack by the local inhabitants. Their horses were then hobbled and put out into the nearby tree-surrounded meadow to graze under the watchful eyes of the men who were soon to be working close at hand. Then Big Foot and Iron Hand took their crosscut saws and axes and went up into the adjacent stands of timber nearest their chosen cabin site. There they felled green trees for the log roof and walls of their soon to be built cabin. As they did, Crooked Hand and Old Potts gathered up large rocks from the creek's waters flowing below the wellhead of their vigorously flowing spring, dug out, rocked up and constructed a firepit adjacent their new cabin site for all of their outside cooking that would take place during their nicer weather.

At the end of that first day, out from the forest came Iron Hand and Big Foot, to find a roaring fire in their new outside firepit and a supper of buffalo meat, hot coffee and Dutch oven biscuits. After supper, the men sat around the fire on recently dragged into the area, limbed sitting logs, smoked their pipes and relaxed because they knew the next day would be an even harder one work-wise. Come daylight, the last of the buffalo meat, along with hot coffee was served as the men's breakfast.

Terry Grosz

Then Old Potts with a team of horses, began dragging the previously cut logs from the forest to the cabin site, while Big Foot cut in the notches at the logs' ends for log-upon-log wall placement. Three more days of hard work timber felling, cutting, removing limbs, notching and sectioning the logs, and the men had a pile of roughed-out, green timber ready for constructing their new cabin.

All the next day, with Iron Hand and his great strength and height hefting up and setting the ends of the logs into the pre-cut notches, the walls began taking shape. Soon the walls were raised high enough so that even Iron Hand with his great height, could walk around inside the cabin to be, without having to do so in a stooped-over position for fear of banging his head on the low-hanging ceiling rafters. Then up went the roof's stringers and finally the rest of the smaller logs lining the roof top. Following that, the roof was covered with cut sagebrush filling the cracks between the logs, followed by two feet of dirt placed over the top with brute force-hauled and filled panniers, so no one could start a fire on their roof if the trappers were inside its walls and under attack by Indians.

For the next week, Big Foot, who was a bit of a carpenter in addition to being a blacksmith in his previous life, made tables, sitting chairs for the inside and outside of their cabin, and finally cut the logs needed for a heavy front door and shutters for the window openings in the walls of the cabin. When Big Foot did, he made the front and only entryway to the cabin 'single-man' narrow. Big Foot figured in the hostile country they were in, if anyone ever attacked their cabin and the men inside, they would have to come through the front door single file. If that were the case, he also figured those so entering with killing on their minds, would make easier targets for any of those trapped inside the cabin and fighting for their lives... He also devised a special step just inside the front door that everyone entering the cabin would know about and have to 'step-long' to get over.

133

The Adventurous Life of Tom Warren

Once again, Big Foot figured that if anyone attacked their cabin and came storming inside, they would not be familiar with his >step-long= addition and end up stumbling and falling on their faces in the confusion of any battle. Little did the other trappers complaining about the long step needed upon entering their cabin truly realize just how handy such a 'long-step' device would someday become in the event of a surprise attack. As Big Foot carried on with his backcountry carpentry work, Iron Hand and Old Potts carefully cut out the firing holes along the sides and ends of their 20 by 25-foot cabin for use in case they were ever attacked by those they did not want inside or those who did not want the trappers in their country.

Following the culmination of those works, all four men built a mud and stick fireplace and chimney at the end of their cabin for the heat it would provide and to facilitate indoor cooking when the winter winds howled outside. In building their fireplace and chimney, the men dug out a part of their spring for the mud it supplied and in so doing, made a larger pool to facilitate watering their horses. Lastly, Iron Hand and Big Foot cut down a number of dead pine and fir trees, as Old Potts and Crooked Hand, using several teams of horses, hauled the dry timber down to their cabin and stacked it nearby for their close-in winter's firewood supply when the snow had become too deep for easy firewood access.

Satisfied with their work efforts, the men hauled into their mostly finished cabin all their pack saddles, packs, traps and provisions for safe keeping. Then when Old Potts and Crooked Hand went deer hunting for their 'frontier windows', Iron Hand and Big Foot built a hell-for-stout horse corral next to their cabin, so they could keep a peeled eye out for any horse-stealing Indians who just happened to be wandering by. In order to curtail any such 'horse-thieving', the corral had been built right next to the trappers' cabin for the added protection that closeness afforded. As an added bonus, the wall of the cabin facing the horse corral sported four shooting slots just so positioned to preclude anyone wanting to steal a horse from doing so without 'having one hell of a tough hide'...

When Old Potts and Crooked Hand came back with several packhorses hauling five just-killed mule deer or what were soon to be known and utilized as 'frontier-windows', all four men lent a hand in butchering out the animals. Then Old Potts and Crooked Hand began scraping the loose meat and fat from the deer hides and removing the brains from the skulls of the animals. This they did so the hides could eventually be frontier tanned. Once the deer hides were tanned and scraped very thinly, that allowed their open window areas to be covered with such specially processed hides to keep out most of the cold and yet let in some dim light for those cabin occupants during the daylight hours.

Then Big Foot worked his carpentering skills once again into four log and post bed frames with wooden pegs and ropes for mattress supports for their buffalo robe and bearskin sleeping furs. Following that, he made and pegged several log shelves inside the cabin to hold foodstuffs up off the floor and away from 'the little people' as he called them. Then more pegs were driven into the inside walls of the cabin to hold their cast-iron pots, water pails and three-legged frying pans. Finally the trappers were done with their domestic chores and now the much-anticipated preliminary trapping preparations could begin.

First, all their traps were smoked over a wood fire to get rid of the man smell, and an outside meat smoking and drying rack was built right next to the cabin to preclude bears and magpies from sampling the meat being processed without being discovered in the 'meat-thieving' process. Next, Iron Hand hauled in a mess of cottonwood from along the nearby Poplar River to be used as smoking wood for drying the meat on the racks. Following that, Iron Hand took a packhorse into the adjacent timbered hills and brought back a mess of mountain mahogany, which made for an excellent smoking

The Adventurous Life of Tom Warren

wood for the buffalo, moose and deer meat soon to be harvested and processed.

Then it was out onto the prairie on a buffalo hunt where four cow buffalo were killed, butchered and the meat hauled back to their new cabin. There the majority of the meat was cut into thin strips, some was salted and peppered for later use, some was set aside for their meals and the rest slow smoked for winter jerky for sustenance when the men were out on the beaver or wolf trapping trail.

Throughout the entire cabin building episode and the subsequent buffalo hunt, nary a single Gros Ventre Indian was observed. There were plenty of old, unshod horse tracks showing they were in country but none had been observed, nor had they been discovered during the trappers' many preparations from cabin building to scouting out the best beaver trapping waters...

For the next three days, buffalo meat was smoked, extra wood was hauled for the outside firepit and a small mountain of musket balls were cast for their 'lead-eating' .50 caliber Hawken rifles. Additionally, the men began tanning the hides from the four cow buffalo recently killed, for making into heavier winter wear, new moccasins, and rifle and pistol scabbards and holsters for the extra weapons carried on their packhorses as Old Potts had suggested in case a number of hostile Gros Ventre became a life or death concern...

Early the following morning, the four trappers ventured forth in order to scout out the best beaver waters to trap once those animals' pelts were in their prime. The men had once again decided that Iron Hand and Old Potts would be the ones doing the actual trapping, while the other two trappers remained horsed in order to offer protection to the men setting and tending the traps against unforeseen or surprise Indian or grizzly bear attacks. For the past several weeks, no Indians had been observed and the men were beginning to mentally let their guards down when they were out and about. However, being in the country of the dreaded Gros Ventre, they still stayed

Terry Grosz

pretty observant and made sure their two pack animals brought along on each scouting trip carried the extra rifles and pistols for a spirited self-defense if and when the occasion demanded it.

However, they religiously followed that extra gun tactic because those hostile Indians reportedly met by other trappers usually followed the same methods of attack when attacking inferior numbers of trappers out in the open. That was, once the trappers being attacked had fired their single shots from their rifles or pistols, they had to quickly reload in order to protect themselves from being instantly overrun. The attackers, knowing the required sequence of 'single-shot shooting' and then quickly reloading, would then instantly attack and overwhelm the trappers while they were still in the process of frantically reloading their weapons. That tactic practiced by the Indians usually reduced the number of dead they suffered, and allowed the hated trappers to be overrun, killed and all of their goods and horses taken. Hence Old Potts's tactic of carrying extra rifles and pistols on their pack animals so the men would not be so vulnerable when attacked was strictly adhered to. Or at least that was the thought that they would be able to give a better surprise accounting of themselves with the extra firepower they carried close at hand on their pack animals.

Three weeks later, Old Potts decided it was time to begin trapping for the many beaver the men had been observing on all of their scouting trips along the Poplar River and its many other waterways. He did so because every morning there was a skim of ice on their water bucket, the aspen leaves had turned yellow and orange universally, geese were observed flying south in larger and larger numbers, and the prairie grasses had begun turning many colors of yellow and brown and were drying up. In short, fall and the time to begin trapping for many species of furbearers was upon those who made it their business of trapping for a living.

The Adventurous Life of Tom Warren

In the excitement and anticipation of the next day's endeavors, the men slept little and were up way before dawn. Hot coffee with plenty of sugar, Dutch oven biscuits and staked buffalo steaks headed up the breakfast menu and soon the men were more than raring to go. Horses were saddled, pack animals readied, traps loaded into the panniers, and all their weapons were loaded with a fresh charge of powder and in the case of their pistols, were loaded with buck and ball for any kind of 'in close work'!

With those preparations behind them, the men set off for the nearby Poplar River and their chosen beaver trapping area for the start of the season. With three men standing guard, Iron Hand entered the fall-cooling waters and set his first beaver trap of the season. Seven hours later, Iron Hand tiredly set his last of 40 beaver traps next to an opening he had made in a beaver's dam, allowing the pond's water to exit and flow freely. Knowing the beaver would be there shortly after sensing the loss of water in its pond to fix the break in the dam, Iron Hand figured that trap would yield a beaver for sure by the next day. Then backtracking their trap line past their previous trap sets, the men discovered that 15 of the previously set traps already had large, dead beaver in them!

Finding dead beaver already in their traps the men found somewhat unusual, because beaver were normally nocturnal in their habits. However, those beaver were quickly but carefully skinned on the spot by expert skinner Big Foot and the hides dumped into several panniers being carried by the men's two packhorses. Those traps were then re-scented with castoreum (a pungent fluid removed from a beaver's caster glands) and set once again in the same spot to be checked the following day. Keeping three of the larger beaver carcasses for their supper that evening, the men retreated back to their cabin via another route, so a well-used trail of shod horse tracks would not be a dead giveaway that white men trappers were in the area and lead any cold trackers with evil intent on their minds and in their hearts directly into those trappers' campsite.

Terry Grosz

That evening after their first successful day of trapping, fleshing out their catches and hooping the same, supper made from fresh beaver meat was then specially prepared by Iron Hand and Big Foot. After supper, the men retreated to their homemade chairs around their outside fire site previously constructed by Big Foot, smoked their pipes and drank a cup of rum to celebrate their first day's successes. Realizing the next day would be another busy one, especially with 40 traps set out in such heavily populated beaver and muskrat-infested waters, the men retired to bed early after bringing in their grazing livestock, hobbling and placing them into their corral for safekeeping.

Iron Hand was glad to stretch his huge frame out on his bed and drew his bearskin sleeping furs up around his neck. It had been a long day, immersed in the cooling fall waters throughout, setting beaver traps and retrieving dead animals that had been caught the day before. In fact, his feet and lower legs from his knees on down had not warmed up as of yet because of all the cold water immersion and felt just like ice to the touch. Soon, the loud snoring from Old Potts filled the cabin but sleep came easily to Iron Hand in spite of his noisy sleeping partner, icy legs and cold feet. Iron Hand soon found himself not of the waking world.

About an hour into his sleep, Iron Hand's eyes suddenly flew wide open and having been raised and living so long on the dangerous frontier, he did not move or give away the fact that he was now wide awake and alert to what had just awakened him so abruptly! His right hand slowly moved to the side of his bed and his fingers quietly closed around the comforting cold wood and steel butt of his nearest pistol loaded with buck and ball. Slowly transferring that pistol to his left hand, his right hand once again silently dropped over the side of his bed frame and closed over the comforting handle of his second pistol lying close at hand. Now holding both pistols across his chest and the thick-haired bearskin sleeping robe for

139

The Adventurous Life of Tom Warren

the silence it would provide in what he was about to do, he silently cocked both guns against the silencing blanket of fur!

Listening intently, he once again heard the sounds that had originally had to have been the ones that had awakened him from his deep and tired sleep. The horses outside in their corral were now making all kinds of shifting and shuffling-around noises and other restless sounds in their near at hand corral. Sounds like maybe a corralled horse would make when a grizzly bear was close at hand. But Iron Hand soon realized that the restless animals were not making any alarm sounds like they would upon scenting a bear, but were making horse-sounds like they would generally make when strange humans came into sight or were close at hand to them...

Quietly taking his right hand and using the pistol barrel as an assist, he carefully moved his top bearskin sleeping fur off his long body and laid it off to one side where it would be out of his way in case he had to hurriedly exit his bed. Then ever so slowly, Iron Hand quietly moved his legs over the side of his bed so that he was soon in a sitting position in the coal-black darkness of the cabin and facing their unlocked front door. Controlling his breathing to just slow and shallow quiet breaths as if that would help his hearing and not raise any alarm to other hostile listening ears, he listened intently once again to the mystery sounds their horses were making. It was then that his hearing shifted from the sounds the horses were making to another closer at hand and more sinister sound. THAT SOUND WAS THE SLIGHT SCRAPING NOISE THE WOODEN HANDLE TO THEIR DOOR MADE WHEN SOMEONE WAS SLOWLY SLIDING THE WOODEN LATCH BACK IN ITS HOLDER PRIOR TO OPENING THE DOOR AND ENTERING THEIR CABIN!

Iron Hand was and always had been a light sleeper. As such, he knew none of his fellow trappers had left the cabin earlier to go outside to the bathroom and were now coming back inside to go back to bed. Turning slightly in his sitting position on his bed frame once again so as not to make any

Terry Grosz

noise or alert anyone who might be listening, he was now facing dead-onto their cabin door. THEN IRON HAND HEARD THE DOOR HANDLE FINALLY RELEASE WITH HARDLY A DISCERNIBLE 'CLICK', AND COULD SENSE EVEN IN THE DARKNESS OF NIGHT THAT THEIR DOOR WAS SLOWLY BEING OPENED ON ITS SQUEAKY SOUNDING LEATHER HINGES!

Then all of a sudden, their heavy log door was flung wide open with a loud 'CRUMPING SOUND', as it slammed against the inside wall of their cabin!

"HI–YI–YI–YI!" SCREAMED A LOUD VOICE, which was followed by the sounds of a human rushing through the narrow door opening, only to be followed with another loud 'CRUMPING SOUND' as that individual, not familiar with Big Foot's long-stepping board below the door's sill, tripped and crashed violently onto their hard-packed dirt floor face first, with a loud OOOMPFFF! That was followed by another man yelling "HI–YI–YI–YI–EH!" just as he plowed through the open doorway behind the first man, with what had to be another Indian on the attack!

BOOM! went Iron Hand's pistol full of buck and ball and at ten feet, tore off most of the face from the second man moving through the open doorway, blowing him off to one side and into a crumpled heap just inside their cabin! When Iron Hand shot, the brilliant flash of his pistol being fired for just an instant in the trappers' dark cabin illuminated several more Indians trying to push their ways through the purposely-built narrow doorway all at the same time! However, Big Foot's idea of creating a doorway that only accommodated one man at a time coming through its framing was now bearing survival fruit of the finest kind...

When the third Indian tried pushing himself through the open doorway, he stumbled over the first man through, who had stumbled on Big Foot's cleverly raised door sill and had fallen hard onto the dirt floor of the cabin, stunning him in the process. Down went the third Indian through the open

141

The Adventurous Life of Tom Warren

doorway into a squirming pile on top of the first man through the doorway still lying on the floor as well. The fourth Indian to push his way through the open doorway was shot in the chest with Iron Hand's second pistol shot, the force of which blew him backwards out the open doorway and into the fifth Indian trying to push his way inside the cabin! Upon impact from the backwards-flying dead man, the fifth Indian trying to push his way inside so he could have at the trappers inside, found himself being flung outward and knocked down outside the cabin!

In the flash of Iron Hand's second pistol discharge, he could see more Indians outside the doorway trying to get in at the men inside, as well as seeing his three fellow trappers now making ready to shoot since they were more than wide awake, armed and making ready to get into the battle posthaste!

Yelling in the darkness to voice-identify himself so he would not get shot in the darkness by one of his fellow trappers, Iron Hand grabbed the first Indian through the door who had stumbled over the high door sill, had recovered from his hard fall onto the cabin's hard-packed dirt floor, and was now rising to his feet yelling loudly. His yelling suddenly stopped seconds later, as Iron Hand, now having 'run dry' with his pistols, grabbed the man's head and lower jaw with a two-handed grip that would have stopped a runaway horse, gave a vicious sideways twist and heard the stocky Indian's neck snap like a muffled pistol shot!

The next Indian through the door also died a quick death! The other three trappers all fired at the sound of the yelling man's voice coming through the doorway with their pistols simultaneously! When they did, the man literally blew apart, as the three .69 caliber lead balls and buck crashed into his body, flinging him out the open doorway backwards, in a 'flushing' of blood, pieces of meat, flying snot, and facial pieces of bone and splintered teeth!

Yelling loudly once again for the voice recognition so he would not be shot in the cabin's pitch black and now dense

142

black powder smoke filling the interior, Iron Hand grabbed the next Indian beginning to charge through the open doorway just illuminated by the three just-fired pistol shot flashes and the two of them crashed into the back wall of the cabin with a hard THUMP! Iron Hand could feel a tomahawk in the Indian's upraised hand, grabbed it from the still surprised Indian and split his head clear down to the man's first neck vertebrae! When the head was split open with such force, Iron Hand's face was heavily splashed with blood and brains, as the tomahawk's handle busted off in his hand because such tremendous downward force had been used by the adrenalin-fueled trapper when he struck!

Then the third Indian who had stumbled over the first one through the doorway and had also fallen to the floor, rose and fired his pistol in the darkness at where he figured Iron Hand was standing. Iron Hand felt the burning hot sensation as the bullet churned through his previously damaged cheek skin from his first fight back at the cave with a Blackfoot's tomahawk a year earlier. That same bullet also burned off a sizeable portion of hair from Iron Hand's heavily whiskered face. Feeling the rise of fury he had felt during that first battle with the Blackfoot back at the cave, Iron Hand reached out where he had last seen his Indian adversary in the flash of his firing pistol, and grabbed a portion of the man's buckskin shirt. Jerking the Indian into him, Iron Hand kneed him in the groin and when the severely damaged man bent over in pain, another hard knee to the man's head violently snapped it back, breaking his neck and killing him instantly in the process!

BOOM–BOOM–BOOM! went three almost deafening blasts inside the cabin from the other trappers' Hawken rifles, now that they had 'run dry' with their pistols. In the lightning-like brilliance from three rifles going off at the same instant in such close quarters, Iron Hand just saw part of another Indian trying to come through the door, only now flying backwards and out through the door frame! His three friends, not being able to see individual Indians coming through the door in the

The Adventurous Life of Tom Warren

cabin's darkness, knew where its opening was roughly located and were just firing blindly whenever another screaming savage came through the narrow doorway, and being in a 'fatal funnel of fire', that savage also joined the Cloud People in their land of eternity...

Finally, the last of the attackers burst through the open doorway, yelling to the high heavens in the emotion of the moment, realizing he was going into a fierce battle with the trappers based on all the sounds of battle ongoing from within the cabin's walls. His yelling was quickly drowned out, as Iron Hand's quickly retrieved Hawken sent him off to the Happy Hunting Grounds to join the rest of his recent brethren who had been foolish in attacking the trappers in their cabin...

Then utter silence and the acrid smell of burned black powder flowing outward from the open cabin doorway now rent the air from the attacking Indian side of the doorway. "Think we got 'em all?" yelled Crooked Hand.

"Don't know," said Iron Hand, as he fumbled for his possibles bag in the darkness so he could reload his Hawken in case there was more danger yet to come in the form of hostile Indians bursting through the cabin's narrow doorway.

Soon, the other men could also be heard quickly reloading their rifles in the darkness without uttering a word. They jointly figured that to say something would mean any other remaining attackers could echolocate onto anyone inside the cabin by the reloading sounds or their voices, and blindly shooting in that direction. So the reloading quietly continued, as the heavy smell of burned black powder filled the air and made seeing anything of the attackers because of the heavy smoke in their cabin and the surrounding darkness outside the open doorway, impossible! However, the men now took up better defensive positions other than their beds and quietly waited for more attackers to show themselves through the open doorway... Attackers who would soon join the other Cloud People if they decided to unwisely enter through the narrow cabin door frame.

Terry Grosz

Come the dawn when the men could finally safely see, they got a firsthand look at the carnage around them. Inside and out of the cabin lay eight dead Indians! Indians later identified as Gros Ventre by Old Potts because of the clothing they were wearing. However, no more Gros Ventre could be seen anywhere at or near their cabin. Later as the four men cautiously left the protection of their cabin's heavy log walls and took a look around, it appeared they had been attacked by a small party of Gros Ventre, who had apparently either witnessed the trappers setting beaver traps the day before or had smelled their wood smoke from their previous evening's cooking fire. Those Indians had waited until dark, after 'cold tracking' through either sight or smell the men back to their cabin and when they figured the trappers were sound asleep, they had viciously attacked.

Had it not been for one of their kind being a light sleeper, the Indians may have succeeded in killing the lot of them as they slept. Iron Hand later 'cold tracked' their Indian attackers with Crooked Hand as his backup and discovered where they had hidden their horses. There were only eight horses staked out, so the men figured they had killed the entire group of Indians that had found them out and had attacked them from the night before.

When Old Potts heard that news, he was more than pleased. "That means none of them red devils escaped and that also means we will not have to worry about any of their kind that got away, telling the others back in their village and bringing back more of their kind to do us in," he said through his heavily whiskered grin.

"Yeah, but what do we do with all of their bodies? We just can't go out and bury them. If we do, the wolves and grizzly bears will sure as shooting just dig them up and then the rest of the Gros Ventre looking for those eight warriors will eventually discover what happened to them. When that happens and it will, they will then go on the warpath looking for those who did their brethren in," said Iron Hand, as he

145

The Adventurous Life of Tom Warren

nursed the skin on his bullet-creased cheek with the application of some bear grease in order to close up the slight wound.

"You are right," said Old Potts. "We need to find a way to make them just plain disappear so their killing cannot be tracked back to us and our campsite. 'Cause if we don't and those red devils get their hands on us realizing we were the ones doing the killing of their kin, what they will do to us will not be 'purdy'. Maybe if we toss them into the river, they will wash down to the Missouri and then we be rid of the problem as they continue washing downriver to the south," he continued, with just a lilt of hope in the tone and tenor of his voice.

"I got an idea that should keep them off our trail and onto us and what we did," said Iron Hand. With that, he shared his plan with his friends and with a number of heavily whiskered grins from his partners, approved the rather oddball but guaranteed to work plan, if they could pull it off and 'the buffalo would help' with a little cooperation.

An hour later found the four trappers trailing eight Indian horses westward out onto the open prairie. As they went led by Iron Hand, the men passed numerous small herds of feeding and resting buffalo, which figured highly into their plans for disposal of the eight Indian bodies now being carried over the backs of their own horses. After an hour of riding around and still trailing the unshod Indian horses behind the trappers' shod ones to confuse anyone running across their tracks, Iron Hand finally said, "I think this is the spot I am looking for. This small valley lying to the front of us will be perfect for what I want it to do. It forms a perfect funnel out onto the plains below and should work if we can get the buffalo to cooperate. Let us take the eight bodies and scatter them out on a line for about a half-a-mile down through this funnel-like valley like they were killed while hunting buffalo. You know, like a buffalo stampede got the better of them and caused the men to lose their lives. Big Foot, you take these two horses and run them out along the line of dead men we are going to leave all

146

spread out like they were killed while running the buffalo. Then somewhere along the line of bodies, kill both horses and let them lie there with their bridles and saddles in place. That way, it will look like those horses were killed when their riders were trying to shoot some buffalo as well. When we lay these bodies out on a line like they were running the buffalo, we also need to make sure their rifles and everything else they owned are laid out there with them just like it would be in normal life. They need to look like they died while hunting buffalo to anyone who comes to investigate the remains. Because if they look all normal-like, then the searching Gros Ventre will be satisfied that they died naturally like occurs on many an Indian hunt of buffalo and they will not suspect foul play by any white man. Now, let us get this done as quickly as we can so we can cover our tracks and deed like I explained earlier."

After the men had disposed of the bodies as Iron Hand had instructed and Big Foot had killed the two horses like the buffalo had done them in, the men still trailing the rest of the unshod Indian horses behind them to cover their shod horses' tracks, returned to the large buffalo herd Iron Hand had spotted earlier feeding at the head of their small valley. Coming in behind the large herd of quietly feeding buffalo, the four trappers, still trailing the Indians' unshod horses in order to cover their shod horses' hoofprints, sent the buffalo onto a full-blown stampede towards the small funnel-like valley Iron Hand had selected as their cover story holding the Indians' bodies. Soon the large herd of stampeding buffalo had picked up numerous other smaller herds of buffalo and all of them were now running in the direction that Iron Hand had planned. As the buffalo were stampeded along, they were directed down the small funnel-like valley where the dead Gros Ventre and their two horses had been laid out in a line like they had been hunting the shaggy beasts the day before and had run afoul of a buffalo stampede.

Once the four trappers had the buffalo lined out and running like they wanted, the men fired off their guns behind

The Adventurous Life of Tom Warren

the running buffalo, sending them into a full-blown panic. Herding them along down through the small valley, the charging herd ran right over all the dead Indians and their two horses, making it look like the men had been careless in their hunting tactics and the buffalo had gotten the better of their antagonists... All that was left once the buffalo had stampeded through the small valley, was muddy, buffalo hoofprinted soil and human mush where the eight men and two of their horses had been laid to their final rest, just like Iron Hand had planned.

Then still trailing the remaining six Indian horses behind their shod horses, Iron Hand led the men to the Poplar River and walked their horses into the water where there was a strong enough current to erase all evidence of their horses' shod tracks. With that, the six Indian horses were released to run wild on the open prairie like they had escaped the buffalo stampede unlike their masters.

Moving out from the river's current further upstream, the four trappers then returned to their secluded cabin for some much-needed rest and a good supper, topped off with several cups of strong rum to celebrate their continued existence in the world of the living... However, they also celebrated Big Foot's ingenious narrow door-opening design, allowing the entry of just one man at a time and his clever use of an extra high board below the door's sill. A board meant to trip the unwary of its presence and thereby giving any sleeping inhabitants a second chance at survival if they were attacked in their sleep...

For the next two weeks, the trappers ran their extensive trap line without incident. The men now remained extra vigilant suspecting once the eight Gros Ventre members were missed, they would have some friends out looking for them. However, except for the ever-extensive herds of buffalo at every turn on the prairie, all was quiet on the hunting grounds of the dreaded Gros Ventre.

That 'quiet' however did not relate to their beaver trapping successes. The men discovered that they were catching around 20 beaver a day in their traps! As it turned out, way more

Terry Grosz

beaver were being caught than the four men could keep up with when it came to the skinning, fleshing and hooping the skins at the end of each day of trapping. Finally, tiring of just getting only several hours of sleep before the next day's run of the traps occurred after all the beaver hides had been processed, a drastic change in their work efforts was becoming apparent, even in the face of the ever-present Gros Ventre danger always at hand.

Finally, one evening after an exhausting day of running the trap line, followed by the hide processing requiring all the trappers working into the wee morning hours, Old Potts put his foot down. "We need to change our operation. We are all wearing down, working from just before sunup until late the following morning every day. There is no way we all can keep up this hectic pace. Even in light of the high quality 'blanket-sized' beaver we are catching," said Old Potts, as he wiped beaver fat from his fleshing knife against the side of his pant leg by the light of their outside campfire.

"What I am about to suggest is a mite dangerous. But I don't see any other way clear, short of just up and leaving this here beaver-rich country. We kain't watch constantly over our shoulders for them damn Gros Ventre and do all the work required when it comes to processing this mountain of beaver we are a-trapping," he continued.

"I suggest that we pair up and split the trapping and processing duties. I think if I had a say in this matter, I would pair up Iron Hand and Crooked Hand and have them do all the beaver trapping. I say it that-a-way 'cause them two is the best shooters among us. They could load their packhorses heavy with extra firepower, and chance running into some of those damn murdering Gros Ventre devils being the good shooters they are and still survive. If they did with all that firepower and their damn good shooting eyes, they would make them Gros Ventre think twice afore trying to close in on them and 'lift their hair'. Plus, with only two trappers working the beaver grounds, they would sure leave a damn sight fewer

149

The Adventurous Life of Tom Warren

white man-shod horse tracks for them murdering devils to discover and then follow them back to this here cabin."

"Then, that would leave just Old Potts and Big Foot to do all the fleshing and hooping on all the beaver taken the day before. Since we are the poorest shooters among the four of us that would be best, especially if we were a-doing the trapping and got caught out in the open by them murdering savages. That a-way by splitting up all the duties, we could all get more sleep, the beaver trapping numbers would stay the same, and we would be ahead of all the caring for all of our *plus*. Plus, that would leave the two of us back at the cabin to watch over the rest of our valuable horse herd," continued Old Potts, obviously happy with the workings of his new idea.

For the next few minutes, all the men quietly mulled Old Potts's idea over, especially the touchy issue of the trappers being separated deep in Indian country. Then Iron Hand spoke up by saying, "I say if those ideas suit the rest of you men, they suit me just fine as well. I think that Crooked Hand and me, if given just half a chance, could make it a rough go for any Gros Ventre who took us on fair and square. Course if they did not fight fair and came at us like a hundred of them at a time, then them prairie wolves would eat good for a couple of days on a pair of mangy old trappers and a passel of dead Indians once all the smoke had cleared."

"Well, it would get us a little more shut-eye than we been a-getting," said Crooked Hand, as he was obviously mulling over Old Potts's idea from his side of the fence as well.

"Now with that plan out in the open, think about this. We are catching far more beaver than an ordinary trapper by catching about 20 of them critters per day. Soon at that rate, we will have hundreds of *plus*. Additionally, a great number of them are 'blanket size' and that will bring us a pretty penny back at Fort Union...if we live that long that is," said Big Foot quietly. "But if we do and can get out of here with our hair, I say we do it. Do it because with our catch numbers, we will soon have all the *plus* we can carry on our horses, and that

150

Terry Grosz

would allow the four of us to head back to Fort Union maybe weeks earlier than normal...that is once again, if we can keep our hair... Now that I think about it, who the hell would want any of our mops, except for Iron Hands' that is. I can see many an Indian wanting to lop off his mop of hair so he can use it for a horse blanket or a robe in the winter time for him and all of his kids," he said with a large smile over the funny he had just pulled on his larger than life-sized friend...

The next morning right at daylight, with Iron Hand in the lead and Crooked Hand trailing two packhorses, the trappers headed for their distant trap line. However, both packhorses sported an extra rifle and two pistols each, with the pistols being fully loaded with buck and ball for any close-in fighting that might occur if they were discovered and surprised by the Gros Ventre...

Arriving at the first of their 40 trap-sets, Iron Hand discovered a giant-sized beaver drowned and hanging lifeless at the end of the trap chain. What was remarkable however, was that the huge beaver weighed at least 100 pounds! "We have another 'blanket-sized *plus*' in that one," said Iron Hand with a smile on his heavily whiskered and now battle-scarred face, as he waded out to retrieve the drowned animal. Dragging the huge beaver ashore, Crooked Hand, the best skinner of the four trappers, went to work with his flashing but carefully utilized skinning knife, as Iron Hand now stood guard. Somewhat later, the huge skin went into a pannier being carried by one of the packhorses, with the carcass tossed off to one side for the gray wolves and grizzly bears to eat, the trap re-set and then the two men moved on to their next set on their trap line.

At the end of that first day utilizing Old Potts's plan, the two men had successfully trapped another 23 beaver, with nary an Indian seen. What was even more remarkable was the fact that the beaver numbers along the Poplar River and in the adjacent marshes seemed limitless... Relaxing just a little, the two men took another way back to their cabin, so as not to

The Adventurous Life of Tom Warren

leave a well-traveled, white man-shod horse trail leading the way.

For the next eight days, the four trappers executed their new plan without anyone having a 'hitch in their giddy-up'. Come day nine of their new trapping tactic, under a leaden sky portending an early fall storm coming from out of the northwest, found Iron Hand knee deep in icy beaver pond waters and mud, dragging back to shore another very large beaver. "Damn, Crooked Hand, this water gets any colder, I am going to change places with you and let you freeze your butt off instead of keeping it warm in the saddle all day long," said Iron Hand with a happy to be doing what he was doing grin.

Just as he spoke those words and looking up at Crooked Hand for his reaction to his semi-threatening words, he could see that his partner was intently watching the far northwest horizon intently like a 'robin would do when looking at a nearby worm'...

"What you looking at, Partner?" asked Iron Hand, as he waded out from the beaver pond he had been walking in.

"Could have sworn I saw four riders way off on the horizon coming this way. But just as I tried to look more closely, something blew into my right eye and by the time I had cleared it out, whatever I was seeing was gone," said Crooked Hand, as he still intently kept looking to the northwest.

Pausing at the pond's edge, Iron Hand also looked intently at the far horizon but seeing nothing of interest, laid the dead beaver down by Crooked Hand's horse saying, "This beaver is all yours. While you are skinning out that one, I need to stamp my feet around a bit to get the feeling back into them and my near frozen legs as well."

With that, Iron Hand removed his rifle from his horse and took off across the prairie at a slow trot in order to get some life and warm blood back into his near-frozen feet and legs. Moments later after trotting back to where Crooked Hand had almost finished skinning out the beaver, Iron Hand said,

"Crooked Hand, did you see anything else of what you saw earlier?"

"Kinda hard to keep a sharp eye peeled when one is skinning out a beaver," said Crooked Hand, as he tossed the bloody hide up into an almost full pannier of beaver skins on his packhorse. Then he re-mounted his horse and said, "Let us get going, Iron Hand, we have two more traps to run and then we can get the dickens out of here afore that damn storm a-coming our way is upon us. 'Cause as you damn well know, sitting in a saddle riding a horse in a snowstorm is a damn cold proposition."

With that, Iron Hand led his horse the 100 yards or so to the next beaver trap in order to try and get some more feeling into his feet and legs before he waded out into any more icy waters to check his traps. However, he was lucky and faced no more cold water that day, because the remaining two traps were empty. With that, Iron Hand mounted his horse and the two trappers turned their ways towards their far distant cabin's location, just as the first wet snowflake from the oncoming storm flattened out on Crooked Hand's 'hawk's beak'-looking-like nose.

An hour later, riding into the teeth of a full-blown blizzard, the two men and their horses struggled their way towards their cabin. When the flying snow got too heavy to see where they were going, the men gave their horses their heads and let the animals' instincts lead them back to their corral, a mouthful or two of hay and the comfort of the other familiar horses.

Twenty minutes later, Iron Hand heard a sound that made him rein in his horse abruptly! Sitting there tall in his saddle, Iron Hand strained his ears in the face of the flying heavy and wet snowflakes in order to hear the faint mystery sound he had heard just moments earlier. For the longest time, Iron Hand heard nothing but the wind and the soft squishing sounds of the large, wet snowflakes against his huge beard. Then Iron Hand heard Crooked Hand's horse being ridden up alongside his motionless horse as he sat there still carefully listening.

The Adventurous Life of Tom Warren

"What are you listening for, Partner?" said Crooked Hand.

"I don't know but for the last hour, my 'sixth sense' has been running wild with my imagination," said Iron Hand, as he looked intently into the flying snowflakes as if they held the clue to what was sticking in his craw. "Plus, I hear a strange sound being carried out there in the flying snow and deep timber that I could not put my finger on."

Both men sat there quietly on their horses listening for any kind of suspicious sounds for about another ten minutes. As they did, all they did was collect more heavy wet snow on their horses and their winter clothing, before Iron Hand finally pushed on once again. After another ten minutes of stumbling through the wind-driven, heavy wet snow, Iron Hand abruptly drew his horse up short. Once again, peering intently into the heavy, wet, wind-driven snow, he heard the mystery sound once again. That time being closer to the sound, he knew exactly what he had been hearing. HORSES! IRON HAND HAD HEARD A HORSE WHINNYING IN THE TIMBER LYING JUST AHEAD OF THE TWO MEN! A PLACE WHERE HORSES SHOULD NOT BE, ESPECIALLY NEAR THEIR CABIN AND IN FOUL WEATHER...

Once again, Crooked Hand rode up alongside Iron Hand with a questioning look in his now near-frozen and snow-encrusted eyes and totally frosted-over beard. However, Iron Hand, all six-foot, seven-inches of his being, was standing as tall as he could in his stirrups, as his dark eyes peered intently into the gloom of the dense coniferous forest lying directly ahead of their route of travel back to their cabin.

Sitting slowly back down in his now wet saddle with a serious look on his frost-covered face and beard, Iron Hand leaned over and whispered to Crooked Hand, "I just heard a horse whinny! I think there are some horses tied off in those trees to our front and when they smelled or heard our horses coming their way, one of them whinnied. My guess is we are not far from our cabin and those sounds are not coming from our horses in the corral back at the cabin! My guess is they are

154

horses from an Indian raiding party who somehow located our cabin, probably from the wood smoke smell coming from our fireplace being blown about in this storm. I think you did see something on the horizon before we left the beaver trapping waters and I will bet what you saw were Gros Ventre. They may have been looking for the eight men we killed, when they somehow stumbled upon our cabin's secluded location. That being the case, they must have tied their horses off in that grove of trees ahead and since I don't hear any shooting, I figure they are now sneaking up to the cabin. Or they have already surprised Old Potts and Big Foot in this damn snowstorm, killed them off and are now looting our cabin and getting ready to steal our horses back in the corral..."

Without another word, the men rode into a nearby grove of aspens, tied off their horses and made a 'beeline' to where they figured the mystery horses were tied off. Once discovered, Iron Hand's suspicions were validated upon viewing four Indian horses tied off, with their owners' sets of moccasin tracks in the fresh snow leading up towards where he suspicioned their cabin was located.

Checking to make sure their rifles and pistols were primed and ready to go, both men began sneaking along those sets of moccasin tracks left by the Indians heading in the direction of the trappers' cabin. Finally the trappers' horse corral came into view in the swirling snows and Crooked Hand noticed there were two Gros Ventre hidden in the trees nearby watching the front door of the cabin. Then Iron Hand noticed that two sets of the Indians' tracks led over to their stack of logs of their winter woodpile. Ensconced behind the log pile where they could also watch the front door of the cabin and not be seen, were two other crouching Indians with their flintlock rifles held at the ready.

Motioning with his hand, Iron Hand sent Crooked Hand after the two Indians hidden at the end of the trappers' horse corral, as he began his sneak over towards the winter woodpile of stacked logs hiding the other two Indians of the now-

The Adventurous Life of Tom Warren

discovered raiding party. Once in place where he wanted to be and out of sight, Iron Hand just waited in the heavily falling wet snow knowing what was soon to come once Crooked Hand got into position near the other two Indians. Iron Hand did not have to wait long... All of a sudden, Iron Hand heard an Indian yell out in discovery and then a quick rifle shot cut short his surprised shouting! That shot was then quickly followed with another of that being a pistol fired and then all was quiet over at the horse corrals, except for those animals' nervous stamping of their feet and their milling around in confusion in their corral over what had just occurred so close to them.

In that same instant right after Crooked Hand had fired, the cabin door flew open and backlighted in the faint candle light from within, stood Old Potts with his rifle at the ready. When he appeared, one of the Gros Ventre hiding behind the stack of logs, ignoring the shooting heard at the corral, rose up, shouldered his rifle, and then his head exploded throwing a brilliant red, 'fan-shaped' bloody 'spew' onto the white snow lying in front of him! The second Indian waiting in ambush by the log pile, realizing his partner had just been shot from behind, whirled and seeing Iron Hand standing there, fired a snap shot from the hip with his flintlock rifle, missing the trapper 'eyeballing' him by at least ten feet. However, Iron Hand did not miss with his pistol loaded with buck and ball from just ten feet away... When Iron Hand fired, once again the snow was spewed in the front of the man with a bright red smirch!

"What the hell is going on?" yelled Old Potts in surprise at what was occurring in front of the cabin. Then he lowered his rifle, as he saw Iron Hand moving around the woodpile of logs towards the cabin. Old Potts was then even further surprised with the close at hand appearance of Crooked Hand from the horse corral side of their cabin. "What the hell are you two damn fools doing shooting off your guns next to our cabin this time of afternoon? Don't the two of you realize powder and lead is hard to come by way out here?" yelled Old Potts at his fellow trappers.

Then he spotted Iron Hand reaching down and picking up the dead Indian's flintlock he had just shot with his pistol and then moving around their log pile to another dead Indian lying on the ground. Once again, without a word being spoken, Iron Hand picked up another rifle from the hands of a head-shot Indian and began walking towards their cabin. Walking up to a still very surprised Old Potts, Iron Hand handed him both of the Indian's flintlocks saying, "Here you go, Old Potts, have a couple of rifles, a present from a couple dead Gros Ventre who almost shot the hell out of you standing there backlit in that doorway."

About then, Crooked Hand approached a still surprised Old Potts and handed him another two flintlock rifles from the Gros Ventre he had just killed over by the corrals. For once, Old Potts had nothing to say as he fumbled with the four rifles and his own all at the same time. "Can we come in now and get warmed up a bit?" said Iron Hand with a semi-frozen-faced grin. "How about me, too?" asked Crooked Hand.

Once inside and standing by the roaring fire in their fireplace, all the while dripping from the now-melting snow off their heavy bearskin coats, Iron Hand and Crooked Hand told Big Foot and Old Potts just how close they came to being wolf bait. For the next few minutes the two trappers related their story of finding the Gros Ventre's horses tied off in a grove of trees and how they had stalked their four riders to the very edge of their cabin. What events followed were quickly related to the two still very surprised trappers in the cabin, and then Iron Hand and Crooked Hand realized they still had work to do because their day was not yet done.

Moments later, both men exited the cabin and walked back to their still tied-off riding and packhorses down in the far aspen grove. Gathering up those horses, the men then walked over to where the Indians' horses had been tied off. Leading all eight horses back to their horse corral, they proceeded to

The Adventurous Life of Tom Warren

unload and unsaddle the mounts and placed them into the getting-crowded corral.

As Crooked Hand struggled over to their cabin with the panniers full of fresh beaver hides, Iron Hand wiped off the snow and secured their saddles, pack saddles and riding gear from the Indians' horses under the lean-to next to their cabin, built for just such items to get them out of the weather. Then Iron Hand fed all their horses from the hay pile gathered earlier in the summer and then hightailed it for the warmth and dry of their cabin.

Once inside their cabin, the two men filled in Big Foot and Old Potts as to the day's adventures. Then spying a cast-iron pot cooking away something smelling wonderful in the fireplace coals, Iron Hand and Crooked Hand filled their bowls with hot buffalo stew, as the other two trappers went to work on the latest batch of semi-frozen beaver hides, fleshing and hooping the same.

Finished with their supper, Iron Hand and Crooked Hand went back out into the storm, gathered up the four dead Indians and hauled them off a short distance from their cabin into a small grove of aspen trees. There they left the cooling carcasses in that small stand of trees, figuring they would make the final disposition of the same the next day. Then the trappers realized they were once again faced with what to do with the bodies in order that the rest of their band of Gros Ventre, once they realized their four warriors were missing, went looking for them.

Dawn the following morning showed a clear blue sky, temperatures at or near freezing and the four men just finishing their breakfast of roasted buffalo steaks and Dutch oven biscuits. "Well, Iron Hand, how do we hide these four dead Indians so we ain't found out?" asked Old Potts, as he was having trouble mouthing a piping hot biscuit, fresh from the Dutch oven.

"Well, I have been thinking about that and I am stumped. Trying the old buffalo trick probably wouldn't work this time.

158

Terry Grosz

We would be leaving too many of our shod horse tracks in this latest snow. Tracks that would be easily read and then followed quickly back to our cabin. Maybe we could dump one of those bodies into the Poplar River and let him drift off before freeze up and the rest scattered off out across the prairie away from our cabin and let the gray wolves and grizzly bears have at them. After all, many of the grizzly bears are still out and about gorging themselves on just about anything they can find to eat in preparation for their next five months of hibernation without eating. I would imagine any grizzly bear that would come across any Indian body we dumped out on the plains, would be very happy with just such a meal. Other than that, I am just not sure what to do in order to remain undiscovered. However, Crooked Hand and I need to get that done and fast in case any Gros Ventre just happens to come wandering by our cabin and gets nosy. I figure we can take care of that chore this morning. We can load the bodies on their horses and take them with us when we go to check our traps. Along the way, we can scatter out three of their bodies in gullies and the like and leave them to the critters out on the plains to eat. The fourth one will be slipped into the river stark naked, so the birds can have at him as he drifts towards the Missouri River. By the time the body gets to that point, he should be more than picked clean by the crows, ravens and magpies. Then the crawdads can have at whatever is left," concluded Iron Hand.

Since there were no objections from the rest of the trappers, Iron Hand and Crooked Hand dressed for the winter cold, left their cabin and saddled up their horses for the day's work of running their trap line and Indian body disposal. Finished with saddling up their stock animals, all four trappers headed for the small grove of trees away from their cabin trailing the four Indians' horses onto which they were to load the bodies for disposal.

Approaching the grove of trees where the dead Indians had been placed the evening before, the four trappers got one hell

The Adventurous Life of Tom Warren

of a surprise. The snow in the area holding the dead was smeared blood red in every direction, and all the men could find of the four dead Indians was one heavily chewed upon pelvis and two badly chewed-all-to-hell skulls! That was all that was left! The rest of the Indians' bodies had been cleaned up and eaten cleaner than all get-out! Looking all around, the trappers discovered the tracks of one large grizzly and those from three smaller grizzly bears. From all the signs left behind, a sow grizzly with three two-year-old cubs had discovered the bodies sometime during the night and in their extreme hunger just before hibernation, had consumed everything of the four dead Indians except three pieces of their skeletons and a few badly chewed long leg bones...

"Well," said Old Potts with a grin, "those bears sure made it a darn sight easier when it comes to getting rid of all those four bodies. That plus the fact, come deep winter time, most bands of Indians stay close to their tepees unless they need to get out and hunt some buffalo for food. That in mind, maybe they will not be out looking so hard for their lost kin until springtime. By then, the wolves will have chewed up the rest of these body parts left behind and what they don't in the way of the little pieces, the gray jays will tote off, so no one, even them good at cold tracking 'war-hoops', will be able to find the rest of their kin."

Pleased over the previous night's grisly events, Iron Hand and Crooked Hand headed off to continue checking their trap line, as Old Potts and Big Foot still had a passel of fresh beaver skins from the day before to tend to in fleshing out and hooping for the drying to take place inside their warm cabin.

Come noontime, Iron Hand had been breaking thin ice when out in the water, as he checked all of his beaver traps. Each time he brought in a dead beaver for Crooked Hand to skin out, he would stomp around in the snow trying to warm up his soaked feet and wet legs. As Crooked Hand deftly skinned out the beaver carcasses, he would throw their bodies up onto the nearby shore for the wolves and remaining grizzly

160

bears yet out on the prowl and not in hibernation, to consume. By so doing, that reduced the carcass weight the packhorses would have to tote back to their distant cabin. If they did not do so, the trappers would have to find a spot to toss the beavers' carcasses near their camp and that would bring in unwanted predators. So the carcasses were left shoreside for the critters to pack off and consume out along adjacent the marshes where they had been trapped.

Long come the afternoon near the end of running their trap line and in a very dense stand of willows, Iron Hand dismounted and began walking out into the shallow pond water to retrieve another dead beaver floating at the end of the trap chain. Splashing through the water towards the dead beaver, Iron Hand heard VERY CLOSE IN THE ADJACENT STAND OF WILLOWS, A LOUD "OOOOH-OOOOOHHH-UMPH! UMPH!"

Before he could react to the terrifying sounds, there was a very loud and close at hand crashing of a heavy-bodied 'something' smashing through the willows, and THAT 'SOMETHING' WAS COMING RIGHT TOWARDS IRON HAND! Dropping the dead beaver he had been holding, Iron Hand went for his pistol in case he had to defend himself, only to have the 'something' come smashing out from the willow thicket some ten feet away! The next second, Iron Hand found himself facing a monster-sized bull moose, deep in rut in a head down, full charge!

Trying to run and shoot all at the same time at the sex-crazed and testosterone-fueled full of fury, bull moose, Iron Hand was able to take two steps before he felt the antlers of the huge moose smash into his backside, lift him upward and toss him some 20 feet away into the deeper water of the pond! Stunned from the terrific impact dealt out by a 1,000-pound moose in full charge, Iron Hand just lay in the water motionless trying to get his wits and wind about him!

Crooked Hand on the other hand, who was supposed to be providing protection for Iron Hand 'against all comers', was in

The Adventurous Life of Tom Warren

the middle of a damn good and violent rodeo with his horse and the two packhorses once they had seen and smelled the enraged moose! One thing Crooked Hand learned right then and there was that all horses were deathly afraid of a moose! Especially when it came to a bull moose in rut looking for something to 'mount', like anything 'moose-looking'-sized like that of a horse... So as a result, when the moose burst from the cover of the willows in a full blown charge upon hearing and thinking Iron Hand was another competitor bull moose, the horses seeing such a threat to their 'manhood' began doing their best to get the 'hell out of Dodge'! And if that meant tossing everything they were carrying on their backs to hell and gone, they were now in the process of doing just that! That included all packs and riders unfortunate enough to be 'in country' and that meant one unfortunate Crooked Hand in the process of being violently unhorsed into the 'wild blue yonder' as well!

Bucked off forthwith, Crooked Hand landed hard on the back of his shoulders! When he did, a frightened out of his mind packhorse stomped the hell out of him as it tried running away from the huge, noisy, smelly and dangerous, close at hand moose in rut!

Meantime, back in the middle of the beaver pond, Iron Hand had finally regained his wind and upon rising shakily out of from the water, was instantly lifted up off his feet with a 'face' full of moose antlers and hurled once again airborne and into a nearby patch of willows! That time, Iron Hand did not move when he landed. When the moose had hit him in his second charge, one antler tip had hit Iron Hand squarely in the center of his forehead and the 'lights went out'! Fortunately, Iron Hand had landed flat on his back, otherwise he would have drowned right then and there in the pond's deeper waters.

Meanwhile back at the 'rodeo', Crooked Hand was trying to stand up, grab his rifle and come to the aid of his 'high-flying' friend. Instead, he found himself puking up his breakfast once he stood up, after having his stomach being

162

violently stepped upon by his 1,000-pound packhorse! Getting rid of two buffalo steaks and four Dutch oven biscuits took a bit of a doing and when Crooked Hand had finished feeding 'the little people', he had another problem. HERE CAME THAT DAMN SEX-CRAZED, RED-EYED MOOSE RIGHT AT HIM, THINKING HE WAS FACING ANOTHER OPEN CHALLENGE TO HIS 'MOOSE-HOOD'!

Fortunately for Crooked Hand, Iron Hand was beginning to come around after being knocked silly, and seeing his partner in terrible trouble, drew his pistol, swung it at the head of the now charging and enraged bull moose in full tilt, and pulled the trigger. "POOOFF" went the now thoroughly wetted pistol, as its black powder charge failed to ignite!

Then it was Crooked Hand's turn to see how high he could be tossed on a set of moose antlers being 'used' by an enraged critter going full tilt. As it turned out, Crooked Hand, being a damn-sight smaller than Iron Hand, made it almost to the 'clouds' in tossed height before he fell back to the ground with a hard, BONE-RATTLING CRUNCH!

Then the moose, having had his 'druthers', began ambling off back into the willow patch looking for the female moose he had been so closely trailing and courting before the trappers had the audacity to 'rudely' interrupt his amorous hunt for the close at hand female in heat. However, that was the last 'lady in heat' he ever chased... BOOOM! went Crooked Hand's Hawken, as a .50 caliber lead ball ripped through the moose's neck, just below the juncture of his head! With that, the moose dropped into three feet of water and remained stock still except for the ripples flowing away from his now stilled body...

That was after Iron Hand, whose head was ringing like a church bell on a Sunday and with his insides trying to figure what day it was, had made it to shore, retrieved Crooked Hand's dropped rifle and had put an end to their moose and his little 'having a bad day' escapade...

Reaching down, Iron Hand helped Crooked Hand to his knees, as the man continued puking up the remains of his

The Adventurous Life of Tom Warren

'horse-stomped-in-the-guts' breakfast. That was when Iron Hand told him that would be the last time he would make him Dutch oven biscuits for breakfast if that was what he was going to do with them...

Upon hearing those words, Crooked Hand, finding the humor in Iron Hand's statement, tried to laugh and that only brought on more puking, only this time the eruption of his vomit even squirted through Crooked Hand's nose in the process because he was puking so hard...

Finally, the two trappers 'gathered up their skirts' and took stock of their situation. Here they both hurt like hell, had a dead 1,000-pound moose lying out in the beaver pond in three feet of water and nary a horse in sight! Finally, after a few more moments recovering, the two men got enough 'wind in their sails' to realize they were still among the living. With that, Crooked Hand staggered off after his lost horses, and Iron Hand, now thoroughly soaking wet from being made a 'plaything in the marsh for a moose', staggered off in three feet of icy cold water to put his sheath knife to the belly of a moose, whose best parts were shortly to be destined for the trappers' supper...

Half-an-hour later, Crooked Hand returned with their horses and taking a rope, helped Iron Hand drag his freshly gutted moose out from the beaver pond. Leaving the field-dressed moose where he lay after being dragged up on the ground at the edge of the beaver pond, the two trappers finished checking out the rest of their trap line. Upon finishing, they returned to the moose, now covered with what seemed to be a hundred of the ever-hungry magpies, ran the birds off the trappers' supper to be, and began butchering. What a sight the two trappers presented. One was soaking wet and could hardly walk and the other, covered in his own puke, could only stagger around after his run-in with a moose at the end of his charge subsequent to him having his horse stepping all over the softest part of his carcass...

164

Terry Grosz

One hour later of moose-butchering, the two bedraggled and stove-up trappers packed the meat upon their horses, called it quits for the day and headed for their cabin. Both packhorses were groaning under a huge load of moose meat, so the trappers took their time crossing the plains and heading for their secluded cabin hidden in the pines. Upon their arrival, both men were greeted by their partners, who were soon amazed over the story to be told of the two trap-line trappers and their tales of adventure. In fact, the 'storytelling' started when Big Foot and Old Potts had to physically help both Iron Hand and Crooked Hand down from their saddles. That was because both men had since their 'moose dance in the marsh' physically stiffened up from their 'run-in' with the 'supper' their packhorses were groaning under, and could not safely dismount under their own power...

Suffice to say it took more than two cups of their fiery high proof rum and an hour sitting in front of a hot fireplace fire to get the two 'moose hunters' loosened up in body and tongue in order to hear their whole story on how that evening's great tasting 'supper' came to be.

But when it came to supper that evening, both 'moose hunters' had managed to find their appetites. One because he had not eaten all day, in reality, after puking up his breakfast after the moose's and packhorses's impact with his carcass, and the other one over being antler-tossed not once but twice for good measure... By bedtime, both 'moose-men' had more than enough of their high proof rum under their belts that the pain from their lumps and bruises was no longer an issue. In fact, if the Gros Ventre had attacked their cabin that night, both 'moose-men', so loaded with rum, would have died happily in their beds without a concern in the world...

The next day, Old Potts and Big Foot took a turn running their trap line, while Iron Hand and Crooked Hand stayed back recuperating from their previous day's adventure. In so doing, they found themselves doing all of the fleshing and hooping of the beaver skins trapped the day before. As for the rest of their day, they happily made 'revenge-jerky' from one ill-fated, sex-crazed moose over their smoking and drying racks.

The following morning, both Crooked Hand and Iron Hand felt good enough to go back to running their trap line as Old Potts had originally planned. Both men were still so stiff and sore, that it took them some time to get up into their saddles but by noon, they had wrung out the kinks in their sore bodies and were moving right along from one trap-set to the next. However, Iron Hand's 'sixth sense', which had deserted him prior to the previous day's moose attack, was acting up as he ran the traps and removed the dead beaver for Crooked Hand to skin. As such and from experience over such innermost feelings, he kept a jaundiced eye on any and all willows in case the first moose had a mean-assed brother of the same temperament.

By late afternoon, storm clouds had once again gathered in the northwest and a cold and ominous wind was softly blowing and was heavy with foreboding moisture. Finally heading home with both panniers loaded with 31 skinned-out beaver hides, the best day to date numbers-wise, the men noticed that the clouds now seemed to be hanging unusually lower than normal when such a storm blew out from the northwest. Additionally, those clouds were an ugly blue-black in nature and occasional thunder could be heard coming from them off in a distance. As it turned out, a rare snow and thunderstorm was in the wind, and the men soon found themselves dangerously exposed and out in the middle of the open prairie as the storm roared in their direction! It did not take long for the men to realize that being out in the middle of the prairie during a lightning storm was not the best place to be while on horseback. With that, the men hustled their horses off in the direction of a set of nearby timbered hills with deep gullies in which to seek shelter from the coming tempest!

Then all of a sudden, there were four or five very close lightning strikes less than a mile or so distant to the northwest! Kicking up the pace of their horses, Iron Hand realized his 'sixth sense' of danger was almost raging inside him, and so much so that he began looking all around for any sign of hostile Indians. Seeing none, the men now began experiencing what is called 'thunder-snow' as small balls of snow pellets began heavily pelting them and their pack string. That was when Iron Hand noticed a strange noise. A strange rumbling sound like faraway thunder was coming from the northwest. It was almost like the constant sound that distant low rumbling thunder would be making. Still out on the prairie with lightning strikes being seen all around them, the men continued lying low in their saddles and spurring their mounts for the distant timbered hills and what they hoped would be sheltering cover.

Then it happened! One of the packhorses broke loose from its hard-jerking lead rope and bolted away in terror from the sights and sounds of the now violent storm lying almost low overhead of the fleeing trappers. Then, BAM! Both men found themselves riding madly bucking horses, as a lightning bolt hit so close to them that they could smell the acrid odor of ozone hanging heavy in the air, feel the concussion from the close at hand strike, and feel the long hairs of their beards standing straight out from their faces from all the static electricity in the air! Finally getting their madly bucking horses back under control, Iron Hand looked back and saw where the lightning strike had just occurred. Their packhorse that had broken its lead rope and had bolted away in fright now lay back behind them on the prairie, as a steaming and smoking mess! That packhorse had just been struck by lightning and killed deader than a stone...

Stopping and turning around, the two men stared in disbelief at the smoking ruin of what had once been a damn fine packhorse that was fully loaded with their beaver catch for the day! Now the horse was nothing more than partially cooked wolf bait! Realizing that had been the horse packing most of their prime beaver pelts, the men quickly rode back to see what they could recover. Dismounting and looking over

The Adventurous Life of Tom Warren

the smoking ruin, both men paused and then looked over at each other still in head-shaking wonder.

It was then that they could feel the ground shaking under their feet! "What the hell?" said Crooked Hand, as the sounds of rolling thunder and the acrid smells of ozone filled the air. It was then that Iron Hand said, "Crooked Hand! Mount up! That sound we are hearing and the feeling we are getting through our feet is not thunder! It is a buffalo stampede and from the sounds of it, we are right in its path and it is a big one! Let's get the hell out of here and I do mean now!"

Then the two men saw what had to be an apparition from hell bearing down on them! Over the next ridge flowed a vast sheet of brown heaving bodies of thousands of stampeding buffalo heading right their way! But that was not what scared the hell out of the two men, even though that alone was more than enough! All across the huge herd of stampeding buffalo bodies was what they knew from past experience and tales being told by other trappers to be the seldom-seen phenomena called "Blue Lightning"! Blue lightning that was glowingly affixed to hundreds of horns of the now terrified and madly charging buffalo. For the briefest of moments, both men stood there transfixed over the rare phenomenon of what they were seeing quickly bearing down on them! Literally hundreds of charging buffalo with blue-sparkling, static electricity-covered horns, all caused by the energy from the monster rare winter electrical storm they were now standing directly under!

"Let's get the hell out of here! Leave the dead horse and its beaver-filled panniers, we need to go now!" yelled Iron Hand in the increasing wind, crashing bolts of lightning and ear-splitting sounds of thunder directly overhead in the low-hanging, ugly blue-black looking clouds.

With that, both men vaulted into their saddles and with Crooked Hand leading their last packhorse, bolted for the nearby hills and the hoped-for safety they represented. By now, the men found themselves in a fierce whirlwind of flying dust, airborne frozen ice crystals, pellets of snow, and bolts of

Terry Grosz

lightning striking the ground all around them at random and then racing along the tops of the prairie grasses! Then the worst of the worst from the violent, overhead weather event occurred!

Huge stabbing bolts of lightning would strike the ground and then in an explosion of dust and flying snow, out would come bouncing blue balls of static electricity racing along the prairie in every direction! Then some of those racing balls of lightning slammed and bounced into the oncoming herd of panicked buffalo! When those balls of lightning slammed into the racing brown carpet of terrified buffalo, there would be a loud explosion and numerous animals could be seen flying through the darkened sky with their dark legs all askew like jackstraws!

Seeing all of that, the two men and their three horses found that terror-from-hell lent wings to their feet as they raced across the prairie, all the while lying low over their saddles to avoid being hit with the bolts of chain lightning now streaking all around them! By now, the charging and blue lightning-covered glowing buffalo were fast closing. Racing as fast as their horses could carry them, the men and their lone packhorse just made the timber, as the edge of the herd of fast-closing buffalo swerved to avoid running madly into the stands of trees...

Not slowing, the two men fled deeply into the dark timber, as the thundering herd of buffalo streamed by glowing, showering devilishly blue sparks and showing nothing but the whites of their eyes in the terror of the moment. Then it was all over... At least the fear from being crushed alive by the madly fleeing, highly electrified buffalo. But the lightning and the tall pines they were in turned out to be a 'horse of another color'. Trees around them were being struck right and left by bolts of lightning as the air once again smelled strongly and acridly of ozone, and now the new smell of burned pine needles filled their nostrils as well! Bailing off their horses in a small draw, both men laid out on the ground, as their horses stood

169

The Adventurous Life of Tom Warren

there trembling from their exhausting run across the prairie and in fright from the weather events unfolding loudly all around them!

About an hour later, the extreme weather event had moved on and now the trappers were faced with the utter silence of softly falling snow. Both men rose from the ground and then Crooked Hand said, "Damn, I ain't never seen such a thing. I swear we just saw a little bit of what hell looked like and after all that, I am going straight to my knees and saying my prayers like my beloved mother taught me to do every night at the foot of my bed."

After a few more moments of silence, Iron Hand said in a very measured tone and tenor of voice, "I saw hell once when my wife of just two years and my young son both died of smallpox in my arms. I say "hell" because I knew I didn't think that I could ever live a normal life or love again after suffering such a loss. That is why I came to the frontier to forget, figuring I could do so by experiencing and looking into the many faces of death on a daily basis. But like you, I think we both saw the Devil open his gates of hell for us to look into today and I for one, don't ever need to be reminded of what awaits all of us sinners if we continue to stray..." For the longest time Iron Hand just stood there 'drinking' in the emotion of the moment surrounding him and then said, "Crooked Hand, I never want to see another short period of time as we have just experienced these last few days! From Indian fights to fights with the Lord have we seen and experienced these last few days, you and me. Hopefully we two never again experience such wild events as we have just lived through."

Without another word being spoken because none were necessary after what they had just experienced, both men mounted up and headed for their cabin and a little quiet and downtime after violently living what they had just gone through. But before they left, they rode back to where they had left their pack animal lying out on the prairie after it had been

Terry Grosz

struck by lightning before the herd of buffalo had charged through the area. All that was left after the huge herd of buffalo had run through the area and over the animal were several heavy thigh bones and a wet spot in the partially frozen soil. "Well, I guess my idea on getting rid of those eight Gros Ventre who stormed into our cabin and tried killing us worked just 'slicker than cow buffalo slobbers' if this is an example of what occurs," said Iron Hand, in a tone of voice that would be associated with those who made their living out on the western frontier and the type of humor that surged through their bodies on a daily basis after facing death at their many turns on the trail...

Arriving later back at their cabin, they saw Old Potts cutting wood for their fireplace and Big Foot skinning out another deer so they could have something for their supper and later, a tanned deerskin to use as an outside covering for a pack of beaver *plus* being transported to Fort Union come the annual summer rendezvous.

As the two of them rode up, Old Potts walked over to the arriving men saying, "Say, did the two of you see that huge storm and all that lightning a couple hours ago?"

"Yep, in fact, we got caught out in the open when 'she' came upon us. When it did, lightning killed our best packhorse and we damned near got turned into mud by a huge herd of storm-caused stampeding buffalo trying to run us down. Don't need to see another such storm as long as I live," said Iron Hand as he tiredly dismounted and began tending to his horse.

"Me neither! Those damn bolts of lightning were smacking the ground all around us every which way, killed our packhorse and I swear, we both got a look at the very face of Lucifer afore it all passed overhead and went on its way," said Crooked Hand as he unsaddled his horse, hobbled it and let it out to graze.

"If we are having mule deer for supper, do I need to make my brand of Dutch oven biscuits?" asked Iron Hand, as he placed his riding gear under the shed meant to protect all their

171

The Adventurous Life of Tom Warren

leather goods from the elements just like nothing unusual had just occurred.

"I was a-hoping you would ask," said Old Potts, "and while you are at it, take some of our dried fruit, set it to soaking and make us one of your famous Dutch oven pies, if you be so kind."

With those words of encouragement and a heavily whiskered smile on his face over Old Potts's sweet tooth, Iron Hand took his rifle, lumbered over to their nearby spring, broke some of the ice off the pool of water and washed off his face and hands. As he walked back to their cabin, Big Foot handed Iron Hand a slab of hindquarter from the mule deer he was butchering saying, "Since you drew the biscuit-making detail, you might as well as throw a mess of steaks over the coals in a frying pan or on some metal cooking stakes as well. You know where the bear grease is and I suggest you make up several batches of biscuits. From the looks on your face and that of Crooked Hand, I would think a big supper and several cups of rum are in order. Most especially since Lucifer apparently got a good look at you two men's faces. That being the case, your last supper if he comes a-looking for you two mug-faces here at the cabin, might just as well be a 'good-un'." Thusly passed another 'fun-filled' winter week out on the frontier in the world of beaver trappers...

With winter fast upon them and ice that was getting too thick to easily trap through, the four men spent their daylight hours hunting mule deer for their needed tanned hides to be used as covering over the beaver packs while being transported to Fort Union for sale. Additionally when the weather allowed in their northern clime, they trapped wolves, hunted buffalo-- their favorite source of meat, chopped firewood on a never-ending basis, and tended to their stock when they were allowed to graze out on the open prairie under the watchful eyes of two of the men at all times to avoid any problems caused by horse-stealing Indians.

Terry Grosz

Come spring and ice out, the two-man team of Iron Hand and Crooked Hand continued their customary beaver trapping regimen from the fall before. As for Old Potts and Big Foot, they returned to their fleshing and hooping duties, so their small mountain of pelts would be ready for travel to Fort Union come late spring. Additionally, it was up to Old Potts and Big Foot to fold the dried *plus* fur side in and make compressed packs of 60 skins per bundle. This they did so each packhorse could comfortably carry two such 90-pound bundles while en route to Fort Union. As it now stood, the four trappers had amassed for sale or trade at Fort Union, 467 beaver *plus*, 32 gray wolf skins, 18 river otter skins, 64 muskrat skins, six buffalo robes, four grizzly bear robes, and a tanned moose hide. That was if the four trappers could manage to get to Fort Union and keep their hair in the process...

As it now stood, the trappers were facing the end of spring beaver trapping and those *plus* would be added to their overall total for transport to Fort Union. Happily, they had lost only one of their packhorses to the storm the men now called "Lucifer". Fortunately, they had gained four horses from the unfortunate Gros Ventre who had decided to sneak up and take out Old Potts and Big Foot earlier in the year, only to be taken out themselves by Iron Hand and Crooked Hand in a timely manner. Those same four now dead Gros Ventre, who in a surprising circumstance became grizzly bear food later that evening, making disposal of their bodies by the trappers, happily unnecessary...

Come springtime when the beaver had finally gone out of 'prime', the trappers pulled their traps and spent the next two days surveying their old and potentially new beaver trapping areas. True, they had trapped out a portion of their trapping grounds, but there was still a lot more new territory for the trappers to successfully run their trap lines for at least another year if they so decided. The question in each man's mind however, was it worth it? Especially in the constant face of the deadly Gros Ventre threat to return in the fall for another

The Adventurous Life of Tom Warren

trapping season, and the very real chance to be discovered by a larger war party and if that occurred, to 'lose one's hair' in the venture...

One morning as Iron Hand made his brand of biscuits for the rest of his party, Old Potts exited the cabin to take care of a call of nature. About ten minutes later, the men inside the cabin heard the loud BOOM! of a rifle being fired close at hand! Racing outside first with rifle in hand since he was up fixing breakfast and close to the front door of the cabin, Iron Hand saw Old Potts with his buckskin pants pulled down around his ankles from going to the bathroom, fighting an Indian with each hand! It was quickly obvious to Iron Hand that Old Potts, tough as he was, had way more than he could handle! It was also starkly obvious that Old Potts was fighting for his life, as both Indians were holding upraised tomahawks and about to strike down the white man trapper they had caught 'with his pants down' while attending to a call of nature!

BOOM! went a quick shot from Iron Hand's Hawken, felling the closest Indian in the fight to him with a spinal hit! Dropping his now empty rifle and sprinting for where Old Potts was still fighting for his life, Iron Hand got surprised. Another Indian rose up from his place of hiding in the elderberry bush just ten feet in front of the hard-charging Iron Hand! That Indian instantly raised his rifle at the oncoming hard-charging trapper and was immediately overwhelmed by a furious six-foot, seven-inch, 250-pound trapper, who now had a hold of the man's upraised rifle with a grip of iron! Jerking the rifle from the Indian's hands, it went off into the ground just inches from Iron Hand's feet! Ignoring the close-in shot at his feet, Iron Hand dropped the Indian's rifle and quickly reached for his antagonist. As that Indian grabbed for his sheath knife to defend himself in the face of such a furious charge by the trapper, Iron Hand's right hand closed around the neck of his assailant and immediately choked and simultaneously jerked the man violently onto the ground! As he did, another Indian materialized from the elderberry bushes just feet away, raised

174

his rifle directly at Iron Hand, who was at that instance in the process of killing the first man in his hands and started to pull the trigger! However, in that same microsecond of time, that Indian's head exploded into a brilliant spew of red and blobs of gray matter! BOOM! went Big Foot's rifle as he took out the Indian about to kill Iron Hand, shooting into him from just a few feet away...

Throwing aside the Indian he had just choked to death with his right hand by crushing his windpipe and internal carotid arteries, Iron Hand then reached out from his kneeling position on the ground, grabbed the leg of the Indian still fighting with Old Potts and jerked him down from behind! When he did, he also jerked Old Potts down as well, who was still locked in the Indian's savage grip. With a quick snap of that Indian's neck, that man never got to face off with his huge assailant before he joined the rest of the Cloud People...

Jumping to his feet so he could continue in the desperate struggle with any other nearby Indians, he felt a sharp pain in his head for just an instant, and then the lights went out!

With a loud ringing in his head and what felt like his whole body crazily spinning, Iron Hand began returning to the real world at hand. "Lay still, Partner. You took one hell of a rap on that big ole ugly noggin of yours," said the familiar voice of Crooked Hand, sounding like he was speaking from way off in a distance somewhere.

Moments later, Iron Hand had regained a better command of his senses and sat up in a still somewhat-spinning world. Then he saw the faces of Old Potts, Big Foot and Crooked Hand all peering down at him with concerned looks on their faces.

"What the hell happened?" asked Iron Hand, as he rubbed the side of his head and in so doing, had his hand come away all covered with his own sticky blood.

"You big dummy, you had just killed the Indian who had a hold of Old Potts by breaking his neck, when another of their party sneaked up behind you and slammed you on the head

The Adventurous Life of Tom Warren

with the butt of his rifle. Don't worry, though. He did not get a chance to go back to his people and brag about knocking you in the head. I saw to it that a lead ball ruined his day," said Big Foot as he continued loading up his just-fired pistol...

"Help me to my feet, Guys. Is everyone else alright?" Iron Hand asked, as his huge frame was being lifted up to his feet with some difficulty by Big Foot and Old Potts.

As Iron Hand leaned his still unsteady body against a small pine tree, he asked, "Old Potts, what the hell happened? Where the hell did all these Indians come from and how did they know we were here in our cabin?"

Then Iron Hand could see Big Foot and Crooked Hand looking over at Old Potts for an explanation as to what had happened, so he continued doing the same.

"Well, I guess it is up to me to explain since I started this whole damn 'hurrah' when I came out to take my morning dump. I had no more than dropped my buckskins, when a dying mule deer burst out from the brush below me, ran right up to where I was squatting and then died right at my feet. Before I realized what had happened and then spying an arrow sticking out from behind the deer's front shoulder, I was surprised by these two Indians running hard after their crippled deer. They was just as surprised as me, especially when I stood up and they saw just how much of a man I was, I guess, standing there all bare naked and all... Then realizing I was the enemy, those two young bucks now lying dead at my feet, ruined my morning's dump by grabbing me with their tomahawks in hand. I was whipping the two of them, when Iron Hand stuck his big nose into my fight and shot hell out of one of them afore I could do him in. Then the rest is a blur. It seemed just as fast as Iron Hand was killing them, they was a-growing up out of them elderberry bushes like flies on a dead buffalo in the July sun and then the killing really began."

"Near as I can tell from all the dead Indians around me, Iron Hand kilt three of them tomahawk-swinging bastards and the rest of you kilt those left standing. Fer as I can tell, we had

Terry Grosz

a party of Indians out on a deer hunt and they shot this big buck with an arrow now lying dead at my feet. That damn deer ran right to me and fell over dead, and then the rest of them 'buggers' dropped in on my morning's dump and you guys can figure out all the rest," continued Old Potts with a big ole sly grin showing through on his heavily whiskered face.

"Now what do we do? We got six dead Gros Ventre and what the hell do we do with all them bodies?" asked Big Foot. "'Cause you know their kin are going to come looking fer them just as sure as Iron Hand is the ugliest among the four of us standing here all agape..."

"I got an idea. We dump all these buggers in the Poplar River, take their horses, load up all our gear and head fer Fort Union fast as we can. We are almost ready to get the hell out of this area fer the season anyway, so what if we leave a week or two earlier than planned? Let's dump all the bodies along the Poplar at intervals as we trail all their unshod ponies behind our pack and ridin' string. That way, we leave the area for a few months, they cannot find us, we get rid of all them bodies and make our ways to Fort Union with all of our hair. By the time we all come back, if we do, them 'war-hoops' will have forgotten all about this killing spree and their disappeared kin. If that plan is alright with you guys, let's find where they tied off their horses, load up our furs and leave the area tomorrow at daylight. Along the way, we can dump their bodies and leave the area with all our furs and our 'God-given' hair, what do you think about them 'apples'?" said Old Potts.

The next mid-morning found all the trappers' horses loaded with their traveling gear and packs of furs. The rest of their gear they did not need for the trip had been cached in the ground behind their cabin for later retrieval, if and when they returned. The Indians' horses, having been located the morning before, were now carrying their dead masters' bodies as the men made ready to leave. This the trappers purposely did because any Indians tracking the dead men's horses could tell by the depth of the horses' hoofprints if they had been

177

The Adventurous Life of Tom Warren

ridden or not. If the hoofprints were deep as normally associated with a ridden horse, the trappers figured that may throw any pursuers off the fleeing trappers' trail. As it just so happened, lashed across one of the Indians' horses was also the 'traveling meat' for the trappers on part of their trip, namely one mule deer, courtesy of the now dead Indian hunters who had 'tangled' with Old Potts and company the day before...

Reaching the west bank of the Poplar River not far from their cabin, the trappers turned south trailing the six Indian horses behind their pack string in order to hide their obvious white men's horses' shod tracks. Reaching the south end of the rolling and timbered hills leading them out onto the open expanses of the prairie, the trappers dumped the bodies of the now identified by their dress, Gros Ventre, from the battle the day before into the strong current of the Poplar River.

After making sure the bodies were drifting southward on the Poplar towards the Missouri River, the trappers continued heading south as well. They had only traveled about five miles when Crooked Hand spotted about 20 Indian riders riding in a loose group on the far horizon. Realizing that remaining out in the open expanses of the prairie would make them easy targets since they were so outnumbered, the men scampered their pack string into a large grove of nearby aspens, rode their stock into the middle of the trees where they were more than out of sight and dismounted. Walking their horses over into a deep gully in the grove, the men secreted the animals out of sight and then took up their battle positions at the top lip of the gully, in case their pack string's tracks were discovered by the distant Indian riders and then they were ridden down and attacked where they lay in hiding.

Come nightfall and still undiscovered by the distant Indian riders, the trappers re-emerged from their place of hiding and continued riding south along the Poplar River. After a three-hour ride in the darkness, the trappers once again pulled into another grove of aspens along the Poplar River and made a cold camp. However as a matter of precaution, Big Foot and

Iron Hand hobbled all their horses and remained with them as they grazed into the night. Come daylight, the horses had been re-packed with their bundles of valuable furs and after a breakfast of jerky and a long drink of water from a spring, the trappers once again hit the trail. Around noontime, the trappers and their pack string arrived at the shallow ford in the Poplar River, crossed over and by nightfall, had reached the north bank of the Missouri River.

Making camp in a dense stand of trees and brush along the Missouri River bottoms, the trappers finally chanced a fire. There they prepared a staked dinner of fresh venison from the buck deer the Indians had killed earlier in the fight with the trappers and relaxed as best as they could while their horses quietly grazed deep in the river bottoms. Early the next morning, the trappers set out once again heading for Fort Union with their heavily loaded pack string and with luck, would soon have a chance to relax under the fort's protective walls from the chance of hostile Indian discovery.

Two days later, the four Free Trappers were once again met at the fort's front gate by their old friend and Factor for the American Fur Company at Fort Union, Kenneth McKenzie. Once again, McKenzie was amazed over the sight of their trapping successes, the size and quality of their *plus* and another again-increased horse herd comprised in part by somehow acquired Indian ponies!

The following day after being hosted by McKenzie the evening before to a dinner fit for the class of Free Trappers they represented, the usual ritual for fresh from the field trappers with their furs began. While Old Potts and Big Foot oversaw the counting and grading of their trapping successes, Crooked Hand and Iron Hand visited the fort's storehouses and began compiling the next year's stocks of needed provisions, since the four men had finally decided on spending at least one more year on the frontier as trappers, since it suited their fancy just fine.

The Adventurous Life of Tom Warren

The following evening, two Free Trappers sporting a long pack string of heavily loaded mules and horses moved into view of Old Potts's camp located along the Missouri River bottoms off to one side of Fort Union's log walls. Suddenly, those riders stopped cold in their tracks upon seeing Iron Hand cooking around Old Potts's campfire and intensely scrutinized him for the longest time like something was the matter with what they were seeing! As they did, Iron Hand could see that one of the men moved ominously closer to his closest pack animal carrying a rifle as if to be able to quickly withdraw it for immediate use! Then the intense looks from the two newcomers shifted from the camp cook onto Old Potts, Crooked Hand and Big Foot who were working around the campsite, with the same intense stare they had originally focused upon Iron Hand just moments earlier. *Strange behavior indeed for the two new arrivals to the general Free Trappers' campsite to say the least,* thought Iron Hand.

Finally satisfied over what they were seeing or not seeing around Old Potts's campsite, the two strangers casually brought their pack string down the riverbank, pitched their tents nearby, unpacked their tired animals and let them out to graze in the lush river bottom grasses adjacent the Missouri River. As they moved around establishing their campsite, Iron Hand noticed that both men were strapping, strong-looking individuals, weighing at least 250 pounds apiece and who were at least six-and-a-half feet tall. Finally seeing someone almost as tall and large as he was and interested in meeting the new arrivals, Iron Hand ventured over to their campsite once they were all set up and introduced himself to the two strangers.

"Evening. I am Tom Warren from Missouri but my friends here call me "Iron Hand", and you two are…?"

"Good evening, yourself. I am Joshua Dent and this here runt of the litter is my younger brother, Gabriel," he said with an obvious 'smile' in his voice, unlike the serious and strange initial scrutiny they had just given Iron Hand and his camp mates an hour earlier. "Our friends just call us Josh and Gabe.

Terry Grosz

If you don't mind me asking, Tom, how the dickens did you come by that moniker of "Iron Hand"?" asked Joshua.

"Oh, that comes with a long story. But in short, while being attacked by some Blackfeet warriors up around the Medicine Lake area, I managed to kill one of the attackers with my hands during the battle," replied a slightly embarrassed Iron Hand over what he considered a rather personal question. Then to quickly change the subject over the genesis of his unique frontier name among his friends and the rest of the knowing trapping fraternity and looking over at the two trappers' beautiful string of horses, Iron Hand said, "Those six matched buckskins with the coal black fetlocks are some of the finest looking animals I have seen in a long time. How did you two come by such fine and beautifully matched horses way out here on the frontier? I ask, because most of the stock we see out here are rather rangy and not anyway near the fine quality of those animals."

"They at one time belonged to our uncle back in Missouri. Uncle Jack was killed, as was our aunt, by four Missouri ruffians or "Bushwhackers" as they are called locally. Those killers are led by an evil son-of-a-bitch from the next county over from our old farm, named "Black Bill Jenkins". Those horses are in part, the reason why my brother and I are here in this country today as trappers. You see, Black Bill and his three brothers, named Clio, Stilt and Lem, also killed our Ma and Pa earlier that same year on our farm, while my brother and me were out gathering several tubs of wild honey. Then because the law was hot on their trails for all the killing and robbing they were doing around our county, they fled to the frontier and became trappers to avoid a damn good hanging back home in Missouri," surprisingly offered Joshua in more detail than Iron Hand would have normally expected from having just met the two men.

"When that bunch of killers left the county, we sold Uncle Jack's farm since we were his last of kin and became trappers for a damned good reason as well. In so doing, me and my

181

The Adventurous Life of Tom Warren

brother here dedicated our lives to hunting down Black Bill and his kin and have been doing so now for several years. We know and have also heard from a number of other trappers, that they have been trapping and trouble-making up on the Yellowstone, the Musselshell, the Big Horn, and now along the western reaches of the Missouri. After hearing they may be in this neck of the woods, we came to Fort Union in the hopes we would find the four of them selling their furs and getting new provisions so they could continue on the run out here where there is no law. If we run across their trails and the four of them, we intend to kill the lot for what they did to our folks and our aunt and uncle," continued Joshua with a degree of iron in the tone and tenor of his voice.

With those words of extreme family loss, Iron Hand's mind once again flashed back to his painful past and the loss of his young wife and beautiful, first-born young son to a deadly 'killer' as well. A 'killer' to his way of thinking, which was just as evil as Black Bill was purported to be, but one which moved silently and had favorable consideration for none...

Then Joshua broke into Iron Hand's thinking back on the darker earlier days in his life by saying, "So, Tom, back to our 'horse' story. When we left Missouri hunting Black Bill and his kin, we needed good horses for what we planned on doing in tracking down that bunch across the untamed west. Being that Uncle Jack had some of the finest horseflesh around that neck of the woods, we came by them naturally since we were the last of his kin. Those buckskins are the only ones like them, and my brother nor I will never, ever part with them unless someone kills the two of us and steals them from our cold, dead hands. They mean that much to us, because that is all of what we have left from our past to remind us of the loved ones we lost in such a violent manner. And we aim to keep them unto death," continued Joshua with a look of grim determination flashing across his dark eyes. Surprisingly, the same dark look

182

Iron Hand had when he was deep in battle with the Blackfoot that had attacked their camp earlier...

Then Gabriel got a big grin on his face as did his brother, when the following was revealed regarding their earlier behavior upon riding up onto Iron Hand's campsite. "When we rode up onto your campsite, we damned near fell off our horses after all these years of chasing Black Bill and his kin. You see, Iron Hand, you are built just like what we have heard tell about Black Bill, beard and all! He is at least six-and-a-half feet tall, weighs about 250 pounds and has a large, dark beard just like you are sporting. That is why we paused at the top of the trail leading to this camping area in amazement when we spotted you tending the campfire. On first look, we figured you were Black Bill and we were almost ready to start shooting and kill you on the spot, before we realized you might not be him because of the friendly appearance of your fellow trappers at the campsite! You see, Black Bill has three brothers who are running with him that will never leave his side, and are all reported to have flaming red hair and beards. Just as we were ready to come down and see if you were Black Bill and if so, send you off to hell, Joshua here cautioned me not to shoot unless we were certain you were him. But then when it came to looking over your friends, we could see that none of them sported red hair and beards, so we backed off knowing you were not the man we were after. However, we came down to check and make doubly sure who you were anyway, up close and personal like," said Gabriel with a wide and winning smile...

With the story of that discovery out and in the open, all three men had a good laugh over what might had happened had Gabriel unlimbered his 'smoke-pole' and had exhibited an itching trigger finger. That and a keen eye for details, like a mess of flaming red-haired and bearded cohorts accompanying a large and tall man with a dark in color and full beard not being in the company of the rather large man the two brothers were at that moment observing...

The Adventurous Life of Tom Warren

It was then that Iron Hand's eyes took a closer look at the different type of rifles the two brothers were carrying. Rifles that were not exactly like the Hawkens he and his fellow trappers were carrying but appeared to possibly be their rifle's forerunners.

"Say, what manner of rifles are you men carrying? They kind of look like the Hawkens me and my friends are carrying but are not exactly the same," asked Iron Hand, out of his curiosity for all manner and type of firearms, being an ex-Army man.

"They are an old style, 1803 U.S. Military rifle, caliber .50 in nature. When we left St. Louis, other trappers who had been to the frontier told us to get heavier caliber rifles than our old Pennsylvania rifles, which were .40 caliber in nature. They advised that a rifle shooting at least a .50 caliber ball would be needed for such large-bodied animals found on the frontier like the moose, buffalo and such mean-assed critters like the much-feared grizzly bear."

"So my brother and I went to Samuel Hawken, a local gunsmith in St. Louis, to see if he might have and we could buy from him, such a heavy caliber rifle. He did not have one but said one like we wanted was in his mind for making in the near future. However, he did have a shipment of the 1803 short barrel, half-forestock rifles coming and would be willing to sell us those. When we heard that, we bought out his stock of four such rifles and have used them ever since and very successfully, I might add," advised Joshua.

Walking over to look at Gabriel's rifle lying against a nearby sitting log by their fire site, Iron Hand could see that the rifle had been heavily engraved and had carved into the stock the wording, "Gabe's Rifle".

"That sure is some fancy 'smoke-pole'," said an admiring Iron Hand, as he looked down at the customized rifle lying propped against the sitting log.

"Like our matched buckskins, that rifle will never leave my hands unless someone kills me and removes it from my

Terry Grosz

dying finger," said Gabriel quietly, but with a touch of 'cold' in his voice. "That rifle has saved my life three times in just this last year alone. As such, it will remain with me until my dying day," he continued...

"I hear what you are saying about your rifle. My Hawken is just as special to me as is your rifle to you. I too have not only saved my life with my rifle but that of some of my friends as well," quietly said Iron Hand.

"I be doing all the talking since we met; what about you guys?" suddenly asked Joshua.

"Well, some years back, the four of us signed on with Kenneth McKenzie, Factor for the American Fur Company when we were in St. Louis. We signed on as Free Trappers but agreed to accompany the expedition to the Missouri headwaters and help them in building a trading post and fort for trade with Free Trappers, Company Trappers and the local tribes of Indians. Then upon the building of Fort Union and the end of our contracts with the American Fur Company, we four struck out on our own as Free Trappers. There is not much left to say except that we have been very successful and still have all of our hair," said Iron Hand with his classic, heavily bearded grin.

For the next month, Old Potts's trappers remained in their camp along the Missouri River bottoms visiting with old friends; purchasing new handmade and gaily decorated clothing items from the numerous groups of Indians visiting the fort that befitted the flashy dress of Free Trappers; hunting buffalo; topping off their supplies, especially more high grade rum, powder, lead, beans, rice, dried fruit, and sacks of coffee; trading off four of the Indian horses from the earlier fight with the Gros Ventre while keeping the rest as pack animals; and having all of their animals's hooves trimmed and re-shod by the fort's blacksmiths for the coming year's travels.

During those times, Old Potts's group of Free Trappers and the "Brothers Dent" spent many an easy evening visiting, telling tall tales, drinking rum and eating together as good

The Adventurous Life of Tom Warren

friends will do. About a month into their stay, the Dent brothers received word from several other freshly arriving Free Trappers at Fort Union that they had a violent run-in with four trappers working the Missouri further west, who had caused them grief over ownership of overlapping trapping territories. The descriptions given by those trappers fitted Black Bill and his three red-headed and heavily bearded kinfolk to a "T"! With that current information in hand, the Dent brothers topped off their needed supplies for the coming year and headed out west along the Missouri River looking for the killers of their kin. But before the Dent brothers had left, Iron Hand and his partners made sure they had a small mountain of freshly cast rifle and pistol bullets in their inventory in case they were able to hunt down and kill those four Missouri bushwhackers who were long past needing more than just a damned good killing, from all they had been told around the men's campfires over the last month!

Then Iron Hand's peaceful and quiet times at the fort drastically came skidding to an unexpected halt. Walking out one morning from one of the fort's storehouses carrying bags of cooking spices forgotten in his first go-around of re-provisioning, Iron Hand walked right into the rather large and imposing 300-pound figure of John Pierre entering the same storehouse!

"Well, if it ain't you and the rest of your worthless self," bellowed John Pierre, equally surprised at the unexpected meeting of his old antagonist from the days of little Sinopa!

Surprised over seeing such a disgusting individual up close and personal like, one who brought back ugly memories regarding his treatment of the young Blackfoot women named Sinopa, Iron Hand quickly recovered his composure, said nothing and kept walking away with his armload of supplies. It was then that his 'sixth sense' came roaring back into his inner being, causing Iron Hand to tense up!

That was when the day turned ugly! John Pierre, not wanting to be ignored, reached out and grabbed Iron Hand's

186

arm, causing him to spill to the ground the various bags of much-needed spices. Without missing a heartbeat, Iron Hand whirled and with one punch, knocked John Pierre clear off his feet and in the process, sending him airborne into a nearby horse trough, where the man landed unceremoniously with a huge splash in front of God and everyone else looking on!

Seeing that quick reflex action on the part of Iron Hand and the splash of their friend going into a horse trough, brought John Pierre's nearby standing cohorts into the arena of battle at a dead run! Seeing that he was outnumbered, Iron Hand faced off with the crowd of onrushing friends of John Pierre with the same calm and demeanor that one does when it is realized there is a snake at hand that needed a damn good and quick killing...

The first of John Pierre's friends and fellow trappers to reach Iron Hand, a man known only as "La Rochelle", got a thoroughly smashed flat, broken and bloodied nose for sticking it into places where it did not belong! That thunderous blow to the nose sent La Rochelle to the ground like a sack of spuds off a wagon... The second running man to reach Iron Hand with an upraised tomahawk and blood in his eyes was a smallish French-Canadian man whose last name was "Galipeau". He in turn, had his tomahawk ripped out from his upraised right hand, his arm broken and his entire carcass tossed off to one side like a rag doll into a loudly howling-in-pain heap! The third and fourth trapper friends of John Pierre to reach Iron Hand received, simultaneously in concert, a busted mouth and the other, an eye that would not be looked out from for at least a week until the blood in the damaged eye socket had dissipated into its surrounding and now swollen tissues... Those two men then whimpered off and out of range of Iron Hand's flashing right hand like a just-kicked cur dog...

Then it was a soaking wet from his surprising immersion into a horse trough John Pierre, who upon coming back into the fight, ran into a 'whirlwind' of hammer-like fists at that point in the battle that quickly covered his face and head with

The Adventurous Life of Tom Warren

soon-rising knots the size of ducks' eggs... It took exactly three seconds of Iron Hand's flying fists to lay out John Pierre cold as a long dead beaver for the next ten minutes!

As for John Pierre's trapper buddies numbers five, six and seven, whose names Iron Hand did not know, they drew up short of the physical combat arena after seeing what had just violently happened to the rest of their unfortunate cohorts...

Then those three remaining cohorts from John Pierre's group of American Fur Company Trappers simultaneously drew their pistols and leveled them at a standing still and quietly fuming Iron Hand!

"Now, you big bastard, you will pay for what you just did to our friends and for stealing our little Indian 'play-pretty', Sinopa, when you should have been minding your own business," said an evil-looking man, who smelled even worse than he looked!

"Shoot 'em! Shoot 'em, Monk, afore that bastard McKenzie puts a stop to our little surprise party," said a tubercular-looking stick of a man who came to be known as "Jacques Du Mont" and a close friend of John Pierre.

"First trapper who pulls a trigger will be dead before he hits the ground," said a strident mystery voice being emitted from the corner of one of the adjoining storehouses.

Du Mont, his two still-standing cohorts and Iron Hand spun around in concert to see who the threatening mystery voice belonged to. There stood Spotted Eagle of Chief Mingan's band of Blackfoot from the Medicine Lake area and seven of his warriors! They had just left one of the fort's storehouses with several armloads of supplies and upon seeing their friend Iron Hand in trouble, acted accordingly! Their supplies now lay strewn about on the ground in disarray, as the eight warriors now held their rifles on John Pierre's trappers with a look in their eyes meant to convince one of what was to follow if a single man of the three holding leveled pistols on their friend Iron Hand, even dared breathing too deeply, much less, if they unwisely pulled a single trigger...

Seeing that they were now outnumbered by eight very serious-looking Blackfoot warriors, the trappers slowly laid their pistols down on the ground, just as a very concerned and previously alerted to the fight McKenzie hurriedly rounded another building and entered the battle arena.

"All of you lower your weapons! What the hell is going on?" McKenzie bellowed, as he stepped in between the two warring groups of men. "Iron Hand, what the hell is the meaning of all of this," said a still-concerned McKenzie over the possibility of a small war being carried out within the confines of his fort's wall between his trappers and his Indian trading partners.

"Ask him," quietly said Iron Hand, as he nodded to a still out cold John Pierre lying on the ground off to one side.

"I might have known," said McKenzie in a disgusted sounding voice, once he recognized the identity of the heavyset man lying stretched out on the ground with a badly swollen face.

"It is not going to do me any good asking John Pierre until he comes around from what I suspect is something he started and could not finish," said McKenzie, as he looked over at the towering and still quietly standing at the ready for whatever came his way, figure of a very determined looking Iron Hand.

About that time, several well-armed Company Clerks ran up to their boss McKenzie, and looked at him for instructions relative as to what he wanted them to do.

"Throw some water from that horse trough onto that bastard lying on the ground over there and wake him up," said McKenzie, as he pointed to an inert John Pierre lying a few feet away.

Following their boss's orders, they soon had John Pierre brought around, up on his feet and trying to explain away his earlier behavior through a set of badly split and swollen lips recently received from the fists of Iron Hand. Once McKenzie got the story straight as to what had caused all the uproar, he ordered John Pierre to settle up with the American Fur

189

The Adventurous Life of Tom Warren

Company as to their wages. Then he ordered the men to get their needed supplies, leave the fort and never return because every one of John Pierre's group was now released from their contracts as American Fur Company Trappers and were now out and on their own...

As John Pierre's group of 'slightly damaged' and still mad as hell and fuming trappers staggered off out of sight, McKenzie turned and told Spotted Eagle and Iron Hand that he apologized for the behavior of some of his Company Trappers. Then he said in a contrite tone of voice, "All of you men need to follow me." With that and not another word spoken, the men adjourned to another storehouse, where all the really valuable items of trade, like kegs of gunpowder, firearms and casks of his best rum, were stored.

For the rest of that afternoon, the men were treated to all the rum they could drink, as a gesture of McKenzie's friendship and as an apology for his Company Trappers' bad behavior. During a moment of that time, Iron Hand went over to Spotted Eagle, shook his hand and told him that he appreciated his getting involved during a rather tense moment when the three trappers had the drop on him and had him covered with their three pistols.

Spotted Eagle just smiled back and said, "Remember when we met last at your camp in the winter and had our discussion about Sinopa. I told you after that meeting that we would be like brothers forever, just like you and Chief Mingan. I only did for my brother what he would have done for me under the same circumstances."

The rest of that afternoon was spent in enjoying each other's company and when Iron Hand walked back to his camp with his armload of needed spices, he felt pretty damn good, considering the amount of rum he had consumed and all... Plus, there was additional good news that came to light as the men enjoyed each other's company over the free rum. Spotted Eagle and Sinopa were now husband and wife and happily so, with one very young son and another child on the way!

190

Terry Grosz

The next day, McKenzie, Fort Union Factor, good to his word, ordered Company Trappers John Pierre and his seven other trapping cohorts off the grounds of Fort Union for their previous day's bad behavior. That was also followed with a stern warning never to return, either to sell their furs or to seek re-provisioning of their annually needed supplies. With that and escorting guards leading the now-released fur trappers from the fort, John Pierre and his slightly worse for wear cohorts left Fort Union and disappeared. As that group of dejected men sullenly rode off, Iron Hand could not get the stirring 'sixth sense' out from his being that those trappers and Iron Hand would meet again, someday, somewhere and when they did, the outcome would be so very different.

The Adventurous Life of Tom Warren

Terry Grosz

Chapter 9: The Return, The Gros Ventre and A Fortune Is Made

COME DAYLIGHT ONE WEEK LATER, Old Potts and his trappers had their pack string loaded and ready to go. They had told all their trapper friends to 'hang onto their hair', had breakfasted with Factor McKenzie as Free Trappers were accustomed to doing one last time before they left for the next trapping season, and then began moving out from their old campsite along the Missouri River bottoms heading west. Traveling along the north bank of the Missouri, the four trappers made good time, as they headed for their cabin in the land of the dreaded Gros Ventre and the streams and marshes full of blanket-sized beaver in their old and familiar Poplar River country.

Several days later, they crossed the Big Muddy River, then continued westerly along the Missouri, en route to their trapping grounds on the Poplar River. Along the way, the men passed numerous small herds of peacefully feeding or resting buffalo, antelope, elk and a few bighorn sheep on the prairie. On the morning of their sixth day, the men crossed the Poplar River at their usual fording location, than turned northerly. Finally arriving at the southern end of their familiar line of

193

The Adventurous Life of Tom Warren

timbered foothills lying to their west, the four trappers became even more vigilant, as they continued passing numerous trails of unshod Indian horses and *travois* drag marks heading north in the same direction of travel as that of the four trappers.

Stopping along the way one afternoon, Old Potts gathered up his trappers for a 'confab'. "It appears we are trailing an entire band of more than likely Gros Ventre on the move. They probably are moving their main camp because they ran out of horse feed or for sanitary reasons. I don't think it wise for us to just follow blindly along behind them for fear of stumbling into one of their campsites. I would suggest that Iron Hand and Crooked Hand ride out ahead of us and make sure we are not riding into an ambush of sorts or stumbling into one of their camps."

Without another word, Iron Hand handed off the lead rope of his pack string to Big Foot as did Crooked Hand his to Old Potts. Then checking the priming on their rifles and pistols, the two lone riders spurred their horses on ahead as they followed the *travois* drag marks and the hundreds of unshod Indian pony hoofprints, all the while keeping hidden from sight by using the features of their terrain as much as possible. Iron Hand and Crooked Hand rode until the edge of darkness overcame them and not finding any Indians or being discovered themselves, they stopped and waited along the way to their cabin site. About three hours later, Old Potts and Big Foot rode up trailing their pack strings to their two waiting partners for another hurried conference.

"We are only about an hour out from our cabin," said Old Potts. "What say we continue on and see if we even have a cabin left or not? Besides, the quicker we get out of sight and not wandering around here out in the open with a valuable string of horseflesh, the longer we get to keep our hair."

Agreeing, each trapper once again took hold of a lead rope to an individual pack string and hurried on towards the direction of their cabin under the cover of darkness. As they did, they continued spooking off along their route of travel

Terry Grosz

numerous herds of resting buffalo and elk. Finally arriving at a darkened point of trees coming down from the adjacent foothills that led up to their old cabin, the men quietly moved in that direction hoping for the best when they got there. When they arrived at their familiar geographic turning point, they left the trail of travel taken earlier by the band of Gros Ventre on the move and headed in the opposite direction into the timbered foothills leading up to their fairly secluded cabin site.

Under a full moon, the four trappers quietly rode into the immediate area of their cabin and horse corral. At first glance in the moonlight, the cabin appeared to be intact and undisturbed as did the horse corral. Iron Hand dismounted and remembering the 'welcoming' they had received by a grizzly bear who had taken up residence in their old cave campsite by Medicine Lake, moved cautiously forward to the cabin's front door which was slightly ajar with his rifle in hand and at the ready.

Slowly opening the front door wider with the end of his rifle barrel so he could cautiously look inside before he entered, Iron Hand was met with a sharp-smelling stench from some animal that had occupied their cabin throughout the summer during the trappers' absence. However, further examination inside the cabin's dim light revealed that whatever had left the strong smell was now long gone and it was safe to enter and occupy their old home.

Turning to let the rest of his party know the cabin was now free of any still smelly critters, Iron Hand took two steps and then felt a sharp, stabbing pain in the toe area of his right moccasin! Jumping in pain, he let out one hell of a loud howl like a grizzly bear in heat, as his big toe exploded in intense pain like something had just taken a rather large bite from the luckless appendage!

"What the hell is going on in there!" yelled Crooked Hand, as he bailed off his horse and raced toward the cabin to aid his big friend who was still inside and howling in pain like a banshee.

195

The Adventurous Life of Tom Warren

Rounding the front door's frame, Crooked Hand ran straight into a madly exiting Iron Hand, who was bailing out from their cabin like a scared cottontail rabbit just discovered by a red fox at close quarters...

WHOOM went Iron Hand's big body into the lesser-sized Crooked Hand and a loud "OOOOFF" was heard to follow, as both men now exploded out from the cabin's doorway like they had been shot from the barrel of a rather large cannon!

By now with all the howling and the loud crashing sounds made by two men impacting into each other at a high rate of speed, the hair on the necks of both Old Potts and Big Foot was standing straight up and both men now had their rifles at the ready. Had their rifles at the ready for whatever had obviously gotten hold of Iron Hand, the biggest and strongest man in their group and from all the sounds of it, the mysterious 'thing' causing all the commotion was winning! Something that had him now howling in pain and had put him on the run like there was no tomorrow! Put him on the run and in so doing, had caused him to plow into his friend Crooked Hand who had been coming to the rescue, causing one hell of a wreck! When the crash of the two bodies occurred, it spilled both men out onto the ground in gay profusion in front of God and everybody into an embarrassing heap!

Finally, peace and quiet once again reigned, as Old Potts got a small fire going in their old outside firepit, which cast some light on the issue. As it turned out, Crooked Hand had a bloody nose and blackened eye from having the larger, stout as a bull buffalo sized-trapper, namely Iron Hand, plowing into a rather smaller, soft as the 'fluff' on a cottontail rabbit, man, namely Crooked Hand, at a very high rate of speed!

As for Iron Hand, all six-foot, seven-inches of his huge frame now found him hopping all around on one foot like a grizzly bear had just taken a chomp out of his last part over the fence... Finally, Old Potts got all the wild dancing dust settled down and with the light from his fire, the problem was solved.

196

Terry Grosz

While walking around in the dark of their cabin checking things out, Iron Hand had picked up a porcupine quill from a previous visitor in the cabin with the front of his moccasin, had jammed it clear up and almost out of sight, under the toenail of his big toe and two inches into the flesh beyond! Thinking a grizzly bear had taken a chomp out from his foot, Iron Hand had hightailed it in intense pain, out the open doorway from the cabin at a high rate of speed and had plowed directly into his friend Crooked Hand, who was coming inside to rescue him. Well, when a 250-pound man going faster than a 'bull buffalo in rut to meet a cow in estrus', runs into a 150-pound man, who is cautiously creeping along in the dark, violent physical things could and did happen...

Not having enough light to adequately address the 'porcupine quill' issue that evening and not wanting to sleep inside their cabin because of the intense smell from a porcupine living therein all summer, the men bedded down outside and after having a long day, drifted off to sleep. That was all except for Iron Hand. His big toe was throbbing so much, that he sat up all night tending the fire and soaking his sore appendage in a Dutch oven full of warm water and a pinch of their precious salt from one of their packs.

The next morning, Big Foot took his knife and dug out the troublesome quill from Iron Hand's toe with a lot of howling and squirming by the rather large fellow in the process. Then Big Foot doused it with a dash of rum on the injured appendage and then took a swig himself, after having to timidly operate on a man twice his size, possessing the killing power and strength of an ox if one was to rile him! Then the men cleaned out the porcupine droppings left inside the cabin and threw back the deer hide window coverings so the place could air out. Following that, Old Potts dug up their cache placed behind the cabin when they had left to go to Fort Union earlier in the spring and retrieved the items previously cached. As he did, the other men unloaded all of their packs, putting their supplies inside their cabin for safer keeping and made it homey once

The Adventurous Life of Tom Warren

again, notwithstanding the residual strong porcupine smell left behind after its uninvited summer's use.

As Big Foot hobbled all their livestock and then let them out to pasture, Crooked Hand and Old Potts went deer hunting up onto the timbered hillside behind their cabin and an hour or so later, a single rifle shot was heard. An hour later, Crooked Hand and Old Potts returned, dragging a fat and barren mule deer doe back into camp. Hung from their meat pole moments later, she was soon dressed and cooling out in preparation for the many good venison-laced meals to come.

As Iron Hand hobbled about camp on his sore foot, he still tended to his duties as the camp cook. Into a roaring campfire went all of his previously used cast-iron frying pans, Dutch ovens and bean pots to burn off the old built-up residue from cooking times past. A short time later after the cast had 'burned off' and cooled down sufficiently, Iron Hand greased all the cooking surfaces from his stoneware crock of bear grease and put the cooking ware back inside their cabin for later use. However, he kept out a six-quart cast-iron bean pot and soon it was filled with a mess of dried beans and rice, spices, hot pepper flakes and cold water from their spring. Then the large pot was hung off to one end of their campfire on cooking irons so it could slow cook. About an hour into slowly cooking away, Iron Hand cut off the rib meat heavy with fat from the hanging deer carcass, 'chunked' it up and tossed it into the bean pot to cook as well. Then as an afterthought because the men had been 'a-horse' for a number of days and all were having trouble with their lack of bowel movements, tossed in an extra handful of hot pepper flakes designed to make their morning 'clean-outs' just a tad easier...

As he did, he could hear the ringing of axes above their cabin as the other three trappers began dropping dead trees and cutting them into logs with their crosscut saws so the horses could drag them down to their cabin for their winter log pile. Keeping busy since he was no use with the woodcutting detail being as sore-footed as he was, Iron Hand made himself useful by boning out the mule deer doe, cutting most of the meat into thin strips for making jerky, peppering the hell out of it to keep the flies and yellow jackets away, and hanging it upon their meat smoking rack. Then with the addition of a small but smoky fire, he left the meat to dry out and smoke away.

Following those butchering duties, Iron Hand got out two of his Dutch ovens and prepared to make biscuits for the men's dinner. Stirring the previously filled bean pot and satisfied with its slow-cooking process with the beans and rice now mostly cooked all the way through, it was then seated onto some coals for finishing. Then Iron Hand made up the dough for his brand of much-favored sugared, Dutch oven biscuits. Lastly, he got out their weathered old coffee pot, filled it with cold spring water and set it over the center of their fire on cooking rods to boil. Once he could hear the water was in a rolling boil, Iron Hand slipped the coffee pot over to where it was boiling with less vigor, added six handfuls of coffee grounds, since his fellow trappers liked their brew stronger than an angry mule's kick, and let it simmer away on the cooler side of the fire on higher-racked cooking irons.

A shout from the hillside near their cabin and the sounds of horses struggling as they pulled a number of logs downhill, was Iron Hand's signal he had better have supper ready and lots of it... Into his last Dutch oven that had been soaking away for the last hour loaded with dried fruit, went several handfuls of sugar, a heavy sprinkling of cinnamon and nutmeg, and then his cobbler dough was laid over the now fire-heated thickened filling. That was then set over a mess of raked-out coals from their firepit with a shovel full of coals sprinkled over the Dutch oven's heavy cast-iron lid just retrieved from the open fire and left to bake.

As the men stacked the logs just dragged into their winter woodpile area, unhooked their horses and watered the same, Iron Hand let everyone know that supper was ready when they were. Old Potts, Crooked Hand and Big Foot, after a hard day of logging, did not need any further urging, as they washed up

The Adventurous Life of Tom Warren

in their spring and then hustled over to their outdoor cooking fire and sat down on their sitting logs. Hobbling around the firepit, Iron Hand made sure everyone got a large bowl of his special spiced-up bean, rice and deer meat stew mixture, a hot cup of coffee, and two Iron Hand-sized biscuits hot from a Dutch oven for starters. With his crew plated-up, Iron Hand served himself and sat down with his friends to the 'noisy silence' that comes from a group of hungry men slopping down and eating what they loved, in a place they loved to be living... When supper was done and everyone had eaten his fill, Old Potts said, "Iron Hand, you outdid yourself. Them victuals were 'plate-lickin' good'."

The next morning and days following were a repeat of the days before, as the four trappers prepared for the winter to come and the industry that had to follow when one was trapping beaver out on the frontier. Coffee, Dutch oven biscuits and staked slabs of venison greeted the men as their main morning meals. Following that, the horses were once again harnessed for the logging that was to follow, as the winter woodcutters went off to work once again. However, with Iron Hand still hobbling around although not as badly as he had the days right after his porcupine 'accident', he still commenced with his preparation and cooking duties once again, so as to keep the hardworking trio of his trapper friends fueled up and ready to meet the day and what it offered.

On one of those occasions, out came a six-quart Dutch oven, which was loaded with dried rice, dried raisins, two Iron Hand-sized handfuls of brown sugar cones, filled with cold spring water and set off to one side to long-soak. Then grabbing a shovel, off Iron Hand hobbled onto a nearby rocky hillside, where he spent the next hour looking over his shoulder for any Gros Ventre who might happen along and be sneaking up on him, as he dug out from the rocky ground under a stand of purple flowers, a mess of small in size but tasty and pungent, wild onions. Toting his rifle, the shovel and a leather pouch full of onions back to camp, Iron Hand then checked his

200

soaking pot of dried rice and raisins. They were plumping up perfectly, so the pot was brought to the firepit and set off to one side of the hottest part of the flame to begin slow cooking.

Walking over to their spring, Iron Hand then peeled and washed off the wild onions and returned to his firepit once again. There he checked his previously set aside dried beans soaking in a potful of water and found them to be plumping up as well. Into that bean pot went another handful of red pepper flakes, salt, black pepper, chunks of the last of the deer meat from the day before's kill, and triple handfuls of fresh wild onions. That potful of beans and venison was then set over a mess of coals to slow cook alongside the rice and raisin mixture cooking away in another Dutch oven. Soon, their campsite smelled sweetly of the raisins, melted cones of dark brown sugar and rice mixture, as well as the luxurious smell of cooking venison, beans and onions in the clean air surrounding the prairies...

As Iron Hand's menu of the day cooked away, his partners continued dragging into their winter log pile more dead logs of pine and Douglas fir. But the work carried on without any noontime breaks, because trapping season would soon be fast upon them and they still had to set aside their winter supplies of emergency hay for their horses. Plus they had to gather into their camp needed winter stocks of buffalo meat for smoking and jerking. Those, so they would have the needed provisions to carry them through the hard winter months when moving about and hunting would become problematic because of the harsh winters in their neck of the woods on the Northern Plains.

As expected, when the 'loggers' arrived at their cabin come suppertime, they were famished. Awaiting them was a 'lumberjack' heavy meal of beans heavily loaded with chunks of tender venison, all flavored with the pungent taste of wild onion and spiced with more hot pepper flakes. Accompanying that repast were Iron Hand's style of Dutch oven, sugar-sweetened biscuits and hot trapper's coffee that was stout enough to float a mule's shoe! Then to satisfy everyone's sweet tooth was Iron Hand's concoction of viscous gooey-sweet, cooked rice and raisins, heavily sweetened with ample amounts of dark brown sugar from their sugar cones, nutmeg and cinnamon...

With enough wood now cut and piled to last throughout the worst of the winter months, all the men, now that Iron Hand's foot had healed up sufficiently enough for him to stomp about with the rest of his crew, turned to cutting armloads of high quality hay from the stream and river bottoms, all to be stacked and dried near the men's horse corrals for when the winter snows made caring for their horses out on the open prairie problematic. Following that hot and mosquito-infested work, the men finally found the time to partake in what they loved doing the best, besides eating and drinking that is, namely making the final necessary preparations for partaking of the coming trapping season for beaver, muskrat and river otter...

The following morning after a fast meal of just Dutch oven biscuits and coffee, the men saddled up four packhorses, readied their firearms for hunting and if needed, self-defense, then headed out onto the prairie reaches along the Poplar River. As a matter of precaution and planning, the four men rode the river's watered areas checking out the coming season's beaver trapping opportunities first. But as they did, they also rode northward in the direction taken by the earlier traveling band of Indians they had followed into the area when they first came back to their cabin after a summer's absence. They did so in order to make sure once they began trapping, they would not inadvertently stumble onto the campsite of the dreaded Gros Ventre with the deadly consequences on both sides that would more than likely follow...

After traveling north along the Poplar River drainage for about 12 miles and not finding any evidence of an Indian encampment but lots of beaver sign, the men returned to an area of the prairie closer to their campsite. There not terribly concerned about the closeness of any Indians hearing their shooting since they had just scouted out the area, the men got into a small herd of buffalo, killing four fat cows.

Riding up to one of the dead cows, the four men dismounted and keeping their rifles near at hand, watched as Old Potts slit open the side of the buffalo just below the last rib and cut out a large slab of steaming hot buffalo liver. Then the four men gathered around with their knives, cut and smeared the gall bladder juices over the liver, cut off large chunks of the still-quivering organ, and gorged themselves on the bloody, still hot, semi-bitter, mineral-rich tasting liver! Before that frontier feast and celebration of the first buffalo kill of the year was over, the men had consumed most of the animal's raw liver, body-starved as they were for the minerals it contained! Then the work began, as the men removed the most-favored hump ribs, boned out the best portions of each of the four animals they had killed, loaded their groaning-under-the-weight packhorses and then headed for their cabin. Once there, as the men prepared the meat for smoking on their meat racks or for their meals, they gorged on the fresh, raw and still bloody chunks of the rich warm meat, like they had not eaten anything for months on end...

The next few days were repeats of their first day when it came to hunting and killing buffalo. Soon, the men had meat racks groaning under the weight of their bounty, as the warm smoking fires below worked their magic, turning heavy raw meat into a lighter jerked staple. As fast as the meat was properly jerked or dried, the men stored away such valuable foodstuffs in previously tanned deerskin bags and hung them from the ceiling rafters of their cabin to keep the bugs and vermin from getting into such valuable stores and ruining them.

Finishing with their vital buffalo meat-gathering chores, the men turned their attention to the harvesting of the local mule deer population, in order to have enough skins collected and tanned to cover their packs of beaver *plus* when they transported the same to Fort Union come late spring of the coming year. These deerskin procurements were most important because the beaver *plus* were tightly packed within the tanned deerskin bundles to protect them during transport from the elements and getting dirty. For the next three weeks, the men hunted mule deer, jerked the meat from the animals, and brain and urine-tanned their skins as covering for their more valuable, soon to be harvested, beaver *plus*.

As their trapping season moved nearer the cooler fall months, the men began feeling the urgency at hand. They still had to gather in more mountain mahogany meat-smoking wood, dig more wild onions, dry and bag the same for adding variety to some of their meals during the winter months, cutting and splitting up smaller chunk wood for their outdoor firepits and indoor fireplace, mending their clothing for winter wear, smoking their beaver traps to rid them of the man smell, and casting up a small mountain of pistol and rifle balls for hunting and any self-defense needs that would be required.

Then there was another concern facing the men. Lately, they had observed more and more Indian hunting parties roaming the prairies, as they too were hunting buffalo in order to gather and put up their favorite winter meat stores. In so doing, it made it harder for the men to go about their normal chores as they readied themselves for the long and arduous work associated with the beaver trapping to come, for fear of discovery. Come the time for beaver trapping to begin and with the Indians still roaming about or not, the trappers had their trapping hand forced if they wanted to survive economically as trappers.

Around their outside campfire one night, Old Potts said, "I think now is the time for us to begin trapping once again. The beaver should be coming into prime with the days getting shorter, colder water and weather now having an effect on those critters. I think for the short term, until we get buried with beaver *plus* and need to leave some of us back at the cabin for that processing work, we need to all work together on the trapping end of things. That way if discovered, we have more firepower and the Gros Ventre would be less likely to take on the four of us, over just the two of us out and about trapping. So, once again, I suggest all of us form a single trapping party, we load our packhorses heavy with extra rifles and pistols and the best shooters of the group, namely Iron Hand and Crooked Hand, remained 'horsed' in case danger from attack comes, while our trapping is underway. If we do it that-a-way, we will be more secure in what we do best, should be able to catch the beaver as we see fit, and all of us come home each night with all of our 'topknots'," said Old Potts, as he picked up a long stick from the ground and randomly stirred up the coals in their outside firepit.

Hearing no disagreement from his fellow trappers as he looked around at their faces for any comment or disagreement, he said, "Alright. That is what we will do for the start of our trapping season. I will do the trap setting and the three of you can do the watching out for my miserable carcass while I am in the water. Then when I bring a beaver, mink or muskrat ashore, Crooked Hand, being our best and most careful skinner, can work his magic with his skinning knife. That-a-way, that will give me time to warm up, we can get the skinning done in the field and keep our hair all at the same time. Now, I suggest we all get some shut-eye, for tomorrow will be a long day because if we set all 40 of our beaver traps, that will take some doing," remarked Old Potts. "Any other thoughts concerning what I just said?" he asked as he rose to head for his sleeping furs. Hearing none, he walked off in his usual shuffling gait and into the cabin. He was followed shortly thereafter by the rest of the trapping crew after they had put out the pipes they had been smoking.

Yeah, little did Old Potts know that tomorrow would be a long day in the marsh.

Come pre-dawn the following morning, found everyone a-horse with saddle bags stuffed with jerky, possibles bags topped off, and leading two packhorses loaded with jangling beaver traps and extra rifles and pistols attached in such a

The Adventurous Life of Tom Warren

manner to the animals for easy retrieval if attacked or any other need for their use was to arise.

With Old Potts leading the string of trappers out from their camp since he was to be the lead trapper and decider when it came to trap placement, he was followed closely by Crooked Hand, who was an excellent shooter and the group's main skinner. Next in line trailing the packhorses came Big Foot, who was trailed by Iron Hand, who was the most accurate shooter in the bunch. The reason Old Potts had placed Iron Hand at the end was because in any surprise attacks, the Gros Ventre, when historically given the option and choice of their mode of attack, usually attacked trappers from the rear of their pack strings. This they did because coming in from behind usually caused the greatest amount of confusion within the ranks of those being attacked. Also, the greatest amount of damage could be done in so doing, with the least amount of loss of life or limb to the attackers.

Out from their secluded campsite the trappers streamed, moving in and around small herds of buffalo and elk (Author's Note: Elk were originally plains animals. That was until they were hunted to the point that they fled into the mountains and remain there to this day.), as they headed for the Poplar River, its marshes, backwaters and numerous beaver dams and ponds. About an hour later found Old Potts knee deep in the water and mud of a beaver pond driving in the trap's deep water stake. As he did, Iron Hand and Crooked Hand sat 'horsed' and alert to their surroundings, as Big Foot tended to the pack string and the handing of beaver traps to Old Potts as needed. The four trappers, with time-honored and practiced movements, made good headway in their trap site selection and placement. As they moved forward, the trappers saw that the waters being selected as trap sites were alive with not only signs of lots of beaver but numerous large adults swimming around as well. *The start of their day had all the looks of another good trapping season,* thought Iron Hand, with a smile of appreciation for the goodness of the day and the beauty he was enjoying in his wild surroundings.

It was then that the 'color' of the trappers' day turned dark... With a howl, Old Potts went down as he was driving in another end of a trap's chain-stake in deeper water. Instantly, Big Foot was off his horse and into the water to aid Old Potts. But before Big Foot could get to his partner, Old Potts was back up and standing, albeit somewhat wobbling in his now unsteady stance! When Big Foot got to Old Potts to see what had happened, he was advised that he, Old Potts, had stepped into an unseen muskrat hole and had twisted and sprained his knee!

With that injury, the best laid plans that Old Potts had suggested the evening before, had just dissipated off into the prairie winds... Aiding Old Potts in moving out from the usual clutching deep mud of the beaver pond, Big Foot helped the badly limping man up onto firm ground and then after looking his knee over, helped him onto his horse so he wouldn't have to put any more weight than normal on his now bad leg. Handing Old Potts his Hawken, Iron Hand dismounted, finished setting the beaver trap and avoided the muskrat hole hidden by the muddy pond water that Old Potts had stepped into.

The rest of that day, Iron Hand selected the trap sites, set the traps and did so until they had set all 40 of their beaver traps. Then the four men scouted out their trapping area even further to the north, in case the trapping grounds just selected did not pan out as they had hoped. However, on their return riding along their recently set trap line, their hopes were more than realized. Before they had backtracked along their entire trap line, the trappers discovered they had already caught 16 beaver in their earlier sets! To a man, that was a good omen and in keeping with their tradition for good luck, four of the largest beaver caught were not skinned on site as they usually did, leaving the carcasses to nearby land predators, but were brought back to their camp. Once there, while the other three men fleshed and hooped out the skins from the beaver just caught, Iron Hand skinned out the four fresh ones still in the round and soon had the fat oozing from their carcasses as they slow roasted on a spit over their mountain mahogany-fueled campfire.

As the wonderful smell of roasting beaver meat filled the air around the trappers' campsite, Iron Hand took it upon himself to throw together a mess of biscuit dough and soon added another great smell into the cooling evening air. In the meantime, Old Potts had now hobbled over to his sitting log around the firepit and sat there nursing an injured knee and a cup of their high proof rum all at the same time...

That evening, after a dinner of roasted beaver and Dutch oven biscuits, the trappers sat around their firepit discussing the day's events, as they all now nursed a cup of rum in celebration of their first day's successful trappings. Even Old Potts, bad knee and all, joined right in because after dismounting when they had arrived back at their cabin, discovered that his knee injury was not as bad as he had originally thought after stepping in a hidden by the muddy water, underground hole made by a muskrat. In fact, the more he carefully moved around on the knee, the looser it seemed to get and the better it felt.

The next morning, Old Potts's knee had stiffened up once again and he found that the warmth from their fire and his moving around was helping a bit. But it was decided that he would not be the main trapper until he got better and that Iron Hand would continue as the main one doing all of the trap setting and beaver, muskrat and river otter removal.

Following breakfast, the four trappers streamed out from their campsite and onto the beaver trapping grounds near the Poplar River's numerous waterways. Arriving at their first set, Iron Hand found it 'occupied' with a very large and dead beaver weighing at least 70 pounds! With some difficulty, he removed the dead animal from the trap and staggered across the pond's muddy bottom over to a waiting Crooked Hand. As

Terry Grosz

Iron Hand once again set the trap next to a heavily used beaver slide area, Crooked Hand skinned and whipped the hide off the monster critter in short order. Once again, the carcass was tossed off to one side and out in the open so the land predators would discover a waiting meal and dispose of the same in short order. That trap checking and beaver removal action occurred for the next 17 traps checked, before a single empty set was discovered! The men were elated over such trapping successes and by the end of the day, 29 of their 40 sets held a dead beaver! Iron Hand set every trap in the same place as he had the day previous along the entire trap line and then the men headed for home to begin their fleshing and hooping duties on the pelts they had trapped that day.

Streaming back into their campsite by late afternoon, the men were looking forward to caring for their pelts and a nice supper of roasted elk that had been killed by Crooked Hand the day before, along with some of Iron Hand's piping hot Dutch oven biscuits. However, that joy of returning to their cabin with a load of fresh beaver skins and hopes for having an elk supper were soon dashed...

Upon turning by the last finger of timber leading into their campsite, THE MEN INSTANTLY NOTICED THAT THEIR CORRAL WAS EMPTY OF ALL OF THEIR PACK ANIMALS AND THEIR EXTRA RIDING HORSES!

Spurring their horses into camp, Iron Hand stepped off his mount and hit the ground running over to the open gate at their corral. Kneeling down, he could plainly see two sets of moccasin footprints around the corral's gate. It was obvious that two Indians had discovered their secluded campsite, had ridden in to investigate, discovered the unguarded pack and riding horses in the corral, and had helped themselves and stolen the lot! As Iron Hand continued examining the ground around the corrals for additional clues left by the horse thieves, Old Potts headed for their cabin to see if anything else of value had been taken, like their valuable stores of furs or gunpowder. Emerging moments later, Old Potts said, "All of our furs and

209

The Adventurous Life of Tom Warren

provisions here in the cabin were not touched. It looks like they were just after our valuable horse herd."

"Crooked Hand, you give me a hand in what I am about to do. In the meantime, Old Potts and Big Foot, you guys care for our fresh beaver pelts so they don't sour and let the fur slip. Don't wait up for us because we may not be back for a while. The two of us are going after our horses right now while the trail is hot and the tracks are fresher than all get-out. If we don't and those two horse thieves make it back to the main Indian camp, we will see more of the same of their kind right at daylight, now that they know we are here in their own backyard. Plus when they do come back, it will be with blood in their eyes and itching to take everything from us that they can lay their thieving hands upon," concluded a deadly serious-looking Iron Hand, as he handed his horse's reins up to Crooked Hand. Then with Iron Hand on foot and Crooked Hand trailing him, he took off at a trot following the fresh trail the two Indian ponies and the rest of their stolen stock had left in the soft dirt.

After about a mile fast tracking their stolen stock by the old trapper's method of running a 100 yards and then walking for 50 and then repeating the same process, it became apparent to Iron Hand that the horse thieves, secure in their theft from the white trappers, were only walking the large herd of stolen horses. It was almost like the horse thieves were novices in what they were doing, or at least very confident and secure in their escape with the valuable animals. The stolen horse herd's tracks, along with those of their new masters, led in a direct line up over a timbered ridge and into a small valley on the reverse side. As Iron Hand continued trotting along the fresh trail on foot, he continued in amazement over the fact that the horse thieves were not running the stolen horses hellbent for leather for home but were just walking them...

THEN THERE THEY WERE! Ahead about a half-mile away, were two Indian horsemen pushing along a now somewhat unruly herd of pack and riding horses that once belonged to the now aggrieved and hot on their trail and closing, trappers.

Mounting up now that the horse thieves were in 'their sights', Iron Hand and Crooked Hand set their course for an interception of the two armed Indians near a finger-line of dense pine trees. As they did, they could see that the two Indian thieves were still having trouble pushing their stolen horses and keeping them together in one bunch as they continued pushing them on down the valley. It was apparent that something was spooking the trappers' stock and in their unruliness, the animals were slowing down the two Indians in their attempt to make their getaway. But as the two trappers made haste to slip into an ambush position, they did not realize that they had been spotted as they slipped their way into the timber to intercept the horse thieves...

About 20 minutes later, Iron Hand and Crooked Hand were what they figured to be in a perfect ambush position if the two Indians kept coming their way. Then as if to complicate matters, a late day, violent thunderstorm was in the making and moving its way down the long valley heading their way. With the loud crashes of thunder and bolts of late summer thunderstorm lightning flashing all around the nearby foothills, the two Indians managed to continue pushing the stolen unruly animals right into the two trappers' ambush site.

Then as the two horse thieves hove into view, they all of a sudden stampeded the horse herd directly at the two trappers lying in wait by a small bluff of rocks. Before the trappers could react, two hard-charging Indians riding right behind the stampeding herd of stolen horses emerged from a cloud of dust stirred up by the horses' hooves, firing at the now-surprised trappers as they came!

ZZZZIPP! WENT A BALL FIRED FROM ONE OF THE INDIAN'S RIFLES, CLIPPING OFF A FIST-SIZED TUFT OF FACIAL HAIR FROM IRON HAND'S FULL BEARD! BEING SURPRISED BY THE STAMPEDING HORSE HERD AND THE NOW AGGRESSIVE ACTION OF THE

The Adventurous Life of Tom Warren

HARD-CHARGING INDIANS, IT WAS ALL IRON HAND COULD DO IN KEEPING HIS NOW NERVOUS HORSE UNDER CONTROL! THE BULLET FIRED BY THE OTHER INDIAN AT CROOKED HAND FLEW BY HIM SO CLOSELY THAT HE FELT THE AIR MOVEMENT FROM THE PROJECTILE, AND HEARD A "WHIRRRRING" SOUND AS IT PASSED CLOSE TO THE SIDE OF HIS HEAD...

Seconds later, BOOM–BOOM! went Crooked Hand's and Iron Hand's Hawkens almost simultaneously, as the two Indian horse thieves now streamed by in plain view behind the stampeding horse herd and within very close rifle range! Neither Indian ever knew what hit him, as the huge .50 caliber lead balls smashed into their chests from about 30 yards away! Sitting on their horses amidst the crashing sounds of thunder and bolts of lightning from the now arriving violent thunderstorm, the two trappers calmly reloaded their rifles. That they did before they moved from their place of ambush and exposed their positions any further, in case they had miscounted the number of horse thieves involved.

Then it was the trappers' turn to round up and get control of their stolen livestock, as the Indians' ponies, now alarmed over the fury of the storm and the shooting so close at hand, bolted off into the vastness of the prairie! In typical horse fashion and behavior, they more than likely headed for home and their familiar Indian horse herd located some miles away before the trappers could get them under control. Finally getting their own stock under control but only after the trappers had moved their horses away from the timber and the recent Indian kill site, did the men chance a return to the scene of the two dead Indian horse thieves. Their stock, now that they were away from that stand of dense timber and whatever was in it causing them such concern, were left to settle down and begin feeding in a long draw, as the late summer rains now came down with a drenching vengeance!

212

Terry Grosz

Now getting soaked to the skin as the center of the storm continued passing overhead, Crooked Hand and Iron Hand rode over to the kill site and got a huge surprise! Both Indian horse thieves were only about 16 or 17-year-old kids, but they were just as dead! That was when Iron Hand realized why the thieves had acted so casually and novice-like in their escape attempt with the trappers' stolen livestock. They either did not realize the inherent danger in what they were doing because they were so arrogant because of their youth, or just inexperienced when it came to stealing horses from white trappers. Either way, it got them a quick trip into the realm of the rest of the Indian Cloud People who had passed before them!

Then both Crooked Hand's and Iron Hand's horses began acting up, a behavior like when they had observed the stolen herd acting all goofy-like when the two Indian boys were attempting to push them through the timber. It was then when a change of the blowing storm's winds brought to the two men's noses a strong smell that told them only one thing, a grizzly bear's den was somewhere close at hand!

That was when Iron Hand got a grin on his face along with an idea. "Crooked Hand, help me carry the bodies over to that bluff of rocks by that point of timber. I think there may be a grizzly bear's den near there because of that terrible smell. To me, it is the same smell we had back at our Medicine Lake cave site when we returned from Fort Union, and the grizzly bear living in our cave 'glommed' onto your left leg when you rode up to the entrance. If we leave the dead boys nearby, maybe the grizzly will find them and dispose of the bodies for us. That way, when their tribal members come looking for their young men, and they will, maybe they will think the bear got them and that will settle the issue of their disappearance. Quick, let's get moving and get the deed done. Then let's move our herd of horses out of here while we still have the rains coming down so heavily. That should help wash away most of the tracks and our involvement if we are successful. Let us get

The Adventurous Life of Tom Warren

going so we can make it look like Mother Nature did them in and not several of the hated white man trappers."

Long minutes later, the deed was done and then the trappers took their horse herd home by an indirect route, including walking them down a small creek to help in erasing their shod 'dead giveaway' tracks. Finally by dark, the men and their horses were back at their cabin. The two trappers were soaked to their skins and cold as the dickens, but they were safely home with all of their valuable livestock and a tall tale to tell.

Met as they came into the cabin site, Big Foot put all their stock into the corral, while Crooked Hand and Iron Hand tended to their wet riding gear and individual horses as well. Then it was into their warm cabin and a change into some dry clothing, as they related their story of horse recapture and the killing of the two young men horse thieves in the process.

When Old Potts heard about the deaths of the two young men, a frown quickly clouded his bronzed face. "You know, those boys' parents will come a-looking fer them. I hope you two managed to hide your tracks and the killing-deed in such a manner that we four are not found out. 'Cause if discovered and they pin them killings on us, we will lose our hair and a hell of a lot more fer sure," he continued in a worried tone of voice.

"Well, we dragged the boys' bodies up near a bluff of rocks where I figured by the terrible smell, that a 'griz' was living nearby and we left them there. Fact be known, when we shot after being shot at first, we did not figure our horse thieves to be anything but horse-stealing Gros Ventre and not young men. In fact, look here. The first shot fired by the young men took off a part of this magnificent beard of mine and made a mess of it on one side." With those words out, Iron Hand turned sideways so his two friends could see the fist-sized clump of beard missing. Big Foot, after taking a look at the missing 'face-fur' remarked, "Made a damn fine improvement on that ugly mug of your-en, if that be possible."

214

Ignoring the funny Big Foot made at Iron Hand's expense, figuring more was a-coming if he made notice, the aggrieved trapper continued with his story. "Then we skedaddled out of there with our stock, under the covering rains of one hell of a rainstorm. With a little luck, our tracks should be more or less washed back into the prairie soil, so that we cannot be cold tracked back to our cabin. Hell, we even walked the horses down that small creek on the other side of the ridge before we turned them for home in order to hide our shod horse hoofprints," said Iron Hand.

"Well, the deed be done and at least we got our horses back. Without them, we would soon be a-foot and being in that manner means you are a dead man out here in the country of the Indians, the beasts and the vastness of this here prairie," said Old Potts. "But now, we kain't take any more chances. We need to leave some of us back here at our cabin in case we are discovered once again. I don't want to lose our stock or any of our provisions, so two of us must stay behind and protect what we have. That means, sore knee or not, we will stay here and you two are now going to have to do all of our trapping. However, when you do, the both of you must make sure your pack animals are sporting an extra rifle apiece and two pistols, so's you can at least have a fighting chance if discovered by them damn killing Gros Ventre and you are out in the open. Enough about what has happened, now how about some grub? We managed to save the two of you some of what we had. It weren't up to your cooking standards, Iron Hand, but it will stick to your ribs," said Old Potts with his characteristic heavily whiskered grin.

The next morning, Crooked Hand and Iron Hand went forth once again to run their trap line. And when they did, between the two trappers, they carried four rifles and eight pistols either on their persons or on their nearby pack animals. As for Old Potts and Big Foot, they also were sporting their own regularly carried firearms and had two extra rifles and four pistols stashed around their cabin in case they were discovered

The Adventurous Life of Tom Warren

and ambushed by those who might be out hunting their two lost young sons who had not returned home from the evening before...

Running their trap line once again in the instant face of danger, the trappers did so in order to prevent the predators of the land from finding the dead beaver in their traps and eating them unless they were regularly checked and removed. Additionally, to leave dead beaver hanging in their traps would just alert any passing Gros Ventre to the likelihood of close at hand and unwanted white man trappers. As they left their cabin site, they could feel the first vestiges of winter coming as there was a nip in the air and on the far distant mountains, there was the first cap of winter's snow ringing the higher peaks.

Throughout that day, Iron Hand was surprised over just how cold wading in the waters and deep mud of the beaver ponds had become almost overnight. However, the traps were full of dead beaver and soon the exhausting work of trap setting and beaver removal overcame his concerns about the cold or the Indian parents out looking for their two lost young men who did not return home the evening before.

That was the two trappers' first mistake! Rounding a finger of aspens en route to their next beaver trap, Iron Hand luckily spotted ten Indian riders moving out across the prairie about half-a-mile away! They were spread out in a long line abreast, apparently looking down at the ground as if looking for any kind of suspect tracks that were telling a tale as to the two boys' disappearance. Quickly dropping low over his saddle, he leaned back and motioned for Crooked Hand to do the same, which he immediately did. Then Iron Hand guided his horse into a nearby stand of aspens, quickly slipped out from his saddle and walked his horse deeper into the gloomy and darkened interior of the stand of trees for the cover offered. Crooked Hand had quickly followed suit and soon, the two men stood by their pack animals looking out from the covering

216

stand of aspens to see if they had been discovered by the band of obviously searching Indians.

Minutes later, the Indian riders disappeared into the adjacent timbered foothills in the direction where Crooked Hand and Iron Hand had intercepted and shot the two young Indians out from their saddles for stealing their horses. For the next hour, the two men saw no further movement out on the prairie, other than several small herds of antelope moving about and feeding nearby. Then off in a distance in the direction the Indians had ridden, the two trappers heard a flurry of shooting and then silence. Fortunately, that flurry of shooting had come from the rough direction where the two young men had been killed and not from the direction of their cabin.

Somewhat later as observed from the cover of their aspen grove, the trappers saw ten Indian riders moving into view once again onto the prairie. Only that time, being dragged behind one of the Indians' horses on a long rope appeared to be a brown, flopping kind of rug-like object of some sort. As the riders finally disappeared in the direction of where their suspected village lay, Iron Hand finally figured out what the 'brown rug-like' thing being dragged behind one of the Indians' horses really was.

"Crooked Hand, I think those Indians discovered where those two boys were killed and like I had hoped, they smelled the grizzly bear and figured as I had. After some looking around, they must have figured that damn old bear had done in their kids because of the 'leavings' of their clothing or bone scraps. That would explain all the shooting we heard earlier. Finding the 'leavings' near the bear den, they must have figured that somehow the critter had done in their kin. When they did, they must have taken out their revenge on what they figured must have been that 'kid-eating' bear. Then when they came out after the shooting had died down, one of the Indians appeared to dragging a brown rug-like thing behind his horse, as they headed back to the north to where we suspected that band of Gros Ventre are lodged. I would bet that brown rug-like thing must have been the hide from that damn poor innocent ole grizzly bear that got blamed for killing those two boys," said a hoping for that fact to be true, Iron Hand.

"I don't know. But just to be on the safe side, I say we stay here this evening and wait and see if they come back looking and poking around some more. If they don't, then we can run the rest of our trap line tomorrow morning and then skedaddle for the cabin. Besides, there appears to be another storm coming down from the northwest. If we wait for that to pass, maybe it will further hide our tracks and that would be good," said Crooked Hand slowly, as if he was really thinking their current situation over very carefully.

That evening, Crooked Hand and Iron Hand spent a cold night in their little grove of aspens. A small fire was eventually built late at night and meat from one of the beaver the men had caught earlier furnished a repast of sorts. The next morning, the storm coming from out of the northwest was upon the two trappers and with it came a light snow. Not being dressed for the colder weather as of yet, the two men finished running their trap line and just loaded their two packhorses' panniers with whole carcass beavers in order to finish what they were doing and reducing their time of being out in the open so much. Then with both packhorses loaded down with heavy dead beaver carcasses, the men finished checking and setting their traps. Then they headed for their cabin by a different route, so as to avoid making a well-used, shod-horse trail leading right back to the site of their cabin.

By the time they got back to their cabin, they were almost blue in color from being exposed to the cold for such a long period of time! Without their heavy winter *capotes* and clothing, they damn near froze to death in the early arriving snowstorm while riding out in the open on their cold as hell horses. As for Iron Hand's legs and feet, they were beyond feeling after wading in all the water and pond mud, checking traps, removing dead beaver and setting the traps anew. In fact

218

he was so cold, when they arrived back at their cabin, he had to be helped from his horse because he had stiffened up from the long and cold horse ride back to their cabin and could not stand and only just barely hobble! Additionally, his wetted buckskin pants had now frozen to his icy legs...

Helped inside their cabin, Iron Hand was set down next to their roaring inside fire in the fireplace, given a steaming cup of coffee laced with a generous helping of rum and both Old Potts and Big Foot knelt on each side, trying to rub the life back into his icy legs... In the meantime Crooked Hand, who had not been immersed in the water most of the day, unsaddled all the horses, fed them up and placed them inside their corral for the protection that offered. Then dragging the dead beaver out from the panniers, he toted them over to the front of their cabin and stacked them into a large pile outside their front door for easy access. Once finished, he went inside and after warming up, Old Potts, Crooked Hand and Big Foot removed the beaver from the outside pile, took them inside their cabin, skinned them out and continued doing so until all the critters had been skinned, fleshed out and the fresh hides hooped with willow limbs for drying.

Finished with the above chore, large chunks of the tender parts from the beaver carcasses were boned out and roasted over their fireplace fire. Staked beaver meat, biscuits and steaming cups of coffee were the trappers' supper that evening. By then, Iron Hand figured he would live and partook of large amounts of tasty and fatty cooked chunks of beaver and Dutch oven biscuits along with the rest of the men. Later that evening to avoid any problems with still roaming grizzly and black bears that had not yet gone into hibernation, Old Potts, Crooked Hand and Big Foot removed all the freshly skinned beaver carcasses from in front of their cabin and hauled them off into an adjacent grove of now almost leafless aspens. As for Iron Hand, he spent that time busily rubbing his now stinging like mad and finally warming up legs. He did so as a result of him almost freezing them solid because of their long

The Adventurous Life of Tom Warren

ride back on their horses in the snowstorm and from being wetted from the waist down from all the wading in the beaver ponds and cold mud while tending the traps. That evening, with the men trying to sleep inside their cabin, they were awakened numerous times over the fighting in the nearby aspen grove between a pack of wolves feasting on the beaver carcasses and what had to have been at least two grizzly bears contesting with the wolves for the fresh beaver meat as well...

The next morning, Iron Hand and Crooked Hand ventured forth once again to run their trap line before the waters froze solid, making trapping very difficult when having to chop through and setting their traps under the ice. Only this time, both men were dressed in their winter's best clothing for the cold horse ride and water immersions when checking the traps. That 'freezing cold' issue was solved by bringing along an extra packhorse carrying a complete change of winter clothing for Iron Hand after all the traps had been checked, beaver removed and then re-set or removed for the winter because of the greater thickness of the now forming daily sheets of ice over the beaver pond waters.

As if that wasn't enough, there was always Crooked Hand's comforting words for Iron Hand advising him that, "The beaver pond mud he was walking in was so sticky that it 'could heal up the crack of dawn or mend a broken heart,' so don't be worrying over just being a little cold..."

As for Old Potts and Big Foot, they still had a small mountain of beaver skins from the day before to flesh out and hoop before their partners returned that evening with their additional catches. That meant not only long hours working in the dim light from the fireplace and beeswax candles in their cabin as they carefully fleshed out the skins, but repeated cold horseback trips to the adjacent Poplar River's willow patches, to cut and bring back to the cabin additional limber willow limbs to be used for hooping the skins so they could dry more easily and not sour.

That night unexpectedly the winter winds howled around the trappers' cabin like there was no tomorrow. When the men awoke the following morning, they discovered a foot of freshly fallen snow now covering the landscape and an outside temperature they estimated to be hovering around 20 degrees below zero! Realizing that an oncoming heavy freeze up was soon to be a major problem with trap retrieval, Iron Hand, Crooked Hand and Big Foot hurriedly dressed for the weather, saddled up their stock and headed out to their trap line in order to remove the remainder of their traps left by the slides for the winter, because thick ice would soon make beaver, muskrat and river otter trapping all but impossible. When they left the cabin site that morning, they left Old Potts behind to make sure if their home site was discovered by any Gros Ventre, there would be someone to watch over their remaining horses, packs of furs and their remaining provisions.

Arriving at their trap line, the trappers discovered a thick layer of ice already covering the watered areas and in the intense cold, getting thicker by the moment. That being the case, Iron Hand spent the rest of the day, alternating with Big Foot, in using a hand ax to break the ice out to the remainder of the traps not removed from the day before, as well as all the dead beaver trapped underneath and removed the same. As they did, Crooked Hand remained on his horse watching over the pack animals and the defenseless men as they moved around in the waters, breaking the ice, removing their traps and previously caught beaver. Soon the three packhorses were loaded to the gills with recovered traps and beaver carcasses still in the round into their panniers.

Finally recovering all of their traps and previously caught beaver, the men changed out from their wet clothing into their dry clothes they had brought along and then headed for their cabin and the work that lay ahead in the skinning, fleshing and hooping of the now frozen beaver just caught. As luck would have it, the men stumbled upon a small herd of buffalo on their

The Adventurous Life of Tom Warren

way home and killed two cows before the herd stampeded off into the vastness of the prairie's white wilderness.

Being that the packhorses were already heavily loaded with all the beaver traps and 29 beaver carcasses, the men were forced to butcher out the buffalo where they fell and using their riding stock, loaded them down with hind leg quarters and hump rib sections for transport back to their cabin. The remaining portions of the buffalo were left for the big prairie wolves, already heard howling just a short distance away, once they had scented the fresh blood from the carcasses.

By now, the men were near frozen as the winds had picked up, reducing the temperature even further. In fact, when they had opened up the buffalo, the men had thrust their near freezing hands into the animals' intestines to warm them up so they could make good and safe use of their knives in the butchering out process that followed. Finished with the loading of the buffalo meat upon their riding horses, the men began trudging off towards their cabin amidst the savage sounds of the wolves tearing at the buffalo remains they had left behind just moments earlier.

Two hours later, the three trappers staggered into their cabin site in an almost frozen condition, as the temperature now approached 30 degrees below zero and the winds had picked up in front of a new storm now rapidly approaching from out of the northwest. Old Potts, seeing the poor condition of his fellow trappers, assisted the near frozen men into their warm cabin, broke out the hot coffee and rum, and helped them to the seating benches near their fireplace. Satisfied he had done all he could do for the moment for his friends, Old Potts then dressed for the cold and tended to the horses still standing out in the cabin's clearing. He first unloaded all the now frozen buffalo meat and stacked it near the front door of the cabin. Then he began unloading the panniers of their now frozen-through beaver carcasses as well, placing them near the cabin's entrance as well. Then all the beaver traps were unloaded and placed in their small cache house near their cabin along with most of the bridles, riding and pack saddles for the protection that shelter offered.

By now, the rest of his trappers had warmed up sufficiently so that they were now back out and helping in the tending to the rest of the unloading of the horses and feeding them armloads of hay previously collected from the river bottoms during the summer months. Once the horses were fed, watered and housed in their sheltered-by-the-cabin corral, the trappers hauled the now frozen beaver carcasses inside so they could thaw, be skinned, fleshed and hooped. As those chores were tended to, Iron Hand, with a hand ax, butchered out a full hindquarter from one of the buffalo cows and brought that stack of frozen meat into their cabin as well. Then with the aid of Big Foot, the rest of the buffalo quarters and rib sections were hoisted up onto their nearby meat pole, so it would be out of reach of hungry predators. When the men had finished with their meat pole duties, they hauled in several more armloads of wood for their fireplace. As they carried in their last armloads of wood for the evening, Iron Hand noticed that the meat pole had now attracted numerous feathered visitors. Looking on, it now seemed that on every exposed piece of meat sat a number of chickadee and gray jays helping themselves to some good buffalo and more importantly, the fat, which would help them in surviving the intense cold. With that, Iron Hand just grinned, walked into the warmness of their cabin and closed the door behind him. As he did, the icy blasts from the oncoming storm arrived in the Poplar River country with a howling vengeance. The fall trapping season for beaver was now a thing of the past, because the thickness of the ice on the now iced-over beaver ponds precluded any further and realistic easy trapping.

However, work for the Poplar River trappers was far from over regardless of the winter's icy winds and the snows that were now steadily assailing the area. Inside the warm and snug cabin, plans were made for the trappers, weather permitting, to

The Adventurous Life of Tom Warren

go forth the next day, kill several buffalo and set their wolf traps around the carcasses.

They realized that wolves were some of the smartest predators on the frontier and extremely trap wise. But they also knew that during severe cold weather events, the wolves had to find food easily and lots of it. That meant they hated to expend any more energy than they had to in order to survive. Hence, the wolf traps being set around an inviting freshly killed buffalo carcass for those wolves hungry enough to walk across a ring of hidden traps. The trappers realized that when hunger struck a wolf, many times caution would be thrown to the wind in favor of a full belly, hence a ring of wolf traps being set around an inviting fresh-killed buffalo carcass...

That evening as the new storm's winds howled around their secure cabin, the men dried out their wetted winter clothing derived from their recent beaver trapping trip, feasted on slabs of roasted buffalo meat, Dutch oven biscuits and numerous tin cups of hot coffee holding just a touch of rum for what ailed the cold in their bones. After such a heavy supper, the men relaxed, smoked their pipes and discussed tomorrow's wolf trapping expedition. Later into the evening, tired from the day's labors and the kind of tiredness that comes from working outside in the extreme cold, found the trappers snug in their sleeping furs, as the storm's winds rattled their tanned deer hide coverings pulled tightly over the cabin's window frames to help in keeping the warmth in and the cold out...

Early the next morning, Iron Hand awoke and having consumed numerous cups of hot coffee the evening before, had to 'see a man about a horse'. Not only that but his 'sixth sense' of impending uneasiness was stirring around so much in his belly, along with an overly filled bladder, that he found further sleep problematic. Pulling on his winter clothing and slipping a loaded pistol under his sash as he always did when initially facing the day while on the many-times dangerous frontier, out the cabin door he cautiously went like he had many days previously.

Terry Grosz

Pausing just outside the cabin's doorstep, he breathed in deeply the frosty cold air and looked skyward in order to 'read' the day's weather. The storm clouds from the night before were now long gone, only about six inches of new snow had fallen and the sky was clear blue, windless and cold. Looking all around one more time as his living on the frontier had taught him to do if he wanted to live to see another day, Iron Hand took two steps and THEN QUICKLY DROPPED TO HIS KNEES AND GRABBED HIS PISTOL FROM ITS SASH! AS HE DID, ALL THE WHILE HE EXPECTED THE SLAMMING OF A MUSKET BALL INTO HIS CHEST AT ANY MOMENT! HE HAD DROPPED TO HIS KNEES AS AN INSTINCTIVE PROTECTIVE REACTION TO REDUCE HIS BEING A TARGET, BECAUSE WHEN HE SWUNG HIS EYES LOOKING ALL AROUND THEIR CABIN AS A PRECAUTION, THEY RANDOMLY TOOK IN THE ADJACENT HORSE CORRAL.

WHEN HIS EYES HAD SWUNG BY THE HORSE CORRAL, HE WAS AMAZED TO SEE THAT IT WAS EMPTY AND THE GATE WAS WIDE OPEN! EVERY ONE OF THEIR 18 RIDING AND PACKHORSES WERE GONE! HENCE THE QUICK DROP TO HIS KNEES EXPECTING WHOEVER HAD TAKEN THEIR HORSES WAS CLOSE AT HAND AND READY TO SHOOT ANYONE EXITING THE CABIN UPON DISCOVERING THEIR GRIEVOUS LOSS OF THEIR LIVE-GIVING HORSES...

For the next few seconds, Iron Hand's eyes quickly swept the entire area around their cabin looking for any sign of danger. Seeing no signs of immediate danger, he quickly rose to his feet as his frontier-practiced eyes kept looking all around. As he did, he yelled out for his fellow trappers and then ran over to the now empty horse corral. Stopping just outside the gated entrance, Iron Hand's eyes quickly scanned the ground by the gate. There in the snow he counted six sets of winter moccasin tracks from Indians and a number of the

The Adventurous Life of Tom Warren

trappers' shod horses' tracks leaving the corral and moving away from the cabin and down towards the Poplar River...!

The next moment, he was immediately surrounded by the rest of his fellow trappers, all partially dressed but armed to the teeth for any kind of danger that was to follow. For the longest time, all the men stood eyeing the empty corral with disbelief, surprise and inner concern all in the same breath! They immediately realized that they were now afoot in hostile Indian country without a chance in hell of making their ways back to Fort Union alive. In essence, being deep in the heart of Gros Ventre Indian country without the means of flight or escape, they now realized that they were 'dead men walking'...

Then Iron Hand quietly said, "Best we make haste and ready ourselves for a long day. The winds of last night's storm covered the sounds of our horse thieves stealing and making off with our only means of survival. To me, that is a killing offense but I have to catch the thieves in order to kill them! So I say we had better get going after our thieves while the trail is still fresh and not terribly drifted over. Otherwise, they and our horses will be gone forever...as will we!"

Then Old Potts walked over to their cache house where they kept all their traps and riding equipment to gather up some bridles so they could bring back any horses they could catch, and then froze in his tracks! "I be damned! Those basterds, whoever they are, took all of our traps as well as our horse gear. They never left a single trap or a piece of leather riding gear in which to give us any chance for survival! Whoever did this not only took our horses but any chance we would of had in the way of making any kind of a livelihood as well... They knew damned well that by taking our horses and traps, we would be forced from this neck of the woods and any and all trapping!"

With those words of their potential deaths in the wind, Big Foot and Crooked Hand headed for their cabin without another word being spoken. They knew what had to be done and were on their way to see to it, as Old Potts soon followed. However,

Iron Hand on another mission of sorts, took off down the trail of the stolen horses, as his 'sixth sense' rolled around in his gut like a great horned owl just run over by a madly stampeding buffalo...

Iron Hand had not gone a 100 yards following the tracks of the stolen herd of horses and the six sets of Indians' moccasin tracks, when he came to a spot in the nearby aspen grove holding a number of new tracks. Kneeling down, Iron Hand discovered two sets of shod horse tracks along with the tracks from the six Indians' unshod ponies! Additionally, he identified the footprints in the freshly fallen snow from two other mystery individuals who were not of the moccasin-wearing type but were wearing what appeared to be sets of boots!

Standing there in the morning's cold, Iron Hand began putting together in his mind what he figured must have occurred. To his way of thinking, there were eight individuals involved in the theft of their horses and traps. Two appeared to be from white men and six from their Indian friends. Those thoughts were further validated by the six sets of unshod horse tracks and the two sets of shod horse tracks normally ridden by white men, all clearly in evidence in the freshly fallen snow in the aspen grove. From there, Iron Hand observed that all of the tracks from those doing the livestock stealing and those of the trappers' horse herd headed south towards the distant Missouri River, and its comforting dense river breaks of vegetation in which to hide from any pursuit by the aggrieved trappers and would allow one to get out from any adverse weather. Standing there in anger, Iron Hand did not know where the horse thieves would be going but he knew of four trappers who would quickly be hot on their trail! That was unless the thieves were quickly caught by the trappers, or their four trackers following behind were killed outright by other groups of Indians catching them out on the open plains without the built-in defenses of their horses...

The Adventurous Life of Tom Warren

Trotting back to the cabin, Iron Hand loaded up his possibles bag with extra caps, balls and wadding. Around his now-dressed in his heaviest winter garb-covered shoulders went two powder horns filled with powder enough for a long fight and tucked into his belt went his ever-present tomahawk. Grabbing his carry bag, he loaded it up with jerky, as his partners were simultaneously doing the same. They jointly figured that they would be limited to cold tracking and cold camping along the way in order to not alert the horse thieves that they were being hotly pursued. Then Iron Hand rolled up his bearskin rug as the others did the same, realizing they would be sleeping along the trail and would need some kind of comfort to keep them alive in the intense winter cold and up off the snow-covered ground. Those rugs were hastily placed into slings so they could be 'shoulder-carried' once the men hit the trail of the horse thieves.

Once the rest of the trappers had finished dressing into their heaviest cold weather gear, they all gathered up their rifles and two horse pistols apiece. Finally, on went their powder horns as well as their rifles and pistols, knives and tomahawks, possibles bags and carry bags full of jerky. Without further delay, out the door of their cabin they went into winter's cold with Iron Hand leading the way. Being the best tracker of the group, Iron Hand led the men at a time-proven way of travel or a 'trapper's trot' following the herd of stolen horses and those who had done the stealing. (Author's Note: A 'trapper's trot' was a way many trappers covered great distances of ground when an emergency or drastic need arose for an extreme amount of ground to be covered. That ground-eating formula of travel consisted of a trapper trotting a 100 yards and then walking 50 while he rested. Then the whole process was repeated time and time again as needed. It has been documented in many history books relating to those storied Mountain Men, that many miles a day could be covered by using such methods when the need for such necessary emergency travel was called for or existed.)

Initially, the stolen horse herd's tracks and those of the thieves made a beeline for the Poplar River. Then the tracks turned southward following the Poplar River towards the Missouri River. As they did, Iron Hand was amazed that the horse thieves were not hard-pushing the stolen livestock in order to get the hell out of the country. That told him that the horses had been stolen during the deep of night when the winds had covered any sounds made by the thieves. Figuring they had made a timely escape and the snow would soon cover their tracks, the thieves had not hurried up in their travels. They were right in one aspect though. The newly falling and drifting snows had pretty much covered up their tracks once they were out on the open, wind-swept prairie. But there were still indentations in the fresh snow from so many horses and it was those marks that Iron Hand was able to follow.

As Iron Hand trotted along in his typical, long-legged trapper's trot, he was still easily able to follow the trail the horses' hoofprint indentations. By so doing, he was even able to cut across country and intercept the trail at different points and many junctures, thereby shortening their route of pursuit. Come nightfall, the four tired men retired into a small stand of timber along the Poplar River and made their cold camp. No fire was lit for fear of giving away their presence to the horse thieves if they were close at hand and watching their back trail. Cold jerky was their supper, snow was dissolved in their mouths for water, and then the exhausted men rolled up in their sleeping furs that had been carried on their backs and when they did, they found that sleep came easily despite the snow and intense cold...

Come daylight, the four trappers were once again hot on the trail left by the horse thieves. As they continued cold tracking the horse-hoof indentations, Iron Hand finally realized who might be riding the shod horses normally ridden by the white men in that neck of the country. The thought of "Hudson Bay Company men" crossed his mind many times as he continued along the trail, now getting harder to follow due

The Adventurous Life of Tom Warren

to the ever-shifting prairie winds and the snow now being wind-drifted over the indentations. *Why else would someone steal all of their traps as well as their horses?* Iron Hand kept asking himself. He knew there were Hudson Bay men trapping further north of where they had been trapping, but how did they know where he and his three friends would be trapping?

Then he remembered back about the six sets of Indian moccasin tracks clustered around the corral gate. Somehow the Indians had previously discovered the trappers' camp and had gone to the Hudson Bay men and had so advised of their find. Since the Hudson Bay Company offered rewards, which included alcohol to any Indians for information on any American trappers in country trapping and competing with their own operations, Iron Hand figured he had now put two and two together. With the Indians' fresh information regarding American Fur Company Trappers or Free Trappers in country, the Hudson Bay men figured they would accompany the finders back to the cabin, steal the horses and traps, and thereby severely cripple any other trapping competition in the area, if not killing it and the aggrieved trappers outright. The thought of anyone doing such a deadly, far-reaching thing over beaver *plus* did nothing more than just harden Iron Hand's thoughts and determination on the coming retribution if and when the thieves could be run down and confronted...

On day three of the four trappers' pursuit of the horse thieves, Iron Hand and company had finally arrived on the north bank of the Missouri River. There the stolen horse trail turned and led westward along the Missouri River breaks and even deeper into the dreaded Gros Ventre Indian country! But by now, with the intense cold, lack of warm food, cold sleeping conditions and the basic wrong of stealing a man's way of survival out on the frontier, all four of the trappers resolve had hardened like river stones! If they ever found the horse thieves and their stolen horses, there was now little thought of leniency or forgiveness in the minds of each man...

Come day four of their pursuit, the trappers ate the last of their jerky! Now they were on their own, as if they had not been for the last three days of their pursuit. Long about noon on day four of their pursuit of their stolen horses, Iron Hand spotted a group of about ten Indians out hunting buffalo a short ways off. Being that the four trappers were moving along the edge of the Missouri River breaks, they quickly 'dissolved' into the brush along the waterway for their own protection since they were a-foot and outnumbered. There they were forced to lie low for four hours in the snow, until the Indians had left their buffalo kill site. When they left, Old Potts and Iron Hand ventured across the prairie to the kill site and discovered the Indians had killed 13 buffalo, taken what they could and had left the rest for the wolves. That night, the four trappers gorged themselves on still warm and raw buffalo meat for the first time in many such days! Sleep came much easier that evening with a more than full belly of rich and energy-boosting buffalo meat.

Dawn on day five found Iron Hand and his fellow trappers hard on the cold trail once again and now very much aware that the tracks from the stolen horse herd they had been following were getting fresher! Come nightfall on day five, found the four trappers sneaking towards a small flicker of light being emitted from a campfire down along the Missouri River bottoms out of the way from the ever-blowing and chilling winter winds. Another hour of sneaking towards the light of a campfire and one could hear human voices laughing and having a pleasurable good time. Sneaking even closer to the large campfire, Iron Hand and his fellow trappers discovered their horse herd inside a rope corral and eight individuals located a short distance away around a campfire roasting great slabs of buffalo meat. In fact, the four trappers were close enough that they could even smell the rich smells of roasting buffalo meat and found their bodies making lots of quiet noises from deep inside their now empty stomachs...

The Adventurous Life of Tom Warren

It was obvious from all the happy laughter and loud talking, that the two white men in the group of eight horse thieves had broken out the rum now that they figured they were out from harm's way when it came to any form of pursuit. It was also obvious to the quiet observers that everyone was 'deep into their cups'. Without any words being spoken and with just the use of hand signals, the four trappers quietly spread out alongside the eight men's camp in the darkness for what was to violently come if they had their druthers and if the good Lord would just look away...

Moments later, Iron Hand quietly stepped out from the river bottom's dense leafless brush and just silently stood there at the edge of the light of the horse thieves' campfire. For the longest time, Iron Hand just stood there stone-cold still and unnoticed by the 'happy eight' celebrating around their campfire over their horse-stealing successes. Then one of the white men of the thieves' group noticed the giant of a man standing silently and boldly at the edge of the light of their fire! When he did, HE INSTANTLY REALIZED THAT THE MOUNTAIN OF A MAN FACING THEM WITH A DEADLY LOOK IN HIS DARK EYES WITH HIS RIFLE HELD AT THE READY WASN'T ONE OF THEIRS!

Quickly reaching for the pistol in his belt realizing he was now possibly staring death in the eye from one of the stolen horses' owners, he drew it and that was his last worldly deed! Iron Hand shot him between the eyes from 30 feet away with his .50 caliber Hawken and that man's head exploded into a 'spew' of bright red blood and grayish looking mush! For an instant, there was utter surprised silence coming from those horse-stealing celebrants around the campfire. THEN, THE RIVER BOTTOM BRUSH FURTHER ERUPTED SIMULTANEOUSLY WITH THE ROARING SOUNDS OF MORE HAWKEN RIFLES SPEWING DEADLY HOT LEAD INTO THREE OTHER NEARBY MEN STANDING AROUND THE CAMPFIRE! THAT DEADLY MOMENT WAS QUICKLY FOLLOWED BY THE LIGHTER

SOUNDS OF FOUR PISTOLS BEING FIRED ALMOST IN UNISON! AS PLANNED, THOSE PISTOLS LOADED WITH BUCK AND BALL WERE DEVASTATING AT SUCH A CLOSE RANGE... THEN STONE-COLD SILENCE REIGNED EXCEPT FOR THE SOUNDS OF A CRACKLING CAMPFIRE AND THE SNORTING AND NERVOUS STAMPING OF FEET FROM THE NEARBY, ALARMED OVER WHAT HAD JUST HAPPENED, HORSES! EVEN MORE SO DID THE HORSES BECOME ALARMED WHEN THE SMELL OF BLACK POWDER SMOKE DRIFTED THEIR WAY, AS DID THE FRESH SCENT OF HUMAN BLOOD!

For the next several moments, the air around the campfire was faintly clouded with the white smoke and acrid smells of freshly burned black powder... Seeing no other threats, the river bottom brush parted, producing three more determined-looking trappers from the cover in which they had been hiding. The next sounds heard were from four men reloading their rifles and pistols, in case they had overlooked the number of men in the horse thieves' camp and there was still danger close at hand. Finding they had 'harvested' their 'crop' of horse thieves, the four trappers walked over to the campfire to see exactly what their death and destruction had wrought.

An hour later, the eight horse thieves' bodies had been dumped out onto the ice of the Missouri River for the wolves and bald eagles to find and devour. As for the remaining trappers, they warmed themselves around a roaring campfire, drank rum supplied by the Hudson Bay Fur Company horse thieves, and ate all they could hold in previously roasting buffalo meat meant for the eight deceased men. Men who were now cooling out on the ice of the nearby frozen-over river... Later that evening, the four exhausted trappers slept soundly around their warming campfire, the first they had experienced in five days of wintering out on the open plains in the dead of winter. Now what had earlier been 'dead men walking', were now just four tired fur trappers quietly and warmly sleeping in the comfort of knowing they were once again 'a-horse'...

The next morning, Old Potts and Iron Hand, while Crooked Hand and Big Foot cooked breakfast, went through the dead horse thieves' belongings. As suspected, Old Potts, by the cut of their clothing, had identified the six dead Indians as from a band of Gros Ventre Indians. As for the two white men, according to their personals dug out from their saddle bags, was one Wayne La Due and the other was a William Hamilton, both employees from the suspected Hudson Bay Company. In fact, Hamilton was a *Bourgeois* for the Hudson Bay Company, from one of their forts located up on the northern reaches of the Wolf River according to the "Crown" papers carried in his saddle bags. A Hudson Bay trading post which was, according to Old Potts, about another long day's ride to the northwest from where they now stood!

Old Potts went on to figuring aloud that the Hudson Bay men had probably trapped out the Wolf River area and had been casting their eyes further to the east along the Poplar River, when they received information from their now dead Indian consorts regarding the discovery of the four trappers' cabin in the same area. Deciding the trapping grounds along the nearby Poplar River belonged to their company, Hamilton probably figured they could eliminate such competition by stealing the American Fur Company's trappers' horses and traps (not realizing the men were Free Trappers and were not in the employ of that company), and then without any form of transportation, leave those trappers out on the vast and unforgiving plains in the dead of winter to die a slow but certain death.

The four trappers decided they would stay another night in camp to rest and then leave the following morning at the crack of dawn. However, aside from the fresh buffalo meat consumed and the warm campfire, that extra day's stay to rest up turned problematic. The wolves had found the eight bodies out on the river's ice and had made many great noises all night

Terry Grosz

long, as they feasted and fought over the remains of the horse thieves. When they did, that made any planned sleep on the part of the trappers back at their camp, almost impossible...

Dawn the following morning, after another breakfast of fresh buffalo staked out over their fire, the men hit the trail back to their cabin. Trailing behind the four men were now all of their riding and pack animals that had been previously stolen, plus the eight riding and two packhorses from the Hudson Bay men. Men who by all accounts, were now more than likely wolf or bald eagle droppings along the Missouri River...

Still mindful of the fact that the four of them were deep in Gros Ventre country trailing a large and valuable herd of horses, the men stayed to the river and creek bottoms for cover, as they traveled back along the Missouri and then northward along the Poplar River. As they did, Iron Hand had to smile. Their pack string now carried an extra eight rifles, five pistols, two full kegs of rum, 40 beaver and 24 wolf traps, and one small keg of a high-grade French powder, all retrieved from the dead horse thieves, not to mention an additional eight riding and two fine packhorses plus all their gear and provisions.

That also suited Old Potts just fine, as he remembered all the packs of furs they had back at their cabin and the additional ones they would soon be harvesting come spring after ice out. As such, he had often wondered how they would get all their packs of furs back to Fort Union, seeing that they had more packs than their original herd of horses could carry. Now that issue of concern over having more packs of *plus* than their original herd of horses could carry had been solved with the welcome addition of ten more horses previously belonging to their now-dead horse thieves. Then he had to smile broadly once again. Had to smile, because for the last two years when they had returned to Fort Union, they had arrived with more valuable horses than they had left with. Now, if they kept their hair and were not discovered by the horse-stealing Gros

235

The Adventurous Life of Tom Warren

Ventre, they would be going back to Fort Union a third time with more horses that they had started out with! *That would surely make McKenzie smile over his four Free Trappers' successes once again,* he thought with another even broader heavily whiskered smile...

Arriving back at their cabin, the men found it undisturbed and unloaded all their gear plus their previously stolen traps. Then it was back to work as normal. There was wood to cut up from their winter woodpile for their fireplace; buffalo to hunt for their meals; wolves to trap, skin, flesh and dress out; meals to cook; and an enlarged horse herd to look out for after they had been turned out onto the nearby prairie to graze on a daily basis.

Finally came a welcome spring and the much-awaited first day of beaver trapping for the four trappers. For the two days previous to that 'jump-off' date, the four trappers made ready for what awaited them. Firearms were checked, horse gear gone over and those items time-worn or torn were repaired, the traps were smoked to reduce the man smells, shoes on the horses were checked to make sure none were loose, a small mountain of rifle and pistol balls were cast, and all the men's knives were sharpened. To date, the men had 20 bound and ready to go beaver packs, 18 wolf pelts, 16 of river otter, 20 deerskins, 64 muskrat pelts, and four black bear rugs. That represented a small fortune in furs, plus they still had the spring beaver trapping before them before the animals went out of prime.

Once again, it was decided all the men would go forth initially and trap beaver for the safety in numbers it offered until they got behind on the skinning, fleshing and hooping. Then they would split up, leaving Iron Hand and Crooked Hand to do all the trapping and Old Potts and Big Foot to do all the fur processing and watching over the remaining horseflesh left back at their cabin.

Come the day decided upon to begin spring beaver trapping, the men were eagerly up before the crack of dawn

Terry Grosz

and as the horses and pack animals were made ready, Iron Hand was hard at work making a hearty breakfast of venison steak, Dutch oven biscuits, a Dutch oven pie made from the last of their dried apples and raisins, and the usual 'thick as beaver pond mud' and as strong as an angry mule's kick, loaded with the last of their sugar, trapper's coffee.

Finishing breakfast and heading out, found the trappers' string of horses and pack animals loaded for 'bear'. Each of the two pack animals carried two panniers full of beaver traps, axes, a shovel, two extra rifles mounted on the pack saddles and two additional pistols loaded with buck and ball just in case they had unwanted company, all riding side-saddle in holsters on each horse for quick retrieval.

Since he was to do all of the actual trapping, Iron Hand led the string of horses and trappers from their cabin as they headed for their usual trapping area along the Poplar River, its tributaries and adjacent marsh lands. Behind him came Big Foot, leading a packhorse and behind him came Old Potts doing the same. Last but not least rode Crooked Hand, along with his excellent shooting eye and steady hand, in case someone came along and wanted to try and 'ruin' the trappers' first day of spring trapping...

Approaching the initial starting point of their old trap line, Iron Hand noticed that there seemed to be less beaver in their regular watered areas than they had seen in previous years. Additionally, a goodly number of the old dams seemed to be in disrepair, as well as did a number of the old beaver houses. Those were not good signs, as Iron Hand quietly figured maybe disease or the winter had been so hard and the ice so thick, that a lot of the animals had starved and died. Those suspicions were confirmed when Iron Hand rode up on two feeding beaver on shore that were thin and their fur looked rough and poorly kept, which was unusual for any normally healthy animal. All were signs indicating a possible winter die-off due to starvation as a possible culprit in the beaver's reduced numbers and their overall physical condition.

237

The Adventurous Life of Tom Warren

However, Iron Hand pushed on and began setting out his traps for about four miles up the waterways on their old familiar trapping grounds. When finished with the setting out of 40 beaver traps, the trappers rode another four miles up along the waterways looking for newer and better trapping grounds. They were disappointed in what they saw. All along the way, they discovered beaver dams in disrepair, few swimming animals and those seen were in poor body and fur conditions. Even worse, no young beaver were seen swimming in the waters and Iron Hand did not see one immature beaver track in the mud next to his sets or along the usual slides.

On the way back in the afternoon, as was usual because beaver are so territorial, they checked and discovered ten curious beaver already in their traps. Those beaver had smelled the unfamiliar-smelling castoreum used on the bait sticks, came over to investigate the new arrival's scent and had been trapped. Upon closer inspection, all the beaver were found to be thin and their fur poorly kept and of low grade in quality. All of which was a sure sign of a major winter kill beaver die-off situation on their old trapping grounds, possibly because of the trapping area's shallower waters.

Then Crooked Hand gave Iron Hand a low warning whistle and looking in the direction in which he was pointing, the trappers saw two Indians riding by on a far hilltop. Not wanting to be discovered, the men slipped off into a grove of aspens, dismounted and watched the two Indians as they rode along the ridge line, seemingly unaware of the trappers working their trap line far below. The men waited until the Indians had disappeared off into the distant foothills. Satisfied they had not been seen, they finished riding their trap line, removing four more beaver in the process.

Then the men scouted out another tributary on the Poplar and discovered a number of deeper beaver ponds, numerous well-kept houses and evidence of currently used slides and fresh cuttings in the nearby aspen and willow patches.

Heartened over what they were seeing, the men returned to their old trap line, removed a dozen previously set traps and brought them over to the new beaver area just discovered. There Iron Hand set the 12 new traps in likely looking places and then the men headed back to their cabin to flesh out and hoop the few beaver they had caught earlier in the day.

On the way back to their cabin, Crooked Hand killed a fat cow buffalo and the men feasted on warm, bile-covered, raw buffalo liver right out of the animal. Then they cut out the hump ribs, back straps and one hindquarter, loaded up their pack animals and headed for their cabin.

Walking their pack string and riding horses into the cabin site, the men dismounted and then Old Potts noticed that their cabin door was ajar! Then Big Foot noticed the number of horses in their corral seemed to be a little light in number! A quick head count by Iron Hand who had noticed the same lessened numbers of animals, showed that there were eight horses missing! Jumping off his horse, Iron Hand ran over to the corral gate and noticed that a different kind of knot in the gate rope had been tied other than the one he had tied earlier that morning... Looking down at the ground, Iron Hand noticed one set of moccasins footprints!

Then he heard Old Potts shout. "Those damn devils not only made off with eight of our horses, but 16 packs of our beaver *plus* we had ready for transport to Fort Union! Plus they helped themselves to our few precious brown sugar cones we had left, eating most of them," he yelled in a 'badger'-mad tone of voice!

"Come on, Crooked Hand. I bet it was those two damn Indians we saw earlier in the day before we reset our beaver traps in that new tributary, that were our thieves," said Iron Hand as he leapt back into his saddle and following the hoofprints made in the soft and damp spring soil of the stolen horses, rode off after them. Moments later, the two riders were out of sight as Old Potts and Big Foot surveyed their cabin for anything else that had been taken. When it was all said and

The Adventurous Life of Tom Warren

done, only the eight horses and 16 packs of valuable beaver *plus* had been taken. *No two ways about it, a major loss, especially in light of the heavy winter kill of beaver in their area and the trappers' reduced chances to rebuild their* plus *numbers with their possibly limited spring catches!* thought Big Foot.

Daylight the following morning, found Big Foot and Old Potts still alone! Crooked Hand and Iron Hand had not returned from the night before and now the two remaining trappers faced a dilemma. They had two of their number missing and they still had a trap line to run with 40 traps out! Doing what they had to do, Big Foot and Old Potts rode forth and ran the trap line. Bringing back only 13 beaver out of the 40 traps set, the men were concerned. If they didn't do better in the beaver-catching department, they could not survive in that area as successful trappers. Even more importantly, they were still missing two of their friends! However, they still had work to do, so they set about their fleshing and hooping duties involving the 13 beaver they had caught that morning.

The next morning, Crooked Hand and Iron Hand had yet to return from their chase after the horse and *plus* thieves! Once again, Big Foot and Old Potts ran their trap line, brought home their catch of 21 beaver, fleshed out and hooped the same. By then, it was deep into the night and still their two friends had yet to return! By then, both trappers figured their friends had run afoul of a larger number of Gros Ventre, been overrun and killed. Figuring that, they also realized if that were the case, it would be a simple matter for their killers to backtrack the two trappers to their cabin and then...

That evening, the two trappers made sure that a rifle and several pistols adorned every shooting port along the inside walls of their cabin. Then as an added precaution, they stacked a dozen packs of furs against the inside of the front door in case a number of hostile Indians decided to make a run at the cabin's only door during an attack. They figured if attacked, the Indians would not try and burn them out because they would want any valuable *plus* the trappers had in the cabin, plus there were other goods the white man trappers would have that they would want as well. So the thinking of the two trappers was, if the attacking Indians wanted all of that, they knew where it was, so come and get it...

The next morning before the crack of dawn, Old Potts while lying in his sleeping furs, heard a number of horses approaching their cabin! "Big Foot, get up," he whispered. "I think those Indians that killed Iron Hand and Crooked Hand have backtracked them and they are now here!" Both men got up and stumbled around in the dark of their cabin, as they hustled over to their firearms and made ready for whatever came there way.

"Don't you two old bastards start shooting," yelled Iron Hand. "Open the front door so we will know the two of you are still alright," he continued in a louder tone of voice, to make identification of the one doing the yelling easy to be identified by those inside the cabin.

With that, the cabin door flew open and there stood Big Foot and Old Potts, naked as jay birds having just jumped out from their sleeping furs and now with their rifles in hand.

"Damn, Iron Hand! Damned good thing you yelled out or we was prepared to blow your miserable butts clear out of them saddles in the dim light," grumbled a damn happy Old Potts upon seeing his two friends were safe and sound. "Did you get them damn thieving horse and *plus*-stealing bastards?" he continued in a tone of voice still coming down in the emotion of the moment.

"Can we have some coffee and hot grub and then we can talk? We haven't eaten in two days and are hungry enough to eat our horses and two more from out of the corral," said Crooked Hand.

"Well, you two damn knotheads kain't eat standing out there in the dark and in the middle of our yard. I suggest if you want some grub, you get your tail-ends in here afore we decide to shoot the two of you anyway for making us worry so much,"

The Adventurous Life of Tom Warren

said Old Potts, now getting the 'happy' back into his naked being and tone of voice.

With that, the two tired men dismounted and took care of their horses, while the happy smell of cooking meat now began emanating from inside their cabin and graced their 'smellers'. Later once inside, the two travelers sat down tiredly and had cups of weak, lukewarm coffee thrust into their happy outstretched hands, as their big pot was just beginning to boil over the fire in their fireplace.

"Well, what the damn hell happened out there? I see that you two did not bring back any of our horses or *plus*, so there must be a story on what happened somewhere," said Old Potts.

Gulping down his second cup of now getting hotter and stronger coffee, Iron Hand began their tale of woe. "Well, we stayed hot on the trail of our horse and *plus* thieves until dark overcame us that first day. It was very obvious by the way and speed in which they were traveling, that they did not wait around to see if we would be mad over what they took that was ours and were hot on their trail. They fair moved out on those stolen horses and never slowed down one wit. Crooked Hand and I slept in an aspen grove along the Poplar River that first night and at first light the next day, we were back on their trail. We found their cold camp about two miles away from where we had bedded down that first evening but they were already long gone when we arrived. It was like they now knew we would be hot on their trail. They just kept pushing those horses hard, as they made for the Missouri. Once there, they turned west and followed the river until the second night. We stayed on their trail until dark overtook us once again and then we bedded down along the Missouri and let our tired horses out to graze. The next day, we were up early once again but they pulled out ahead of us before we got there. I don't think they even took the time to unpack those horses or even feed them that time, they were in such a hurry. About midday, we caught a glimpse of them just as they rode into a camp along the river that was clear full of Indians. There must have been 10 or 12 of their brothers waiting for them in that camp, so we backed off and hid along the river until dark. Then we staked our horses out of the way so they would not be discovered and then sneaked up to their campsite to look and see what we could do about our furs and horses."

"Those horse thieves were not fools. They had guards posted all night so Crooked Hand and I could not sneak in, grab our horses and packs and then scoot. There just wasn't any way of getting back what was ours, without getting our hair lifted in the process if we tried anything. So we backtracked out of there and made our way back here. There was no two ways about it. They were a mess of Indian renegades, apparently sweeping the country for anything and everything from other trappers that they could lay their thieving hands upon. That is about it. We came home by another route so no one could track us if we had been seen. In watching our back trail, it was obvious they did not know we were so close around them. Now, how about some chow before my big guts eat my little ones?" said Iron Hand in his characteristic 'I am glad to be home and alive' heavily whiskered grin. Whiskers that had now long since grown back after one of the two young Indian horse thieves had shot off a patch of them in their shoot-out. However, Big Foot had remarked upon their most recent return, that he felt Iron Hand was still not the prettiest among their lot...

The following day to avoid any 'horse-stealing event' repeats, Iron Hand and Crooked Hand commenced with their trapping duties, while Old Potts and Big Foot guarded the remaining *plus* and horse herd. Still finding beaver trapping on their old trapping grounds somewhat skimpy at best because of the huge winter die-off, the two men shifted their entire trap line over to the newer trapping grounds discovered the day the two Indians had run off with part of their horse herd and 16 packs of *plus*. As near as the four trappers could figure, those two Indians who had stolen their eight horses only took that

The Adventurous Life of Tom Warren

many because that was all they could handle at one time and make haste in their escape...

For the next month or so, the trapping remained good on the new trapping grounds and the furs collected were of prime quality. Then when Old Potts and Big Foot, who were doing all of the fur preparations, noticed that some of the beaver trapped were shedding their winter coats, especially their long guard hairs, they advised the team doing all of the trapping, to pull their traps and call it quits.

That first evening as the four men sat around their outside firepit smoking their pipes and swatted at the hordes of mosquitoes, they discussed their plans for the future. It was apparent that they had pretty well trapped out most of the beaver in their neck of the woods along the Poplar River and its waterways. With that realization out in the open, the men looked to Old Potts for some guidance when it came to their future as trappers in the country heavily utilized by numerous bands of the dreaded Gros Ventre.

"Well," said Old Potts once the future plans as trappers was broached, "I guess we can call her quits, go back to Fort Union with what we have and sell our furs from this season. Then we can head downstream to St. Louis, collect our $40,000 in credit that the American Fur Company owes us from our past successes, split the money and go our ways," he said slowly, as he expelled a large cloud of pipe smoke into the cooling evening air and momentarily scattering the horde of mosquitoes flying around him.

For the longest time after those words were spoken, all one could hear were the sounds of the buzzing clouds of mosquitoes hovering around the four men, just out of reach of their clouds of pipe smoke and the wood smoke boiling up from their firepit. "Course then," continued Old Potts, "I hear tell that the beaver trapping further west along the Missouri and up along the Porcupine River is pretty much untouched, except by a few of them Indian trappers and maybe a Free Trapper or two. And if I recollect the lay of that particular part

244

of the country, that river is just about four days' ride north and west of here. I ain't never seed it but hear tell from "Griz Johnson", late of the New York State area, that he caught so many beaver when he trapped there, that he had to pull some of his traps because they was every time filling every one he set! Then another good friend and one hell of a shooter named Harlan Waugh, told me that he trapped the Porcupine for one season and caught so many 'blanket-sized' beaver, that his horses could not pack all of them back to the summer Rendezvous and he had to cache a mess of them, according to Harlan, behind his cabin by a huge fir tree, which was located near where the Milk River joined the Porcupine. But his cache is still up there somewhere because after his last son was kilt by a renegade Mountain Man and some of his Indian friends during a Rendezvous, he pulled up stakes and went over to trap in the Wasatch Mountains. And I hear tell from Griz Johnson that when Harlan did, he jest up and disappeared. But both of them men were good friends and they would not lie to me about the trappings in a good place to go if I was to have a notion to go there."

For the longest time after Old Potts had spoken, quiet reigned around their outside firepit. Then Iron Hand spoke up and asked Old Potts if that neck of the woods around the Porcupine River wasn't crawling heavy with Gros Ventre same as the area they were now in.

"No more than here, I suppose," quietly replied Old Potts.

Once again, quiet reigned around the firepit as the men continued thoughtfully pulling on their pipes, pondering their options of continuing life on a dangerous frontier or the quiet of the city life back in the St. Louis area...

"Well," said Iron Hand slowly, "I came out here after I lost the only family I truly loved to forget my past. Since then, the four of us have become like family and all of you are all that I now have. Won't be the first time I have been shot at or had my life threatened and more than likely, it won't be the last if we go into that Gros Ventre-thick country. If you three want

The Adventurous Life of Tom Warren

to see what lies further west on the Porcupine and take your chances with those damn meddlesome Gros Ventre, count me in. Count me in because I buried my life back in Missouri and traded it for what I have here and now."

"Well, I don-no. As much as Iron Hand eats, I am not sure there are enough buffalo to keep him fed. And if I have to lug around a small mountain of bullets for my lead-eating Hawken just to keep him fed or protected from the damn Indians, I am not sure my body will hold up to such rigors. Hell, them Gros Ventre are a small concern compared to feeding my huge friend here. But wherever he goes, so goes I. Someone has to keep an eye on him, else he will slip and slide into more trouble than he can handle and will need me there to bail him out. Nothing like being dragged 'back-ards' into 'hell hath no fury', but count me in if that is what the rest of you want to do," said Crooked Hand, as he grinned over at his big friend like a hungry pig in a corn crib.

"Well, damned if I will be left behind! I ain't got much hair to lose in the hands of a murderous Gros Ventre, but he can have it if he can come through the hail of bullets I plan on sending his way and laying down in the air around him when that time comes. Count me in 'cause I don't plan on being left behind on my own in this damn country," said Big Foot, with a huge knowing grin of realization of just what he was possibly stepping into.

"As I seed it, that leaves me out here all by my lonesome if I chose to go it alone elsewhere. I kinda figured all you 'gal-loots' would fall for what lies over the horizon, be it arrows, a rifle bullet or a damn good 'buffler' steak. Well, Old Potts kain't live with all of you knotheads and he kain't live without you, so count me in as well and may the good Lord have mercy on all of our black-hearted souls. It looks like the Porcupine may well be our next place to lite-down and take a good look-see..."

For the next week, the four men made ready to leave their old beaver trapping grounds behind and head for Fort Union to

Terry Grosz

trade in their furs, re-provision and eventually, after a lot of drinking and celebrating with old friends at the fort, head for newer pastures and trapping grounds yet unknown, such as the Porcupine River country...

The day of departure found the four trappers up before daylight and loading all their horses heavily with their packs of *plus*, sleeping furs, cooking ware and all the rest of their 'possibles' for the long trip to Fort Union. Upon departure, Old Potts led their string of 18 riding and packhorses with Big Foot trailing one string of nine animals. Riding as outliers on either side of the pack strings rode Crooked Hand and Iron Hand. Both of those men were packing their rifles in hand, two pistols in their sashes and two additional pistols per man holstered on the sides of their saddles.

Moving slowly because all the animals were so heavily loaded, the trappers did not arrive at the Missouri River until late on their second day of travel. The sun had long since set as the men finally picked their ways to the north bank of the Missouri River, only to be brought up short shy of the actual river itself.

Old Potts had brought the group of trappers and their pack strings up short of the river and then sat in his saddle just looking intently towards the distant Missouri River breaks. Seeing Old Potts reined up short in the early evening's low light, Crooked Hand and Iron Hand quietly rode up and paused alongside Old Potts. No one said anything until Big Foot brought his string of animals alongside, then Old Potts stood up in his stirrups as if to get a better look at what had caught his eye and was 'sticking in his craw'. As they sat there, Iron Hand, who had been having 'sixth sense' feelings all day along the trail, found them still stirring within his being like there was no tomorrow. Then he heard what was bothering the 'eagle-eyed' and always cautious, Old Potts.

Down in the breaks along the river, Iron Hand heard a lot of faint yelling and celebrating going on. Then a small flickering of light caught his eyes in the early darkness, as a

The Adventurous Life of Tom Warren

campfire was being stoked up by a number of loudly talking men. As that fire grew in size, faint figures of men now moving around in the fast-falling darkness also caught the trapper's eyes looking on from a-far. Something just did not seem right to Iron Hand as well, as his 'sixth sense angels' made like they were trying to fly around inside him...

"I don't like what I am hearing or seeing," said Old Potts quietly. "Something just doesn't look or seem right and I don't know why but that is just how I feel," he continued. "Folks just don't advertise their whereabouts so loudly in Indian country unless they be Indians." Turning in his saddle, Old Potts said, "Iron Hand, you sneak down there and see what is bothering me about that bunch of men around that campfire making such a ruckus. I can faintly see one hell of a mess of horses and a large number of men moving all around near that campsite. We should be the earliest of trappers heading to Fort Union I would think. I just cannot understand why there would be such a large party of men and horses down there along the river ahead of us this early in the season unless their beaver all died out like ours did. Something is not right and I don't think they be trappers making such a loud fuss. And before we go down there, I want to see what is really going on before we expose ourselves to more than we can handle or a man should be asked to bear if it were to get ugly. So I want you to go down there, sneak around and see what you can see. In the meantime, we will take our horses back into that draw in the river's breaks behind us and hide them there until you come back."

Without another word, Iron Hand handed the reins from his horse to Old Potts, stepped from the saddle and quickly disappeared into the darkness along the Missouri River breaks just as quietly as a weasel could stalk a field mouse. It always amazed his fellow trappers just how quiet such a large man like Iron Hand could move when he chose to be quiet, and tonight was one of those times.

Eventually sneaking down to the group of men and now their two large campfires, Iron Hand first headed for the larger numbers of horses tied off in the trees to see what he could learn. As he moved closer towards the horses, he got surprised! There in front of him was his favorite packhorse! Stunned in what he was seeing in the faint light from the two huge now-blazing campfires and moving in closer in and around the horses, he was surprised even more. Moving in among the many tied-off horses so he could get closer to the numbers of men now drinking and wildly celebrating around the fires as he tried making sense of the occasion, he got a number of surprises! In fact, seven more surprises! There tied off on a long rope were the other seven horses of the eight taken by the two Indians who had also stolen 16 of their packs holding a number of their beaver *plus* just weeks earlier!

Then Iron Hand realized what he was seeing. It was the same bunch of Indians that he and Crooked Hand had tracked to the Missouri River earlier. The same group of renegade Indians who had stolen eight of their horses and 16 packs of beaver *plus* ready to be taken to Fort Union and sold! Then Iron Hand stumbled onto a HUGE pile of packs holding numerous beaver *plus*! Now the truth of the moment became clearer. The group of noisy men celebrating around the two campfires appeared to be the much-rumored among the trappers, renegade Indians along the Missouri who were catching trappers unawares, stealing their horses, taking their *plus* and more than likely, killing their owners! Discovering his own eight horses bore mute testimony to what Iron Hand had now come to suspect.

WHOOMP! went a heavy fist alongside Iron Hand's head, stunning him in the process and in so doing, he dropped his rifle! The heavy impact to Iron Hand's head of the strike from behind by an unknown assailant knocked him into a nearby cottonwood tree as a result of the full force of the unexpected blow. Quickly recovering his frontier survival instincts and finding the instant fury from deep within rising up into a

The Adventurous Life of Tom Warren

towering rage, Iron Hand rapidly reacted to the attack. A towering rage, because he now realized that if he was discovered, he more than likely would be killed and his three friends eventually discovered and murdered and robbed as well! Instinctively ducking as he recovered from the impact of the strike from behind, he felt the wind and heard the dull 'thunking' sound of a tomahawk hitting the side of the tree when he had recoiled from the initial strike to his head!

Rising up to his full height, Iron Hand saw the darkened form of an Indian trying desperately to retrieve his tomahawk, now deeply stuck in the bark of the cottonwood so it could be used once again. In his towering rage and with the strength befitting his frontier name, Iron Hand reached out, grabbed the Indian by the front of his throat and crushed his windpipe all in one fluid powerful motion! As he crushed the man's neck in his powerful hand, he realized that if the individual was able to yell out, he would be discovered and it would be all over. With that realization, Iron Hand crunched down even more forcefully on the man's neck until he heard and felt a dull-sounding snap!

When Iron Hand had done what he had to do, the Indian still being held up in a standing position, went limp, urinated all over himself and then began spasm-trembling in his death throes... Feeling the man's body finally going limp in his hands, Iron Hand let him down slowly, as the killing fury that had risen up in his body also began subsiding, allowing him to come down from his emotional killing high... Then Iron Hand froze in his tracks! At the very edge of the dancing light from the two nearby blazing campfires he saw something that turned even his stomach! Tied to the cottonwood tree where his attacker had buried his tomahawk in his initial attack, he observed a man dressed in buckskins looking right at him... Iron Hand tensed up figuring he was in for another fight until he realized the man 'looking at him' was doing so through lifeless eyes. The buckskin-clad man, obviously a trapper in life, had been tied to the tree and his body shot clear full of

250

arrows! Staring in disbelief at what he was seeing, Iron Hand turned away in disgust and found his disbelieving eyes looking at three more nearby 'surprises' at the edge of the light from the campfires as well! Three more surprises in that next to the first dead man tied to the tree, were three more trappers tied to adjacent trees, who had died in the same body filled-full of arrows execution-style manner! Then Iron Hand recognized one of the dead men as an old friend and fellow trapper named "Four-Fingered Jack"! Four-Fingered Jack had been an original Fort Union Company Trapper who had accompanied Iron Hand in driving the fort's horse herd up to its present location on the upper Missouri River's maiden upriver trip! Jack was also a close friend of Old Potts, having accompanied him upriver on his first trip into the Frontier in 1807 with a St. Louis businessman named Vasquez. Iron Hand knew when Old Potts discovered what had happened to his old and lifelong friend Four-Fingered Jack, there would be hell to pay on the end of a speeding .50 caliber rifle bullet!

Having seen enough and not wanting to be discovered, Iron Hand lifted up the dead Indian's body, picked up his rifle and quietly carried the carcass off from the activity around the Indians' camp to avoid discovery over what had just occurred. Once far enough away from the suspects' camp and where the body would not be discovered, Iron Hand dropped it into a nearby draw and then quietly headed for where he suspected the rest of his fellow trappers would be waiting for him to return and report his findings.

About 20 minutes later, Iron Hand quietly slipped into the area hiding the rest of his crew. There without mentioning the man he had killed with just the strength in his right hand after the man had tried killing him, Iron Hand reported to the group. "You will not believe what I just discovered. There are about 15 to 20 Indians in that camp. They are the one and same bunch, Crooked Hand, who took our eight horses and the 16 packs of beaver *plus* some time back. The one and same two Indians we could not tackle and try to get our horses and furs

The Adventurous Life of Tom Warren

back because once they got to the Missouri River and joined up with their pals, they outnumbered us. They also have a huge pile of packs and a number of horses I am sure that they have taken from other trappers up and down the line, and more than likely have killed them all in the process. Otherwise, why all the horses, mules and packs in one place like they have near their campsite?"

"I also have some very bad news. Apparently four trappers ran afoul of this bunch and recently paid the price. While snooping around their campsite, I ran across those four trappers. They had been all tied up around a number of trees and then those Indians in camp shot all of them full of arrows execution-style! Each of those men had to have at least 20 arrows in them! There was no two ways about it, they were murdered! Old Potts, one of those murdered trappers was your old friend and mine, Four-Fingered Jack! He too died with having at least 20 arrows shot into his whole front side," said Iron Hand quietly.

For the longest time Old Potts said nothing about what Iron Hand had just said. Then he quietly said, "Alright, I will make damn sure every time I pull the trigger, some son-of-a-bitch is going to feel a hot lead slug ripping through his guts! Shoot my dear friend clear full of arrows will they, well, we shall now see what side the Devil is on…"

"Well, they outnumber us now as well. But if we let them live, they will ambush other returning to Fort Union fur trappers following along the Missouri as many of them are wont to do, kill them and take all of their furs and livestock as well. Right now, they are positioned perfectly along the Missouri River to intercept any other groups of trappers following the river back to Fort Union. When they do, it will be all over for those unsuspecting trappers," continued Iron Hand.

"However, here is what I propose. They have at least one and maybe two kegs of rum that they have opened and are now celebrating their horse-thieving and pack-stealing successes. I

252

Terry Grosz

suggest we let them have their little party and when they get good and liquored up, we swoop in and kill the lot! Since there are so many of them, surprise is the only way we can kill all of them if we decide to take them on. In order to do that, we all will need to be carrying our rifles and at least three pistols apiece fully loaded with buck and ball. If we do it that way, I figure we can kill up to 16 initially when we attack, providing we don't miss killing a single man, which should be pretty easy. The whole damn bunch is now clustered around those two kegs of rum and their campfires, which should make killing the lot of them damn easy. However if we don't get them all killed in a shooting go-around, then it will be rifle butts, knives and our tomahawks if we are to carry the day. There are a lot of them but with their drinking and our element of surprise, I think we four can do it. This is no different than what I learned at the U.S. Military Academy as a cadet years ago. With the element of surprise on our side, even in the face of superior odds, we can carry the day providing all of us shoot straight," continued Iron Hand, now getting worked up internally and mentally once again over what was to come if all went as planned.

"Well, you are the ex-military man and should know what you are talking about. I, like you, feel that the surprise method of attack is the only way we four can succeed. If we let these red devils live, they will kill and plunder among the rest of our kind til their hearts' content. I say we kill the bunch and toss their bodies in the Missouri River for other like in kind ambush-killers along the way to see what awaits them if they so continue with such evil ways against us trappers," said Old Potts, with a lot of 'grit' sounding in the tone and tenor of his voice...

"Well, you know how I feel after our talk the other night around the firepit back at our cabin when it comes to you, Iron Hand. I can't let you go in alone, otherwise you will just get your big ole miserable carcass into more trouble than you can handle. So I plan on carrying four pistols into this fracas just

253

The Adventurous Life of Tom Warren

because those bastards took our horses and *plus*, and made me mash my ass in my saddle those couple of days back trying to catch them and having to go for two long days without anything to eat," said Crooked Hand, with a look in his eyes although unseen in the dark, foretold through the tone and tenor of his voice the seriousness of exactly what he was saying...

"I say we all get to cracking, pull the bullets from our pistols, dump the old powder and recharge them with fresh powder, buck and ball. I don't want to go into a one-sided fight with pistols that may or may not fire when all the chips are down. So while those lads are drinking themselves silly, I say we get cracking. Let's load up, prepare for the worst and by then, most of those 'war-hoops' should be good and liquored up so when we kill them, any critter that eats their miserable carcasses will get a bit loopy as well from all that they drank," said Big Foot, with just a tinge of the thrill of the killing chase to come in his voice as well...

"All right, here is what I propose we do. Since we will be outnumbered, I suggest we come at them from two sides but in such a manner we will not get ourselves caught in our own crossfires in case any of us over-shoot. That way, if any of them try to run away from the direction of the shooting, they can only run in such a manner that will present us the best shooting opportunity at killing every one of them! For to let any get away means they will more than likely come hunting us or other fellow trappers at a later date. Also, by attacking on two fronts, when we 'run dry' with our pistols and rifles, we can still throw our tomahawks and by being spread out like I am proposing, none of us will be in the way of the other when we get to throwing. After that, it will be every man and his sheath knife for himself in case I didn't count right," said Iron Hand, pleased over the fact that some of his previous military training was now coming into play...

After making sure they were more than ready to take on a superior force of Indians, the men stealthfully moved out in two teams towards the Indians' camp. A noisy camp of suspects that was now a madhouse of rum-soaked individuals, yelling, dancing and celebrating. Iron Hand and Crooked Hand sneaked into one side of the encampment and Old Potts and Big Foot took the other so they could set up a deadly crossfire once the shooting started. As they carefully approached the two campfires, each man was armed with a loaded rifle, three pistols, a tomahawk, a long-bladed sheath knife, a mountain of grim determination, and more guts than a boar grizzly.

As was usual in most planned situations of such potential intensity, Iron Hand and Crooked Hand ran into a snag they had not planned on as they stealthfully sneaked into the Indians' noisy camp. With Iron Hand in the lead, they chanced upon an Indian out in the brush urinating. The last worldly thing that man saw through his hazy rum-soaked eyes was the shadow of a huge man in the darkness reaching out, grabbing the top of his head and the bottom of his jaw and giving it a violent twist! His neck snapped with a dull sounding THUD, followed by a low groan, as Iron Hand let his lifeless body slip from his 'bear trap'-like grip soundlessly to the forest floor at his feet... Now, there only appeared to be 18 men from the Indian raiding party left to face their maker and the rest of the Cloud People when they finally 'arrived'!

Positioning themselves off to one side of the wildly celebrating group of Indians, Iron Hand and Crooked Hand quietly waited partially hidden in a clump of elderberry bushes for Old Potts and Big Foot to get into position to shoot as well... Moments later, Iron Hand saw Old Potts and Big Foot slipping into position on the far side of the party of now well-liquored-up celebrants. When the two far side trappers finally stood up unobserved by the happy group around the campfires, Iron Hand and Crooked Hand quickly did the same. As the trappers rose to their feet for what was violently yet to come, a look skyward would have also shown that the Indians' Cloud People were moving in closer to witness the deadly occasion as well...

The Adventurous Life of Tom Warren

BOOM–BOOM–BOOM–BOOM! went the roar of four Hawken rifles being fired into the wildly celebrating, unsuspecting group of Indians! When those four rifles were fired, huge clouds of black powder smoke blew into the area surrounding the dead, dying and the remaining living! Although they had been heavily drinking for a spell, many still had enough of their natural survival instincts that began taking over upon hearing the close at hand shooting and seeing the clouds of black powder smoke come rolling their way from the ends of the four rifle barrels! The entire group of men, seeing four of their number violently spun onto the ground from being shot at such close range with heavy rifles, still with 10,000 years of survival in their genes, tried to drunkenly respond. However, when they did, confusion ruled within their crowded ranks. The Indians had grouped up upon hearing the close at hand shooting and then the next four rapid pistol shots dropped four more Indians to the ground in a literal spew of red clouds and flying gray matter! The combination of the rum, acrid black powder smoke enveloping the Indians at such close ranges, four more of their kind dropping dead at their feet as snot, blood, pieces of bone and intestines blown apart from the buck and ball splattered all over those remaining in a standing position, did nothing but add even more confusion to their fast-thinning ranks! That confusion was further multiplied when four more of their kind spun crazily to the ground as a result of four more shots being fired at them with buck and ball from such close ranges! Then the remaining Indians, in total panic, finally broke and ran for their lives. As they began to scatter in total panic, four more were spun to the ground by the deadly close at hand firing pistols spewing tissue-rendering buck and ball into their depleted ranks once again! The last remaining Indian running from all the close at hand blazing rifle and pistol fire, raised his tomahawk and loaded with too many cups of 'demon run', wobbled right at Iron Hand, who was standing quietly at the edge of a large clump of elderberry bushes with his last empty pistol in hand.

Terry Grosz

Singing his death song at the top of his lungs and fueled by adrenalin, a tall for his race and heavyset Gros Ventre warrior ran right at Iron Hand. Iron Hand dropped his empty pistol, sidestepped the Indian's clumsy lunge with his upraised tomahawk, spun the man around with his own inertia, and snapped his neck with a hand movement that was so quickly executed that Crooked Hand standing alongside, tomahawk in hand to help, exclaimed in amazement, "Damn, Iron Hand! How the hell did you just do that?"

With that, Iron Hand let the lifeless man slip from the grip of his right hand to the ground, looked over at his friend, saying quietly, "I don't know. It just comes from within me and I do as I feel it necessary." In that calm reply to his close friend, Crooked Hand just shook his head over his big friend's quiet demeanor, after he had just killed a man with his bare hands who was trying to kill Iron Hand with a tomahawk...

Then a single shot rang out and Crooked Hand yelled and fell to the ground! Quickly turning, Iron Hand saw a young Indian who had apparently been guarding the huge horse herd, upon hearing all the shooting back at the camp, come charging in with 'blood in his eye'. Seeing Crooked Hand and Iron Hand near a number of his dead compatriots, the young Indian quickly fired his rifle at Crooked Hand, hitting him in his bad leg. Iron Hand, upon hearing the shot being fired, quickly turned and upon seeing the hard-charging Indian coming his way, stopped him in full charge with a quickly thrown tomahawk striking him full in the face with a bone-crushing sounding THWACK!

Then turning to his downed friend, Iron Hand knelt beside him to see how badly he had been hurt. After a quick and worried look, he got to laughing over his friend's mildly superficial leg wound. A leg wound in his bad leg, namely the one the grizzly bear had bitten into when they had surprised the critter living in their cave campsite back near Medicine Lake. Walking over to the now-dead Indian shooter, Iron Hand removed his tomahawk, wiped off its blade on the man's

257

The Adventurous Life of Tom Warren

sleeve, then cut off a piece of the dead Indian's buckskin shirt, walked back to a still-howling Crooked Hand and patched up his painful but not serious leg wound. However, as he did, Iron Hand noticed that by the man's dress just killed, he was a member of the Blackfoot Tribe and was not a Gros Ventre...

With that bit of misfortune out of the way, the remaining three trappers quickly reloaded a single pistol apiece and then scouted out around the Indians' campsite looking for any more of the living. Finding none, they began assessing what had been left by the deceased and got the surprise of their lives.

The Indians had indeed been catching the unwary trappers on their trapping grounds or traveling towards Fort Union, killing and robbing them. The three trappers, aided by a now hobbling Crooked Hand, discovered huge numbers and assortments of rifles, pistols, tomahawks and knives that had apparently been taken from the now dead trappers' hands by the renegade band of dead Indians. Additionally, they counted 36 horses and mules that had been taken from those now dead trappers by the Indians, in addition to the raiding party's 24 riding and packhorses they themselves possessed! Then the surprise of their lives was discovered stacked up under a number of buffalo skins. Before it was all said and done, the trappers discovered in the surprise stack, in addition to their own 16 stolen packs of beaver *plus*, another 56 packs of *plus*, furs and pelts taken from those presumably now dead trappers as well...

Knowing they could never return all those furs to their rightful owners, the four trappers realized they were rich beyond their wildest dreams, because they were now the owners of a small fortune in furs! However, there was little time for celebration. The men quickly reloaded all the rest of their weapons in case other Indians were on the way to the dead Indians' encampment, and then all the dead were dragged off to the shores of the Missouri River and dumped into its currents. That way, if any of the bodies were intercepted later by other Indians as they floated by still fully dressed and they saw all of the bullet holes, they just might be warned away from any white trappers they came across. At least that was the thinking of the four victorious trappers. As for the four dead trappers filled with arrows, they were buried near the campsite and a fire built over their communal grave. That way, Old Potts and his group hoped the fire would remove the human smell and avert having wolves or bears digging up the four trappers' bodies and eating the same.

However, as the four trappers were coming down from their emotional high after the deadly battle, they failed to realize that another pair of dark eyes was now observing what had just transpired. That dark set of eyes, realizing he was too late arriving at the Indians' encampment, slipped off into the darkness with a lasting evil in his heart over what he had just witnessed... Especially in the lasting identification of the one huge trapper-man his Blackfoot brethren called Iron Hand, who had just killed his brother with an accurately thrown tomahawk to his face...

Then Old Potts and Big Foot returned to their hidden pack string and brought all of those horses into the Indians' camp for safekeeping. Following that chore, the men helped themselves to one of the kegs of rum after they had cleaned out and bound up Crooked Hand's leg wound. Soon, they were cautiously celebrating their good luck in surviving their latest ordeal and fortune as well, in all of the packs of furs and valuable livestock they now possessed...

Early the next morning, in case there were more Indians en route to the campsite of death, the men were up early rounding up the horses and mules and loading the packs on the same. By noon, all the animals had been loaded for their trip to Fort Union and then the men slowly started out. However, because of the number of animals entailed, they were not brought down the trail in strings but in a loose herd with Old Potts leading the way with a string of fully loaded mares. That way, the animals could feed up along the way and with the four trappers herding

The Adventurous Life of Tom Warren

them along the confines of the north bank of the Missouri River, progress en route to Fort Union was slowly being made.

Two days later, the fortified structure of Fort Union hove into view. When the huge horse and mule herd numbering over 70 animals ambled into view, upon being sighted by the fort's ever-present look-outs, great human activity could be seen emanating from within and around its log walls. Soon the huge horse and mule herd of fully loaded animals was positioned in front of the fort's front gate, as an amazed Kenneth McKenzie, Fort Union's Factor, stood there in absolute, head-shaking disbelief!

"Potts, what the hell is the meaning of all of this? You guys left last summer with less than 20 head of horses and now you show up in front of my fort with over 70 horses and mules? Hell, your group didn't even have any mules because you told me your three 'mule-headed' trapper buddies were stubborn enough for you to deal with, so you didn't need any mules. Now you show up with a huge herd of such animals. I know all of those critters can't belong to you men, and this is one story I have to hear on how they were acquired. Supper at my place at six and bring your eating and drinking appetites, because I have a feeling this is going to be one hell of a story and a like in kind evening to go along with it!" said an amazed McKenzie.

"Me and my kind would be honored, Mr. McKenzie. But first, we need to run all these animals into your central square inside the fort for safe keeping. There we can unload all of these packs and your Company Clerks can begin grading and counting what we have brought you. At the same time, after we sort out all of the stock we may want for the next trapping season, we need to sell the remainder to you for your Company Trappers who have lost their horses and mules throughout their last trapping season," said Old Potts with a big and damn proud grin on his face...

Still shaking his head over the magnitude of what he was seeing, McKenzie began barking out orders to his now

260

assembled staff, and soon the heavily loaded horses and mules were being led into the fort's inner courtyard for the unload. Once the herd of horses and mules were inside the fort's protective walls, a flurry of human activity around those heavily loaded animals began.

Old Potts and Big Foot watched over the Clerks as they began grading, sorting and counting out all the various pelts, hides and *plus* being removed from the pack animals. Iron Hand on the other hand, met with the fort's buyers of animals and kept records on what was being sold and what was offered for each and every animal the American Fur Company wished to purchase. Crooked Hand, still hobbling around more for sympathy than out of pain, selected out from the entire herd the riding and pack stock he and his fellow trappers would need for the coming trapping season, hobbled the lot for better control and walked them off to one side of all the activity inside the walls of the fort, so they would not 'walk away' in the hands of the many Indians camping around the fort's walls and now watching all the latest activity from the inside.

That evening the four Free Trappers, according to the custom of the times and tradition, were the honored supper guests of Kenneth McKenzie, Fort Union's Factor for the American Fur Company. As the trappers entered the Factor's residence, they were met at the door by a Chinese servant with brim-full cups of high proof rum. Then they were ushered off to the Factor's dining room and seated along a long table with several high level Company Clerks and McKenzie. There they were treated to a supper fit for a king out on the frontier. Fresh roasted buffalo hump ribs dripping with oodles of fat, baked potatoes from the fort's nearby garden slathered in real home-churned butter, pickled beets, Dutch oven biscuits with wild honey, and all the elderberry pie the men could 'cart' away, all supplied by McKenzie's two expert Chinese cooks.

After supper, the men adjourned to the sitting room in the Factor's residence, and were plied with more rum as they discussed the business at hand. There McKenzie advised he

The Adventurous Life of Tom Warren

would buy all of the horses and mules for his Company Trappers, and the 76 packs of furs, pelts, hides and *plus* for a total of $32,500! Additionally, he advised he would supply all of the provisions the four Free Trappers would need for their entire next trapping season as well as part of the pay-out for all of the furs they had brought into the fort for sale!

When the trappers heard that figure and a free allowance for all of their needed provisions, their heads just swam and that wasn't because of all the rum they had been drinking that evening... Here they had over $40,000 in American Fur Company credit waiting for them in St. Louis as a result from a previous year's haul of furs and horses and now this in addition! Once those sums were combined and disbursed, each of the four trappers would be considered a rich man! Especially in light of the fact that a working man's wages in that day and age ran about $300-500 per year!

"Now, as all of you know, I don't keep that kind of money on hand here at the fort, nor would any of you want to pack such a sum of money on your persons back out onto the frontier in the form of gold and silver coin. I will supply you with all the provisions you will need for the coming year, and the $32,500 of what you have coming will have to be supplied by the American Fur Company's fur houses in St. Louis. Then when you four decide you have had enough of this life on the frontier, you can redeem your money at that time with your Certificates of Credit Due," said McKenzie with a smile over the major purchases he had just made for his boss Mr. Astor, and his company.

"We will have to mull that over," said Old Potts. "Right now, we are talking maybe a trip out to the Porcupine River country and see what kind of trappings that neck of the woods will bring us. But that is jest talk at this stage in our lives, however it is purdy serious talk. So, we may have to take you up on your offer and redeem our Certificates of Credit Due at a later date, when all of us age out from this kind of living or get tired of protecting our 'topknots' at every turn in the trail or wading in the damn cold water and mud all of the time."

"That is fine with me and the company. Just let me know what you want to do and when, and then we can settle up. Now, you lads need to tell me the story as to how you came into so many horses, mules and packs of furs. Because it seems every time you four venture forth, you come back with a hatful of furs, a horse herd belonging to someone else and a tall tale to tell that beats all. I somehow suspect that a lot of those animals and furs used to belong to me and the American Fur Company. I suspect that, because I am losing to the frontier and all of its dangers, about 25% of my Company Trappers annually. Thank goodness, every year there are more and more trappers coming upriver and wanting to work for the company and replacing those trappers that I am losing. However, if my trappers are being killed and their furs are being taken by the hostile Indians and then you fellas come by them fairly by whatever means, I guess they belong to the four of you, fair and square. So, let me have all of your cups filled with some more of my high proof rum to loosen your tongues, and tell me how all of your new wealth came about. And damn your grouchy old hide, Potts, don't leave out anything in the telling," said McKenzie with a teasing smile, as he settled down in his huge overstuffed chair, waiting for the storytelling to begin.

The four trappers did not leave the Factor's residence that evening until it read way after midnight on McKenzie's old wind-up clock on the fireplace mantle. As the trappers wobbled out from the Factor's residence, he had them escorted to an empty Clerk's residence inside the fort's walls, so they could sleep in peace and without a cloud of Missouri River mosquitoes swarming overhead all night. After all, those four Free Trappers were bringing thousands of dollars into his company's coffers and that was the very least he could do in trying to keep their favor. Especially in light of the fact, if they chose to do so, they could take their furs to St. Louis

The Adventurous Life of Tom Warren

themselves and make a lot more money than they could at Fort Union...

The next morning, Iron Hand and Big Foot took their horses out from the fort's walls and let them graze adjacent the walls of the fort. As they did, both found it easy to relax, as they sat there under the Fort's nearby guards. Then after the effects of their trying days and the sun's warmth, they finally began drifting off to sleep out on the open ground under the day's warmth and the quietness around their herd. An hour or so later, both tired men were soon soundly sleeping in and among their quietly feeding horse and mule herd...

Shortly thereafter, Iron Hand awoke with a start, as he felt the slight tap of the toe of a moccasin on his side. Opening his eyes and looking upward, he saw standing there quietly beside him was Spotted Eagle and little Sinopa. Both had big grins on their faces, as they found themselves looking down at the biggest trapper on the frontier, previously lying asleep on the ground by a herd of horses like an old buffalo.

Sitting up and before he was even fully awake, he found little Sinopa in his arms, telling him how happy she was to see him. Giving her a big hug in return, he lifted her back to her feet and took hold of Spotted Eagle's extended hand and was helped to his feet. There Iron Hand embraced his "Indian Brother", much to the satisfaction of Spotted Eagle and the happiness of little Sinopa. About then, Big Foot, who had been lying alongside, awoke and was also pleasantly greeted by Spotted Eagle and Sinopa as well.

Then Spotted Eagle's face went from an obvious smiling look to one of more serious concern. Then he said, "My Brother, we need to talk."

Sensing the deadly serious change in his close friend, Iron Hand turned to Big Foot saying, "Big Foot, would you take Sinopa to the fort's warehouse, the one with all the fancy things, glass beads and 'play-pretties', and let her have the 'pick of the litter'?"

Terry Grosz

Big Foot, also sensing the serious change that took place in Spotted Eagle, got a big grin on his face and said, "I would be glad to. It is not often an old geezer like me can escort a pretty lady into the fort on a shopping trip." With that and a puzzled look on Sinopa's face over the quick change in her husband's demeanor, she went off hand in hand with her friend Big Foot, so she could have the shopping trip of her life, all at the expense of the American Fur Company...

Once Sinopa was out of earshot, Spotted Eagle began speaking about what had been bothering him. "My Brother, several months ago my village had a fight with some Hudson Bay men and their Gros Ventre supporters on our buffalo hunting grounds. In that fight, we lost several good warriors because those we were fighting had more and better rifles than we did. One of those warriors lost was a friend of mine and a father to two young men from our village. When our hunting party returned, the wife of my friend lost in battle was so upset over the loss of her husband, that she took her own life. That left her two young sons, who in their grief over the loss of both parents began acting like 'bad Indians'. Soon they were taking from others what was not theirs and being mean to other younger children in the village. Chief Mingan spoke to the two young men about being better members of the village but they continued doing mean things to our people, especially the very young and the very old. Chief Mingan, seeing how they continued behaving badly with my people and not following the teachings of The Great Spirit, banned the two brothers from our village!"

Then pausing, a really serious look splashed across Spotted Eagle's face like the news to follow was very bad, he said, "After Chief Mingan banned the two brothers, word came back to my people that they had wandered off into the lands of the Gros Ventre far to the northwest and were now in a band of people who were hunting down white man trappers, killing them and taking all of their furs and horses. Then the killers of the white man trappers would bring those furs and horses to

265

The Adventurous Life of Tom Warren

Fort Union to sell and buy more guns, powder and liquor. After that, then they would go back and do more killing and stealing. Two moons ago, that band of bad people, along with those two brothers, came onto our lands and stole a number of my village's best horses!"

"When they did, Chief Mingan asked me to form a war party of 20 warriors, hunt down and kill those horse-stealing Gros Ventre and the two now very bad brothers from my own band of people. Since that day, that is what I have been trying to do. But now, information just recently back from part of my war party in the field and whispers from the people in my village, tell me that the bad Gros Ventre and they thought the two brothers were recently ambushed by four trappers along the Missouri River. When that happened, it is reported that there was much killing. But in the killing of that band of bad Gros Ventre, only one of the bad brothers was killed and the other somehow escaped."

Continuing, Spotted Eagle said, "When that happened, the remaining bad brother has since joined another band of Gros Ventre, led by a number of evil white trappers, who were Fort Union Company Trappers at one time. Trappers who just last year were banished from the fort by McKenzie for doing bad things to their own kind. Now all of them are killing some of their own in order to steal the furs and horses to sell for white man's goods at Hudson Bay trading posts, are the whispers now on the lips of my people back in the village. It has also been reported to my people that the older brother of the two, the one with the long scar clear across his forehead that he received after a fall from a running horse, is still alive and involved in all of the killings."

Then Spotted Eagle drew himself up to his full height as if to emphasize what he was next to say. "Other words being quietly whispered by my people is that the bad brother not killed in the battle along the river, had arrived late with a pack string of stolen animals and witnessed one large in size trapper killing his brother with a tomahawk. The bad older brother,

Terry Grosz

the one with the scar running across his forehead, is saying to everyone everywhere, that the big trapper who killed his brother with a tomahawk was you, Iron Hand, and now he and the others in his remaining party of bad white people are hunting especially for you and your group of trappers," said Spotted Eagle with more than just a lilt of concern in his voice! With that information out into the open, Spotted Eagle looked deeply into the eyes of Iron Hand for any sign of worry or deep concern over what could follow should he or his fellow trappers ever cross paths with the bad brother and his group of Gros Ventre and banished from Fort Union white man trapper outlaws.

Iron Hand just smiled back at the information and the way the words were expressed by his obviously concerned Brother, saying, "Don't you worry, Little Brother. Yes, I did kill a man in that battle along the river who was trying to kill me with his tomahawk. And yes, we four trappers were the ones who ambushed that band of bad Gros Ventre camped along the river several days ago. But we only did so because they had stolen eight of our horses and 16 packs of our *plus* from our corral and cabin. Crooked Hand and I tried to overtake the thieves and kill them for stealing our horses and taking our furs, but they escaped to the Missouri River before we caught them and joined up with a much larger band of unknown Indians. Since the two of us could not attack such a large group without being killed ourselves, we had to let them go and ended up losing our horses and 16 packs of furs."

"Days later when we left our beaver trapping grounds and were heading for Fort Union, we spotted the light from a campfire along the Missouri River bottoms. So when we found a large number of Indians camped along the river, I sneaked into their camp to see if they were friendly or not. I did so thinking they might be a band of trappers heading for Fort Union and if so, we would join them for the protection a larger group of us offered. But when I discovered our stolen horses in among their herd, our group later decided we would attack

267

The Adventurous Life of Tom Warren

them in the dark of the night, kill them for what they did to us and recover our stolen property. That we further decided to do when I discovered four trappers that they had recently captured, tied to several trees and were murdered when they were all shot full of arrows! We later attacked this band of bad Indians and killed them all for what evil things they had done," said Iron Hand.

Continuing, Iron Hand said, "We four trappers did that and it was then that I must have killed one of the bad brothers you are talking about, who was the one who tried to kill me with his tomahawk at the end of the fight. However, we never saw the other bad brother. He must have sneaked in on us after the battle had ended, decided not to attack since we now outnumbered him, then sneaked off and left the area. That is too bad because had he attacked to avenge the death of his brother, we would have killed him as well and then your people would not have to worry anymore about the 'bad seed' from your village."

With those words, Spotted Eagle said, "My Brother, he will continue hunting you for as long as he shall live. We will still follow Chief Mingan's orders and continue hunting him and the bad group he is now running with, but if we do not find him before he finds you, you need to be prepared for a fight to the death. For as surely as there is grass for the buffalo to eat, he will pursue you until you kill him or he kills you."

Iron Hand just smiled over his Brother's words of warning and worry saying, "Spotted Eagle, we need to go and see what Sinopa has purchased in the way of 'play pretties'. Otherwise, she may just take everything she can get her hands on and then you will have to buy more horses just to pack all her things back to your village." Both men had a good laugh as they walked over to the fort's warehouses and forgot about the 'bad seed' matter and what he would do if he ever ran across Iron Hand in the future.

The next four weeks found Old Potts and his fellow trappers busy as all get-out. Iron Hand and Crooked Hand spent the better part of three days procuring the provisions needed for their upcoming beaver trapping trip of possibly two years into the unknown in the Porcupine River country. In fact, due to its distance from Fort Union, Old Potts had the two men purchase enough provisions for two trapping seasons in case they were unable to return to the fort after only one year afield. In so doing, the men purchased double provisions of salt, black pepper, sacks of rice, bags of pinto beans, four kegs of gunpowder, four additional kegs of rum, sacks of red pepper flakes, dried apple slices, extra pigs of lead, additional caps for the nipples of their Hawkens and pistols, and triple the amounts of bags of coffee, dried raisins and brown sugar cones.

Additionally, the men purchased one extra riding saddle and two new pack saddles in case any of what they had were lost, busted up or the porcupines chewed the sweat-soaked leathers for the salt it offered while they were stored in a cache house. They just figured if a river was named the Porcupine River, it had been done so for a damn good reason... They also purchased extra bridles, horseshoes, horseshoe nails, files, another ax, and a 100-foot spool of cotton lead rope to replace that which always seemed to get busted when the pack strings got into dense timber and crossways with a mess of trees with a panicked horse on each side thinking the tree had ahold of them. When that happened, there was usually a 'rodeo' and a busted lead rope as the panicked horses went their own ways until recaptured.

As those two men did the 'frontier grocery shopping' for their provisions, Big Foot saw to the needs of the trappers' riding horse and pack string animals. Being a blacksmith himself in the old civilized world, he took their 20 animals to Fort Union's two blacksmith shops. There he had all their old horseshoes removed and new ones custom-made for all of their animals when they were re-shoed. He also had the 'Smithies' custom-make extra sets of shoes for each individual animal in case, as Old Potts had instructed, they stayed two years afield in the Porcupine River country or took their furs to a distant

The Adventurous Life of Tom Warren

Rendezvous instead of trekking all the way back to Fort Union. Big Foot then had the 'Smithies' 'float' the teeth on several of their herd of livestock needing that done to alleviate those problems caused by bad or erupted teeth, causing dangerous teething problems. As the 'Smithies' worked their magic on the trappers' horse herd, Big Foot visited the adjacent leather shop and had the significant and hard-used leather strapping replaced on all of their riding and pack saddles. Additionally, he purchased several bundles of leather strapping in case emergency repairs were needed out along the trail. At the same time, he purchased four new panniers to replace two that were damaged in an earlier horse wreck and two others that were stomped into the mud during the lightning storm that got one of their pack animals killed by a bolt from out of the blue and then run over and pulverized by a large herd of storm-frightened, stampeding buffalo.

As for Old Potts, he spent the better part of a week visiting newly arriving Free Trappers and Company Trappers returning from the frontier, attempting to gain any and all information regarding the Porcupine River area and its reported trappings. He was also 'fishing' for any firsthand information relative to any of the returning trappers' relationships with the local Gros Ventre. He also figured with the cost of a few cups of rum, he could jar loose and loosen up the flow of any information from normally close-mouthed trappers relevant to the actual truth about the Porcupine River country and its beaver resources. Plus it gave him the opportunity to visit with some of his old friends and also see who had lost their hair during the recent trapping season and had become worm food...

Then every evening the four men would return to their camp along the Missouri River and discuss the day's events and the progress being made in their individual preparations for their forthcoming trip into unknown Indian country. During one of those evenings around the campfire after eating another one of Iron Hand's great raisin and apple Dutch oven pies, Old Potts poured himself 'three fingers' of rum, sat back

Terry Grosz

on his sitting log and after a few moments of deep thought, announced to his fellow trappers that he was ready to pack things up and head out.

In the discussion that followed, Old Potts advised, "The four of us have a long ways to go in country unfamiliar to any of us." Then after another long swig of rum from his cup, said, "When we get to where we are a-going, we will need to build a new cabin, especially if Harlan Waugh's old cabin along the Porcupine River is found to be unlivable. If that be the case, that means we will be spending a month building our new cabin, hauling in our winter's woodpile, building a new set of 'hell-for-stout' corrals, constructing a cache house to store our leather goods, saddles and the like out from the weather, get our jerky making, smoking and drying racks built, harvest enough hay from the river bottoms for our horse herd during the worst of winter, hunt buffalo, and build up our winter's supply of good jerky stocks. Then if that ain't enough hard work hanging in the wind, we need to cast a small mountain of balls for our lead-eating Hawken rifles and pistols, and then finally scout out the best beaver trapping waters for us to begin our season. And in so doing, manage to keep our heads down and not lose our hair in case we get nosy neighbors who want nothing to do with our kind or wanting us being there in their backyard..."

After that much 'wind', that necessitated Old Potts taking another deep swig of rum from his ever-handy cup. Then Iron Hand spoke up, saying, "Afore we hit the trail if that be the case, I need to purchase another large three-legged cast-iron frying pan because one of the legs broke off my old one and another 16-quart Dutch oven. I can make better biscuits and pies with a larger Dutch, so I would like to get one. Other than that, maybe another set of cooking irons for a second outside fire, some extra tobacco and I will be set," continued Iron Hand, as he absent-mindedly stirred their campfire with a handy limb.

The Adventurous Life of Tom Warren

"I need to get a couple of tins of some sort of bag balm case we 'gimp' up one of our riding horses or sore-up a pack animal. But other than that, we are now ready to go with our animals any time they are needed," said Big Foot.

"As for our provisions, we have a load to carry in light of your suggestion to prepare for two instead of one trapping season afore we can resupply," said Crooked Hand. "It is a damned good thing we have a goodly number of hell-for-stout pack animals this time for all that we will be carrying. My only suggestion for extra provisions is that we get another keg or two of rum. You never know when I will 'gimp' up my left leg again between a mean-assed grizzly bear's jaws, getting knifed by a Gros Ventre or becoming part of a damned good horse wreck," said Crooked Hand with a big guilty-looking grin over his reasoning for acquiring extra kegs of rum...

"Then it is settled. We should make our final plans to leave in a couple of days. That a-way, we will be ahead of all those pesky Company Trappers as they head out since we have such a fer piece to be a-going. That plus I have a feeling that Harlan's old cabin he told me about last season, may not be livable after being abandoned for over a year and if that be the case, we will need to build another one to suit our tastes. If we do, Big Foot, you need to build another one with a narrow front door and a step-sill like you did last time that saved our bacon when those damn Gros Ventre came a-knocking in the dead of night like they did. Also, I had the blacksmiths build us a hell-for-stout iron chain so's we can really secure our horse corral so's no one can come a-knocking and just open the gate and waltz off with our stock like they did last year. Also, McKenzie gave me one of them fancy 'pad-o-locks' from one of his keelboats that we can use on the corral gate. He did so after I shared our story about them damn horse-stealing Indians that kept running off with our livestock. I think he did so because he is hoping we come back next year with a larger horse herd than we left with like in times past once again. That a-way, he can buy those extra horses from us for some of his

Company Trappers who aren't as lucky or as careful with their stock animals as we be. However, that chain and lock will not do us any good unless we built a hell-for-stout corral with larger logs so's them Indians kain't push them over so easily to get at them horses. So, building that horse corral is just another good reason as to why we need to get a-moving purdy damn soon," concluded Old Potts, as he emptied the last drop of rum from out of his cup.

With those words, the men quietly emptied their cups of rum and shuffled off to their sleeping furs for another night under the stars and clouds of the ever-present mosquitoes along the Missouri River bottoms... Tomorrow would be another good day and the men would need all the daylight they could run under in order to get all of their last minute chores cared for. Little did the men realize that bad news would be coming their way from Spotted Eagle come the morrow...

Standing over his cooking fire tending his Dutch oven biscuits, Iron Hand was surprised to all of a sudden see Spotted Eagle materialize silently from out of the bushes along the riverbank and coming his way.

Moments later, Spotted Eagle said, "Good morning, My Brother."

When he spoke, Iron Hand detected a slight bit of concern in the tone and tenor of Spotted Eagle's voice. Ignoring the sound of concern in Spotted Eagle's voice because he knew he would soon speak his piece, Iron Hand said, "Good morning, My Brother. You are just in time for some breakfast." When he spoke, he did so with a happy smile under his heavy beard over the welcome presence of his close Blackfoot Indian friend and now, Brother.

"I am always ready for some of My Brother's biscuits,@ said Spotted Eagle, as he sat down on a nearby sitting log and in typical Blackfoot custom for a guest, waited to be served.

Without another word, Iron Hand took two of his piping hot biscuits from his Dutch, shoveled them onto a tin plate and then slathered them with a generous helping of honey from a

The Adventurous Life of Tom Warren

nearby jug just recently purchased from Fort Union's stores of provisions. Then as custom dictated, Iron Hand served his Blackfoot Brother.

"UMMPHF," said Spotted Eagle, as the sizzling hot biscuit crackled on his tongue when he shoveled in a generous bite without hesitation over the biscuit's Dutch oven heat!

"Is My Brother like a small buffalo calf when it comes to eating a white man's warm biscuits? Is he finding that he is not yet powerful enough to act like a mature man when it comes to eating something hot served by a white man?" asked Iron Hand with a 'smile' in the tone of his just spoken words to his struggling Brother.

Still mouthing all around in his mouth the piece of a hot biscuit, Spotted Eagle mumbled something that Iron Hand could not understand, so he broke out laughing over his Brother's discomfort.

Soon that 'too hot to eat biscuit' moment passed and with a silly grin on his face, Spotted Eagle said, "I came here this morning to share with My Brother some news from some of my warriors just returning from hunting my village's remaining 'bad seed'. Instead, I am almost poisoned by my Brother's bad cooking!"

Both men had a good laugh over the earlier 'hot biscuit' moment and then Spotted Eagle got down to the business for which he came. "My Brother, my warriors tell me that the 'bad seed' is as elusive as a mountain cat. However, he being from my nation, I can see why he is so hard and clever to catch and kill. My warriors tell me that the word from others out on the plains is that the 'bad seed' is now running with other bad seeds from the Gros Ventre and a small number of white men who have been 'thrown away' by McKenzie. They are now hunting any white trappers they can find, are killing them and taking their furs and livestock. Then they are taking what they steal and often selling those things stolen to their friends from the Hudson Bay Company forts further to the north. In turn, their Canadian brothers are supplying the bad seeds with all the guns and whiskey they wish to have and telling them to bring even more furs and horses from the white man trappers from the American Fur Company. They are also catching lone young Indians from other tribes and taking them captive and selling them to their Canadian brothers as well. That is especially so when selling the women and small children to them. Then the Hudson Bay people are selling those women and children to other trappers of their own kind as slaves and the women to 'pleasure' their lonely trappers."

"My warriors are also telling me that the white man trappers they are finding that have been killed by the bad seeds, have died a long and slow death! That is why I am here this morning. Besides having my tongue burned with a white man's biscuit, I am here to tell you that you and your friends need to be 'mountain cat careful' in everything you do once you leave the comfort of this fort. The last word from my warriors is that the 'bad seed with the forehead scar' is now found along the Missouri River because that is where they are finding the most trappers traveling to and from where they are trapping. Since that is where you wish to go next, I wanted to warn you of what awaits you if you are not careful," continued Spotted Eagle.

Then as an afterthought, he spoke up once again saying, "Just as soon as my warriors rest up and provide food for their families, we all will go out once again and try and follow Chief Mingan's words to find all the bad seeds, both Indian and white, to see that they join the Cloud People and wherever bad white men go when they are dead."

By then, the other three trappers were seated around the pair of men on their sitting logs and listening in on Spotted Eagle's and Iron Hand's conversations.

When Spotted Eagle had finished speaking, Iron Hand thanked him for the information and his outward signs of concern. He then advised Spotted Eagle that he and his band of trappers would be leaving soon for the Porcupine River country to trap beaver and that they would heed his wise words

The Adventurous Life of Tom Warren

and remain alert to the dangers posed by the 'bad seed' and his friends. Following that, the trappers shared breakfast and a cup of rum with their Blackfoot Indian friend.

After Spotted Eagle had left, the four trappers sat around their campfire and discussed the Indian's words of caution. It was then decided that the men would return to the fort's leather worker's shop and reconfigure an addition to each man's riding and first horse in the pack string saddles. By late afternoon that same day, the four trappers found their saddles altered to their satisfaction. Instead of having their extra weaponry on their first packhorse in the strings they would be leading into the Porcupine River country, a change had been made. The first animal in their pack string would still be carrying an extra loaded rifle and two pistols in case the trappers were jumped by hostile Indians. However, to the front of their personal riding saddles had just been added two additional custom-made pistol holders. Henceforth, instead of the two extra pistols being carried by the first horse in their pack string, two extra weapons would also be carried on the front of the saddle of each man's riding horse. That meant that each man now carried a rifle while in the saddle, two pistols in his sash and two additional pistols in custom-made holsters attached to the front of each man's riding saddle. In so doing and in case of extreme danger, each man was now capable of firing five shots at any adversaries, with a spare rifle and two single shot pistols still on his first packhorse. In short, in times of extreme danger, the four trappers would be capable of firing a total of 32 shots before they 'ran dry', if they lived that long that was... No two ways about it, a very deadly fusillade for anyone foolish enough to take the four trappers dead-on in a fair fight! That would be especially so if facing such firepower loaded with buck and ball and two of the men being such accurate shooters at close up and personal ranges with their rifles and pistols...

Terry Grosz

Chapter 10: The Porcupine River, The "Bad Seed" And Saving The Brothers "Dent"

THREE DAYS LATER, after having a final breakfast with Kenneth McKenzie, the four trappers rode out from Fort Union. In front rode Old Potts trailing a string of four heavily loaded pack animals, followed by Big Foot with his four pack animals, followed by Crooked Hand and Iron Hand trailing the same numbers of 'loaded to the gills' packhorses. Being the Free Trappers that they were, each man sported once again recently acquired, beautifully adorned and beaded, fringed buckskin shirts, beautifully beaded moccasins, and each of their riding horses were bedecked with brightly colored 'wearing' finery as well. All befitting the special class of historically high-spending and colorfully individualistic Free Trappers.

Old Potts kept his contingent of trappers heading due west along the north bank of the Missouri for the next several days. As they did, the trappers soon had passed the Big Muddy, Poplar and Wolf Rivers without incident. During those days of travel, the men kept to the river bottoms and at night, after unpacking all the pack and saddle horses, camped in the deepest and most secluded of vegetative cover along the

277

The Adventurous Life of Tom Warren

Missouri. This they did as a result of Spotted Eagle's earlier warning about the "Bad Seed" and his bunch of renegade Gros Ventre and white men outcast trappers rumored to be 'bushwhacking' fellow travelers along the Missouri River because it was a chokepoint of travel, especially for all the heavily loaded trappers moving to and from the Fort Union trading post carrying valuable furs or their annual provisions.

Nine days later found the four trappers and their string of livestock at the confluence of the fabled Porcupine River and the 'gateway' to its rumored fabulous beaver trapping country further to the north. However, for their last two days of travel, Old Potts's trappers had run across numerous herds of buffalo and several dozen trails of unshod horses whose Native American owners were more than likely hunting those buffalo. Camping that evening in the river bottoms at the confluence of the Porcupine River and the Missouri, the men decided to change their plans of travel. Being that the Indians were hunting the buffalo heavily for their winter meat stores, and the chances of discovery by the dreaded Gros Ventre were greater during daylight hours, the men changed their travel plans. Around their small campfire that first evening along the Porcupine River, the men decided for safety's sake they would travel only during the early morning and early evening hours. In between those hours, the men would rest up and let their livestock graze in an out-of-the-way place to avoid discovery. The assumption being the nomadic Indians hunting the buffalo would be most active during the daylight hours, killing and butchering those animals killed. Then come the end of the day, they would be feasting or camping and not as active. So the trappers decided they would try and travel during those early morning and dusk hours of the day they suspected the Indian hunters would be less active, thereby reducing their chances of being inadvertently discovered and attacked.

Even utilizing that strategy, the trappers had to numerous times retire to the dense brush along the Porcupine to hide as numerous bands of Indians passed nearby. Thankfully their

Terry Grosz

shod horse tracks were not detected by the Indian hunters, who were more concerned with the killing of buffalo as opposed to really inspecting a small number of shod horse tracks in and among the numerous buffalo herds moving back and forth across the rolling prairie's lands.

For the next two-and-a-half days as the men rode north on the Porcupine, they scouted out the beaver trapping waters as they went. As they did, it was apparent that the beaver waters on the lower reaches of the Porcupine had already been heavily trapped and most sign was old or almost non-existent. Finally at the end of day two while traveling north on the Porcupine, the men began running into pristine beaver trapping waters. Numerous stick and mud dams dotted with active beaver houses now greeted their eyes at every turn in the river system's waterways. It then became obvious to the trappers, that for whatever reason, other trappers had yet to reach the northernmost reaches of the Porcupine and that suited the four 'Old Potts' men just fine. In fact, camping along the river one evening, the men observed 51 beaver, which are primarily nocturnal, moving about in the ponds hauling clusters of aspen and willow limbs to their lodges and to their underwater food cache and storage sites. They also noticed that a great number of those beaver were of the largest in size that they had ever seen, or 'blanket in size' in the nomenclature of the trapping fraternity! In short, the big money-sized critters were abundant it seemed in just about every waterway...

Then the trappers 'struck gold' in another arena. They now had approached a number of low timbered hills off to the north and west of the river in their travels. Old Potts figured that was probably where Mountain Man Harlan Waugh had reportedly built his cabin out of the way of any main-traveled trails along the river's floodplains years earlier. Pulling their pack strings into the timbered areas and out of sight of any prying eyes, the trappers made a dry camp for that first night.

Unpacking all of their animals and hobbling them, they were then let out nearby to graze under the watchful eyes of

The Adventurous Life of Tom Warren

the trappers as they made camp nearby for the night. Stacking the packs of provisions and necessaries around their campsite in a defensive arrangement, the trappers built a small fire and feasted on almost raw or partially cooked buffalo steak, as they rested their tired knees from the many leg-cramped hours in the saddle riding to their current destination.

After supper, the men huddled around their small campfire, smoked their pipes and discussed plans for the next day. For the longest time Old Potts remained quiet as the other trappers made small talk about the area. Then Old Potts piped up saying, "If it were up to me, I would make my secluded camp somewhere in these hills, especially if I were Harlan. There's plenty of small springs and ground seeps of water in most of the gullies, there is grass a-plenty for our stock, we would be out of the north winds, sight and sound of any winter storms or prying eyes, and we would be closer to some of the finest beaver trapping I have seen in my lifetime. I say tomorrow Crooked Hand and Iron Hand strike out on their horses and scout out this area and see if you can find anything resembling an old cabin. If one is located, check it out and see if it would be something the four of us could live in. However, as you two are moving about, keep your eyes peeled for any location that would make a proper secluded spot for us to build a cabin, if Harlan's old one is not found or found to be unacceptable for all of us and gear. In the meantime, Big Foot and I will let our horses continue putting on the feed bag and guard our supplies a-waiting your return. What say you two to that plan?" asked Old Potts.

Iron Hand looked over at Crooked Hand for visual confirmation of what he was thinking and then said, "Sounds good to us. We can leave at right at daylight and stick to the timbered areas because if it were me, that is where I would build a cabin that was out of sight and out of the way from the worst of the winter winds. If we can find where Harlan built his cabin, we will check it out to see if it would be suitable for

Terry Grosz

all four of us. If not, we will keep looking for a spot of our own."

Having decided the next day's activities, the four men turned to their sleeping furs early on and soon they became part of the night. The next morning, Iron Hand and Crooked Hand were up before dawn, had saddled their horses and were now standing around their small campfire warming themselves. Jerky was the word of the day for breakfast, and soon the two men had disappeared into the sparse timber of the adjacent low lying hills looking for Harlan's old cabin or a spot for their new one where there would be good water, feed, firewood and cover from unwanted prying eyes.

After an hour of fruitless wandering looking for an old trapper's cabin, Iron Hand remembered what had been told to Old Potts by Harlan Waugh, about leaving a cache behind because his horses could not take everything when Harlan had left for that year's Rendezvous being held further to the south. The information Harlan had provided was that his cache was below a large fir tree looking toward the Porcupine River and his beaver trapping grounds. Telling Crooked Hand to follow him, Iron Hand rode his horse to the top of a low lying ridge, then stood up in his stirrups and looked all around at the terrain lying below him. Lying to his north stood a tall stand of fir trees out all by themselves along the edge of a finger of pine trees in which they had been traveling through most of the morning.

"Come On, Crooked Hand. We need to head over to that grove of firs. That may be where Harlan's old cabin sits," said Iron Hand, as he sat back down in his saddle and then spurred his horse in that direction. Twenty minutes later as luck would have it, the two men rode into a secluded timbered glen holding an old trapper's cabin! Riding up to the front of the cabin, the two men dismounted and entered the hell-for-stout looking cabin. Looking all around inside, except for some evidence of small animals like squirrels and mice living therein, the cabin appeared to be pretty sound. Stepping back outside, the two

281

The Adventurous Life of Tom Warren

men moved around the cabin and declared its walls, foundations and roof sound as a Spanish silver dollar. The cabin was well-hidden from any Indian traffic moving along the Porcupine River lying to the east, and a small but active ground seep with adjoining springs provided a plentiful supply of fresh water for man and horse alike. Looking around even further, the men could see there was an abundance of firewood within close distance to the cabin and sported a nearby watered meadow that was knee deep in good grass to a bull buffalo.

However after a more detailed inspection, the men determined that the cabin was too small for the four of them, all of their provisions and later on the stacks of beaver *plus* the men figured they would be accumulating. With that information in hand, the men carefully and always on the lookout for Indians, headed back to their original campsite. About an hour later, Iron Hand and Crooked Hand rode into their original campsite only to find Big Foot skinning out a mule deer from a makeshift meat pole. Upon the two men's arrival, the four trappers sat down around their cook fire and discussed what had been discovered by the two men earlier in the day.

After a thorough discussion regarding Harlan's old cabin, the men had made their decision. They would move their temporary camp to Harlan's old cabin. There the men had decided to make an addition onto Harlan's cabin making the structure like a "T". By so doing, the men figured that would save them a lot of time and hard labor by only building an 'add-on' instead of another fuller-sized cabin. With that, the horses were packed with all the men's provisions and following Iron Hand, they backtracked their earlier trail over to where Harlan's old cabin was located.

Once there, all four men added their input into what needed doing and then the work began. The men set about cleaning up the old cabin first and then moved all of their valuable provisions inside to be out of the weather and away from the temptation of theft from the hands of the 'locals' if their cabin site and stockpile of provisions were inadvertently discovered. Then while Iron Hand began enlarging and rocking in their nearby spring so the trappers and their livestock could water more easily, Crooked Hand began cutting stout poles for the corral from an adjacent aspen grove. As they labored away, Old Potts and Big Foot built and rocked in their permanent outside firepit, set up their pot-hanging cooking irons and began making supper. Prior to supper being served, Iron Hand and Crooked Hand cut and dragged with the aid of their horses sitting logs for use around the outside campfire and set them into place. Once supper was finished, all of the men set to digging and setting the corral's previously cut-to-size and hell-for-stout posts so their horse herd could be safely housed without fear of easy theft by Indians. But for the interim, the horses were just double hobbled and left to graze under the watchful eyes of the close at hand trappers.

After breakfast the following morning, the men set to finish constructing the rest of the horse corral and finally had the work done to their satisfaction by nightfall. With the iron-linked chain and padlock in place around the gate posts, the horses were finally safely ensconced. Then the real work began as Big Foot cut a doorway into the back wall of the cabin with a saw, while the other three men cut the needed green timbers for the addition onto their existing smaller cabin. By week's end, the new addition to the original cabin was up, roofed and ready for storage of all their provisions and the men's sleeping quarters which would now be housed in the great, fresh-smelling pine wood addition. That way, the smaller original one-room cabin could be used for cooking during the winter months and also for the fleshing and hooping room when the fresh beaver, muskrat, river otter, bear and buffalo skins began arriving for processing. Then while Big Foot set to making chairs, tables and shelves for inside their cabin, the other three men cut dead and dry timber and with their horses, dragged the logs next to their cabin to be used as their winter woodpile. By now, things around their cabin were

The Adventurous Life of Tom Warren

beginning to look homey and shipshape. Then Big Foot cut pegs and drilled holes into the outside walls of their cabin so their traps could be hung from the pegs for ready access. Additionally, he built a floor sill below the front stoop just inside the doorway into the front of their cabin as a trap for any persons attacking the cabin. By so doing, anyone making a surprise unwanted entrance and not being familiar with the inside stoop, would stumble over such an extra step and fall. If that occurred, upon rising after falling flat onto the floor, guess what awaited the unwanted intruder once he stood up… Lastly, Big Foot cut out shooting holes in all of the walls of the cabin in case the trappers were attacked and trapped inside their cabin. Then he built shooting-hole plugs that could only be opened from the inside to preclude anyone trying to shoot the trappers by using outside shooting holes to shoot through back inside the cabin at its hiding occupants.

Finally, Iron Hand because of his height, began hanging their perishable food items like the dried raisins, dried apples and the like from pegs driven into the inside rafters to preclude 'the little people' from chewing through the sacks and consuming the contents, especially if they had been left sitting on the now hard-packed dirt floor. Finished with the last of the inside work, Iron Hand and Big Foot set to work building a meat pole, along with drying and smoking racks for any buffalo and venison meat brought in to be made into jerky. As it turned out, Crooked Hand and Old Potts brought into their camp all the buffalo meat their three packhorses could tote and after the men had sliced the meat into thin strips, the drying and smoking racks were put into immediate use in the making of jerky for the winter months.

That evening under the light from their outside campfire, all four men, using chopped off chunks from their pigs of lead, began the time-honored tradition of casting a small mountain of rifle and pistol balls so that once they began trapping and hunting in earnest, they would have an adequate supply for most eventualities. As they sat around the hot fire pouring and

284

casting their bullets, the men made sure their first keg of high proof rum acquired earlier from Fort Union, was tested to make sure it did not contain any 'poison'...

By now, the prairie grasses were starting to turn yellow and brown, the geese were moving south in greater and greater numbers, there was usually a skim of ice now in the morning on the cabin's water bucket, and the aspen trees' leaves in the area were bright yellow and orange in color. Noticing the distinct nip in the air as he hustled around the outside campfire making coals for his Dutch oven biscuits that the men had come to expect every morning for breakfast, Iron Hand also tended to his buffalo meat frying in bear grease in his three three-legged cast-iron frying pans set in a thin layer of hot coals as well. As many great cooking and baking smells filled the air, out tumbled Iron Hand's three partners, as they headed for the coffee pot hanging and bubbling away over the hanging irons. Once all the men had poured themselves steaming cups of boiling hot coffee, they then spent the next few minutes trying to cool down those cups of coffee with lots of blowing. As they did, Iron Hand dished out generous slabs of cooked rare buffalo meat as the men liked it and perched a piping hot biscuit on top of each steak served to his friends. After they had been served and were eating their breakfast, Iron Hand served himself, sat down on a sitting log, and THEN HIS PLATE EXPLODED FROM OFF HIS LAP AS A RIFLE BULLET BLEW CLEAR THROUGH THE EDGE OF THE METAL!

Falling over backwards in surprise, Iron Hand quickly rolled towards his rifle sitting alongside another close at hand sitting log, grabbed it, jumped up and looked for the puff of black powder smoke from the assailant's rifle shot so he knew where to place his shot. Seeing an Indian standing by an aspen near the horse corral hurriedly reloading his rifle, he died as he tried...

Then Iron Hand heard two more shots being fired as he grabbed for his pistol which he always carried in his sash.

The Adventurous Life of Tom Warren

Whirling around, Iron Hand saw another puff of black powder smoke slowly drifting lazily in the air by a single pine tree standing at the end of their cabin. That was when he also saw an Indian crumpling onto the ground with a bright red splash of blood smeared all over his face. To Iron Hand's way of thinking, that Indian had died at the hands of Crooked Hand, the man among them who liked to head shoot those he felt it necessary or ones that he did not like... As for Old Potts and Big Foot, they stood at the ready with their rifles in hand around the cooking fire looking all around for any other signs of danger! Seeing none, that gave Iron Hand and Crooked Hand time to reload both of their rifles. Then the men walked over to the two dead Indians, Gros Ventre by their dress, and made sure they were both dead. As expected, there was no doubt as to their being on their journey to meet the Cloud People... Crooked Hand had shot his man through the head with his rifle appropriately named "Never Miss", and Iron Hand had exploded the heart in the man he had shot...

For the next hour or so, the men scouted out the area around their cabin and finally came to the conclusion that the two Indians had just been out hunting deer. In so doing, they had stumbled upon the secluded trappers' cabin and had taken matters into their own hands upon seeing the trappers quietly eating their breakfasts. Those concerns as to their assailants' deer hunting were further borne out when the men discovered where the Indians had tied off their horses. One of the horses had a previously killed, forked-horn buck tied over the back of the animal. Sensing those two Indian hunters were the only threats, the men continued with the business now at hand.

Bringing the two Indians' horses down off the hills, they were placed into the corral with the trappers' other horses. Then the men once again returned to their breakfast without a word being spoken about the most recent series of deadly events. That was until the men had finished eating and then the conversation turned to the matter at hand. That matter at hand being what to do with the two dead Indians so when others came looking for them, they would not be found nor the trappers found out or discovered for what they had done.

Taking another now slightly browned biscuit since they were not tended to in the Dutch ovens while the ruckus was on going, each of the men took turns pouring honey over them and then made sure they did not last in the morning's cooling air. "Now what do we do with the bodies?" asked Iron Hand, as he finished eating his honey-slathered biscuit and licked the 'sticky' off his fingers.

"Well, when John Coulter was being pursued on foot by a mess of Blackfoot Indians hellbent on killing him, he eventually hid under a log pile of driftwood in a creek," said Old Potts, as he still was wrestling around with the remains of his sticky biscuit. Then licking off his fingers, he said, "Yesterday, I seed jest such a log pile of driftwood down by the bend in the Porcupine, due east of our camp. I say we do the same as did old John Coulter with his self. Let's take the dead down there, wade out into the water and jam their bodies up under that mess of logs out of sight and let them be until eternity takes us all," said Old Potts in a matter of fact tone of voice. Hearing no dissent over his suggestion, Old Potts said, "Guess we best get to cracking afore some of their kin come nosing around and then if they see what happened, there will be hell to pay in this here camp come sundown."

An hour later, Iron Hand stripped down bare naked and waded out into the now getting colder-as-hell Porcupine River dragging Crooked Hand's head-shot Gros Ventre, jammed him up under the log pile to where he was out of sight and in such a manner, that when he rotted he would not float free. Then Iron Hand took the Indian he had heart shot, waded back into the chest high waters gasping for his breath as the cold made his heart skip a beat or two, and jammed his dead Indian way up under the logjam and out of sight in such a manner that he would not float free when he bloated or rotted away, ever.

Iron Hand then swam out from the cold waters and jumped up onto the bank for what little warmth he could get from the

The Adventurous Life of Tom Warren

late fall sun's rays. "Damn, I don't look forward to this fall's trapping. I swear, the water in the Porcupine is one hell of a lot colder than that on the Poplar," grumbled Iron Hand as he shivered violently, while wiping himself off with handfuls of long and now fall-dried grasses from the prairie around him.

"What say we get out of here? Knowing them Indians as I do, if they see us fooling around here by this logjam, they sure as thunder will come and investigate. And if they do, they may see the blood still draining out from that log pile from the one Crooked Hand head shot. Hopefully no damn 'griz' will scent blood drippings from his head and then come looking for a meal. They can be sloppy eaters and many times will leave a head, the hips or some long bones and then if they do, the Indians may be on to us as well. So let's skedaddle afore we get found out and blamed."

That evening before supper, found Iron Hand hammering out his tin plate on a woodpile log with the steel butt on his pistol to close up the bullet hole the Indian had shot through the plate's edge with his poorly placed shot while trying to kill Iron Hand. Later during dinner, he found that his hammered-out plate 'ate' just as good as before, if not better, since the Indian had missed his chance and Iron Hand had not...

Once again, the trappers decided to venture forth as a group of four and initially scout out the beaver trappings. That way, if attacked, they had a better chance of surviving with four 'gunners' instead of just two. At least that was the plan until the trappers got 'logjammed' with too many fresh pelts needing fleshing and hooping in order to prevent them from souring. When that occasion arose, then Old Potts and Big Foot would stay back at the cabin, process the fresh pelts and watch over their horse herd being held in their corral, while Iron Hand and Crooked Hand did all of the trapping.

Trailing two packhorses with panniers full of 40 beaver traps, the men left their camp right at daylight. Then doing as they did before on the Poplar River, Iron Hand did all the setting of traps and the other three men stood guard. As it turned out, the waters were so full of beaver, the men only had to run their trap line for about two miles and then all of their 40 traps had been set. Seeing no use in scouting out any new beaver ground until they had trapped out the waters they were now working, and hoping to avoid discovery, the men headed back to their secluded cabin site. But not before collecting 19 dead beaver in their earlier set traps at the front end of their trap line!

That evening when Iron Hand began preparing their traditional evening meal of roasted beaver from their first day's trappings, Dutch oven biscuits and a raisin and dried apple cobbler in another 16-quart Dutch, the other men fleshed and hooped out their first day's catch. That night after supper, the men sat around their outside campfire, drank a celebratory cup of rum and smoked their pipes. It had been a good first day, 19 beaver had been caught in their first sets of the trapping season, four roast beaver had been consumed in their entirety, and nothing was left of the raisin and apple Dutch oven cobbler except the gas the men were later passing over having eaten such a rich and sugary repast, which was somewhat foreign to their primarily meat, biscuit and coffee-driven digestive systems…!

The next morning was a repeat of their first and Iron Hand, upon running their traps, discovered 29 beaver present out of the 40 they had set the day before! Plus, almost all of the 29 were what was known in the fur trade as 'blanket beaver' they were so big! On the way back to their cabin, Crooked Hand shot and killed a cow buffalo, and soon the packhorses were groaning under the weight of the next mess of suppers and breakfasts, as well as a mound of potential jerky for use later on when the winter winds were howling out on the prairie and fresh, close at hand sources of meat were scarce...

However, things were going to soon take a turn for the worse. The trappers found that they were spending a large amount of their time hidden in the brushy creek and river bottoms, as it seemed more and more Indians were on the

The Adventurous Life of Tom Warren

prowl looking for their two lost comrades. But after a week of lying low, the numbers of Indians out looking apparently had given up and thinned out, and the trappers got back to their normal trapping activities.

But with the increasing numbers of beaver skins arriving back at the cabin on a daily basis, the trappers found themselves splitting their numbers with two men running the trap line and two staying home caring for the pelts, so they would not mold or spoil before they had a chance to dry out. Once again, the trappers by being split up in their numbers, found they were staring danger in the eye from the Indians and discovery almost on a daily basis.

The fall and the early vestiges of winter were soon upon the four men and then all beaver trapping stopped as the thickness of the ice became problematic when it came to easily trapping beaver. Then it was on to buffalo hunting and the trapping of wolves around the buffalo carcasses. Soon, the new addition to Harlan's old cabin began taking on the looks of a St. Louis fur house. The entire back of the add-on was quickly filling with buffalo hides, a small stack of black and grizzly bear rugs, tanned wolf skins, and 16 packs of beaver pelts wrapped with their protective deer hides and ready for the trip to Fort Union come late spring

After breakfast one morning, Iron Hand and Crooked Hand dressed for the cold, saddled up their mounts and one packhorse and left Old Potts and Big Foot back at the cabin casting more bullets, as they went out to check their wolf traps previously set around three buffalo carcasses. As it turned out, there were wolves in a number of their traps at each buffalo carcass. By the time Crooked Hand had skinned out the wolf carcasses for easier transport back to their cabin in the panniers on their one packhorse, it was nearing nightfall.

Iron Hand decided they would take a shortcut back to their cabin since it was nearing dark and with that, the two trappers headed over a low ridge holding a finger of pine trees that ran out onto the nearby prairie. As the two trappers swung by the lower end of the finger of pines, Iron Hand abruptly slid his horse to a stop in the snow and froze river rock-still in his saddle. Crooked Hand, leading their packhorse and seeing Iron Hand quickly skid to a stop, did the same. Both trappers, in the fast-approaching darkness of night, just sat there in their saddles with their eyes cast up towards the crest of the finger of pine trees. As both men intently looked into that finger of pine trees, they both could smell pine wood smoke. Iron Hand knew they were still about a mile from their secluded cabin and that he and Crooked Hand could not possibly smell wood smoke coming from their cabin site! That smell meant one thing and one thing only, and that was possibly the sign of danger near at hand!

Without a single word being spoken between Crooked Hand and Iron Hand, the two trappers quickly turned their horses around and headed downhill towards a nearby gully holding a dense stand of leafless aspen trees. Arriving in the sparse cover the aspens offered, Iron Hand dismounted and walked back to where Crooked Hand sat quietly on his horse, leading their packhorse.

"Crooked Hand, I am going to walk back and sneak up to that source of the wood smoke we just smelled. I need you to stay here and wait for me to return. That way with only one of us sneaking up to that source of the smell of burning wood lessens our chance of being discovered. I need to see if it is Indians or maybe other trappers. Either way, we need to know who else is in country and so close to our hidden cabin. Hopefully I won't be discovered but if I am and you hear shooting, you ride for our cabin like the Devil is on your tail and bring the other two back here to bail me out if I am still alive," quietly uttered Iron Hand.

With those words being spoken, Iron Hand quietly disappeared into the fast-falling darkness following his nose leading him to the smell of a mystery someone's campfire. About 20 minutes later, Iron Hand finally observed the faint flicker of a campfire through the stand of pine trees.

The Adventurous Life of Tom Warren

Continuing to move in for a better look one quiet step at a time, Iron Hand was soon able to see the source of his interest.

Sitting or standing around their campsite were eight Gros Ventre Indians. They were warming themselves up and in the process, cooking up some meat that had been staked out around their now abnormally large fire, not usually of the kind built by Indians. But that is not what really caught his eyes. Another Indian emerged from out of the darkness of the trees and he appeared to be a Blackfoot by his dress. Then Iron Hand spotted it! The Blackfoot Indian had a large scar running clear across his forehead and across the rest of his face on a downward angle that was even apparent in the faint light of their campfire! That Indian appeared to be the older brother and Bad Seed from Chief Mingan's band of Blackfeet that Spotted Eagle and his braves had been hunting for some time! The Indian with the scar was also the one who had pledged to kill Iron Hand for killing his younger brother with a thrown tomahawk to the face back at the fight on the Missouri River, when all four of Old Potts's trappers had killed so many renegade Indians who had been killing and robbing every trapper they had run across...

Just as Iron Hand started to leave since he had seen enough and with the knowledge that he and his fellow trappers had to kill this entire lot of Indians or they would be meeting them violently sooner or later, he heard some strange sounds coming from deeper in the timber behind the group of Indians now gathered around their campfire. Curious as to that strange sound just heard, Iron Hand slipped back further into the timber and then stalked around the men by the campfire into the darkness where he had heard such a strange and muffled sound like someone in despair.

Long moments later, Iron Hand found himself looking at three young white men fur trappers from behind where they had been tied up around three separate pine trees. Each man had been tightly bound with rope and even had a tight strand of rope around each man's necks and wrapped around their tie-

292

down tree so they could not yell out or lower their heads and attempt to chew their ways through the ropes binding the prisoners, allowing them to escape.

With that, Iron Hand had seen enough and being outnumbered, slipped quietly back into the dark gloom of the trees in the darkness, as he headed for his horse tied off back in the aspen grove. Arriving shortly thereafter, Iron Hand instructed Crooked Hand as to what he had observed. Then Iron Hand sent Crooked Hand quietly on his way back to their cabin in order to get some help from the other two trappers. But not before instructing Crooked Hand on who was in the suspect camp, their numbers, who was tied up and that the Bad Seed Blackfoot Indian was among them! Iron Hand then laid out a battle plan and instructed Crooked Hand on how to sneak back with the other trappers to where he would be and what needed to be done once all of them were together again.

With that, Crooked Hand slipped off into the darkness leading their packhorse and headed for their cabin and some help. In turn, Iron Hand turned around and once again headed back into the area of the renegade Indians' camp. Once back at their camp, Iron Hand quietly and without letting the captives know he was in country, silently positioned himself directly behind the tied-up trappers so he could protect them if the Indians around the campfire tried to kill them before help arrived.

About an hour-and-a-half later, Iron Hand became aware of someone quietly sneaking in behind him. Turning, he was happy to see Crooked Hand's ugly mug appearing into the faint backlight from the Indian renegades' campfire. Then behind Crooked Hand, Iron Hand could see two other dark shapes materializing out from the darkness and silently creeping towards him as well.

Iron Hand then slipped back into the darkness, joined his three trapper friends and quietly whispered out a plan of attack. Having been there unknown in the renegades' camp for so long, Iron Hand had time to examine the entire layout of their

The Adventurous Life of Tom Warren

camp and figure out what needed doing when the time came. For the next few minutes, Iron Hand laid out a military-style battle plan to his three trapper friends and then without another word being spoken, the three trappers silently disappeared off into different directions in order to get into their pre-planned positions of attack. Then nothing was heard for the longest time as Iron Hand waited for everyone to get into position.

Finally figuring everyone was in position and seeing that all nine of the renegade Indians had finished their meager supper and were now lying down on their blankets and going to sleep around their roaring campfire, Iron Hand made his move. Quietly slipping up behind one of the tied-up trappers, Iron Hand quietly reached out and firmly clamped his hand on the first trapper's mouth so he could not yell out, and firmly held him still until the man quit wiggling in fear and could see the friendly face of Iron Hand peering at him from just inches away in the fire's faint light. In an instant, the tied-up trapper's eyes flung wide open in amazement! The man tied to the tree was none other than Josh Dent, a fellow trapper and friend Iron Hand and his mates had met earlier back at Fort Union and who now had recognized the face of the huge man, fellow trapper and friend. A man who Iron Hand was still holding his mouth tightly shut so he would not make any noises that would alert his captors. Then Iron Hand could finally see a wave of recognition and relief flooding across Josh's eyes upon seeing and realizing his friend was so close at hand!

Iron Hand in turn, being surprised over seeing his friend, a prisoner of the Indians sleeping back at their campfire, had a face that reflected almost the same degree of surprise as did that of Josh! When Josh Dent had finally settled down from his fright and surprise at being 'taken' from behind and then recognizing that of his friend, physically relaxed in Iron Hand's hand still firmly holding his mouth shut. Iron Hand then quietly cut Josh's ropes holding him against the tree and held his finger to his lips for silence and motioned for him to stand ready in place.

294

As that man's rope fell away, Iron Hand handed the man a pistol that had been fully cocked and laid his index finger over his lips for silence once again. Josh nodded his head in understanding and then standing there by his 'tree of capture', watched as Iron Hand slipped over to the next tied-up trapper in line and from behind in complete surprise, repeated the quieting-the-man-down process first and then cutting him loose as well. Then that man was also slipped a fully loaded and cocked pistol into his hands and motioned to remain silent and stand there as well. Following that, Iron Hand slipped over to the last tied-up trapper who was fast asleep standing up in his tied-to-the-tree position. Reaching out, Iron Hand once again firmly placed his hand against the man's mouth so he would not suddenly yell out in fright as he awoke and arouse the now-sleeping Indians. It was then that Iron Hand recognized the face of Gabriel Dent, younger brother to Josh and also a fellow trapper and friend! In turn, when Gabe's eyes were flung wide open in fear and surprise at being awakened in such a manner, he instantly recognized the heavily bearded face of his friend from Fort Union, namely that of Iron Hand! Iron Hand gestured with his finger to his lips for silence and repeated the same procedure he had performed earlier on the two trappers by cutting his ropes that bound him. He too was slipped a fully loaded and cocked pistol and motioned by Iron Hand to remain there and stand quietly for what was yet to come. Gabe with a look of relief on his face, nodded that he understood and stood as still as a stone awaiting their next move based on Iron Hand's further instructions.

As had been planned earlier, Old Potts had brought extra fully loaded pistols to the event because of the number of hostile Indians that they would be facing, and had slipped two extra pistols into Iron Hand's sash once they were together again. Then as each man had been untied, he was slipped a pistol from Iron Hand's sash for what was soon to come.

Facing the three now-freed trapper prisoners, Iron Hand still faintly bathed in the light from the Indian's campfire,

The Adventurous Life of Tom Warren

made a hand gesture which brought the three now-freed trappers' heads in close to his. When they did, Iron Hand whispered what he wanted each man to do. The three men nodded in concert that they understood when Iron Hand had finished with his instructions. Then quietly forming a line abreast, the four men, three armed with pistols holding buck and ball and Iron Hand carrying his Hawken, began silently walking towards the nine sleeping Indians around the campfire.

As they approached into the brighter light of the campfire, Old Potts and his two other trappers 'melted' into the firelight as well in a line abreast and then all the men stood silently along one side of the group of sleeping Indians for a few seconds gathering themselves for the violent act that was soon to come. As they did, Iron Hand pointed to every one of the men standing there in silence and then pointed downward towards which Indian sleeping on the ground each man was to shoot, so that none of the Indians would escape their final retribution...

Looking around at the men to see if their eyes all 'said' they were ready and getting a silent 'OK' from each man, Iron Hand yelled "HEY!" When he did, the nine Indians previously sleeping on the ground around the campfire erupted upward as if they had each been shot into the air!

In the next microsecond, six pistols and one Hawken rifle erupted flame, smoke and deadly hot lead projectiles directly into nine, now madly scrambling Indians awakening from a dead sleep into what had to be a most terrifying moment of panic! In that fusillade of shooting, seven hearts were instantly stilled as most of the Indians were head shot by the six trappers from such a close range. The remaining two Indian renegades died just as quickly a split-second later, as Old Potts's trappers fired their reserve pistols held in their off hands, simultaneously, into one of the remaining two standing Indians, blowing him into dozens of pieces after being shot with buck and ball from just four feet away!

296

Terry Grosz

The remaining standing Indian, a Blackfoot by his dress with a very distinct scar running clear across his forehead from a previous horse wreck as a younger man, one who had been personally selected by Iron Hand for death that evening, had a tomahawk smashed into the front of his face by Iron Hand with such force, that he snapped his blade from the handle! Not to worry though, the blade had sunken into the Bad Seed's skull so deeply that Iron Hand had to borrow Crooked Hand's tomahawk in order to chop out his tomahawk blade from the man's skull so it could be fixed later with a new handle and still be used at a later date when necessary! Any concerns that Spotted Eagle had for his Brother's well-being regarding what the Bad Seed was going to do to Iron Hand if he caught him off-guard for killing his younger brother, drifted off into the 'forever' of the prairie's ever-blowing winter winds...

Old Potts and his trappers quickly reloaded their pistols and then made a sweep of the Indians' campsite looking for any more of their evil-killing kind. Finding none, all the men hurriedly gathered up their riding and packhorses in camp, loaded them, picked up all the Indians' weapons and walked their horses out of the death-shrouded area to where Old Potts and his trappers had secreted their riding stock. Once Old Potts's trappers were once again a-horse, the entire group headed back to the comfort of their cabin located the next ridge over.

Back at their cabin, Crooked Hand built up their outside fire in their firepit as Iron Hand fetched out a keg of their rum and a handful of cups. Soon the cups were filled with their high proof rum, the men were seated around the fire on their sitting logs and then, at the request of Iron Hand, the Brothers Dent told their story of capture by the Bad Seed's band of renegades.

Josh started off with their story of introducing the new trapper to their gathering, a man named Otis Barnes. Otis had teamed up with the Brothers Dent back at Fort Union and had ridden and trapped with the brothers as they hunted for Black

The Adventurous Life of Tom Warren

Bill Jenkins, who was reported trapping on the further north headwaters of the Porcupine River. Then one evening during a rainstorm, the Dent brothers and Otis had been surprised by the Bad Seed and his bunch of renegades and taken prisoner. As Josh told it, they had been slated for killing the very next day but Iron Hand and his fellow trappers through their timely intervention, had made sure those killing plans had not been carried out by the members of the renegade band of Indians...

Throughout the story being told by Josh, Old Potts had remained silent as if mulling over one of Iron Hand's piping hot Dutch oven biscuits in his mouth without getting his tongue burned in the process. In the moments of silence following Josh's story of capture, Old Potts cleared his throat letting everyone know he wanted to speak next, and said, "Well, as I see it, we trappers gathered around this here fire have two options. You Dent brothers and Otis can continue your quest in trying to run Black Bill Jenkins and his red-headed kin down further north on the Porcupine, or you can join up with us and all of us can work as a team for the protection afforded having the seven of us shooters working the beaver trapping waters together. If we can keep our hair, we can share the profits from our trapping successes here on this stretch of water along the Porcupine. There appears to be beaver a-plenty for all of us, and having seven shooters on the same side in this God-forsaken country full of them damn white man-hating Gros Ventre sure appeals to me over getting a poke in the eye with a sharp stick."

Josh Dent looked over at his younger brother Gabe, and Otis and seeing agreement on their faces over the thought of the possibility of being captured or killed further down the line, nodded his head saying, "Old Potts, the three of us would like to take you up on your 'invite.' We had pretty well searched out the headwaters of the Porcupine looking for any sign of Black Bill but from where we stood, it appeared he and his kin had already left the area for better trapping grounds. That being said, we three still need to make a living and beaver

298

trapping is our choice. So until we hear additional word as to the whereabouts of Black Bill and his kin, we would like to take you and your trappers up on your offer of all of us working together for the protection those numbers of us shooters affords."

Then as an afterthought, Josh said, "We still have all of our traps, livestock and the furs we had accumulated prior to our untimely capture, but those nine 'eager-eating' Indians pretty well cleaned us out of our provisions. That in mind, we can offer little in the way of grub but if you have enough and will have us under those circumstances, we can work for our 'found', if that is pleasing to the four of you? Additionally, now that we have those nine dead Indians' riding and packing stock animals, not to mention all of their firepower, we should be pretty well set in hauling out all the furs we trap, plus be more than able to defend ourselves in the face of most groups of hostile Indians that we might be facing."

"Well, it just so happens that we came with two years' worth of provisions because we would be trapping so far west and with that in mind, were not sure if we could return to Fort Union and resupply ourselves come the summer months. So food should be no problem, especially with all the good eatin' buffalo that we have so close at hand. But we will have a space issue inside our cabin with three more of you chaps living inside when the winter winds howl and the snows get knee deep to that long-legged galloot in our bunch we call Iron Hand. However, with three more hands, we can make another addition to our cabin and make it downright homey after a few days of hard work. What do the three of you say? Do you want to give it a whirl, help us build an addition to our cabin for the additional sleeping space it affords, and then throw in your lot with the four of us for whatever that brings the seven of us in the way of future trappings?" asked Old Potts. Then as an afterthought and before anyone could respond to his earlier question, he brought up the question of what to do with

The Adventurous Life of Tom Warren

having to hide the nine now-dead Indians that all of them had 'placed' among the Cloud People.

Crooked Hand just smiled over Old Potts's worry of body disposal, pointing out to Old Potts that there was a grizzly bear den not far from that place of death, and surmised that the bear would soon scent out and take care of their disposal question for them.

Somewhat later in the early morning hours of that memorable day after a 'joining of forces agreement had been reached', the Dents and Otis threw down their sleeping furs just outside the cabin and after several cups of rum apiece, were fast asleep after a somewhat harrowing several days in the capture of the now-dead Bad Seed and his eight Gros Ventre renegade friends. However, come daylight, Iron Hand and Crooked Hand were up, making breakfast and preparing their stock for another trip down to the beaver waters along the Porcupine in order to run their trap line. As for Old Potts, Big Foot, Otis and the Dents, after breakfast, they adjourned into the timber and began cutting the needed logs in order to build an addition onto the men's existing cabin.

Come the early afternoon, Iron Hand and Crooked Hand returned with 25 fresh beaver hides and then the work really began in earnest. As Big Foot and Old Potts hove to in their fleshing and beaver skin hooping duties, Iron Hand went to work in the 'kitchen' now that he had seven large appetites to satisfy, and Crooked Hand lent a hand in the 'timber-harvesting and log-dragging business'.

Come dusk, Iron Hand had a thick bean, rice and beaver meat stew bubbling away, three 16-quart Dutch ovens turning out biscuits every three minutes or so, and an apple and raisin Dutch oven pie in the making. It was a good thing that he did because soon he had seven almost starving men it seemed, gathered around the campfire looking to store up on some grits before 'their big guts ate the little ones'!

The next day, there were seven men working on the new addition to the cabin-making detail. Seven men now working the cabin-building detail, because Iron Hand and Crooked Hand had not reset their beaver traps from the day before so they could join in on the cabin-making detail in order to get it built before the really heavy snows flew and the temperatures went to that of below zero on a daily basis. By day two of the cabin making, the seven men had the additional space for Otis and the Dents roughed out, walls up and roofed in. All that was left was cutting out a door so the addition led into the main cabin and the work was done!

The following morning found seven hungry men gathered around their outside campfire warming their hands on their metal coffee cups, as Iron Hand put the finishing touches on his Dutch oven biscuit making. After a breakfast of thick brewed trapper's coffee, half-raw buffalo steaks fried in Iron Hand's three-legged cast-iron skillets over a bed of coals and Dutch oven biscuits, the men were ready for work.

The evening before, the men had decided that Old Potts and Big Foot would stay behind and start building an extension on their corral in order to accommodate the 11 horses from the now dead Gros Ventre and Blackfoot Indians, and the nine horses the Dents and Otis had brought into the fold as well. That left Iron Hand, Crooked Hand and the Dent brothers to sally forth and re-set the original trap line and for the Dents to establish an additional trap line further north. Come dusk, the four trappers returned and while Iron Hand made supper, the rest of the men pitched in and finished the corral extension to accommodate the extra riding and pack stock.

Thus began the daily routine of the four trappers venturing forth and running their trap lines, and Old Potts, Big Foot and Otis staying back so they could guard their very valuable expanded horse herd, as well as doing all of the fleshing and hooping of the fresh beaver, muskrat and river otter hides brought back to the cabin on a daily basis.

Two weeks later as Iron Hand kept busy around their campfire making his famous Dutch oven biscuits, a large and wet snowflake plopped on the end of his nose, causing him to

The Adventurous Life of Tom Warren

look up. When he did, he observed what appeared to be dark and ugly low-hanging storm clouds quickly moving down from out of the northwest. Realizing that was the direction from whence came their worst winter storms, Iron Hand shouted for the rest of the men to make haste in their daily preparations. After a quick breakfast amongst now flying snowflakes of the heavy and wet variety, the men had finished eating and laid their day's plans. Old Potts, Otis and Big Foot were to take three packhorses and kill several near at hand buffalo and butcher out the same. Iron Hand and Josh Dent would pull their traps for the winter as Gabe Dent and Crooked Hand stood guard duty, as well as acting as the skinners on any and all beaver trapped that day.

Come an early dusk because of the low-hanging winter storm clouds and the heavy snowfall, the trappers finally returned to their cabin. There the beaver skins were unloaded into their warm cabin and as Big Foot and Old Potts got to the fleshing and hooping duties, the other five men hung the two buffalo carcasses cut into quarters from their meat poles and cut off a mess of steaks for their supper and breakfast the following morning. By the time the men had finished their meat handling and butchering duties, they looked like snowmen they were so blanketed with wet snow from the now fully engaged winter storm that was upon them.

Walking to their cabin, they shook off the snow from their heavy winter garb, entered their now warm cabin, closed their front door on the first heavy winter storm of the season and in so doing, ended that fall's beaver trapping season on the Porcupine...

Come the next morning, found the cabin covered with two feet of freshly fallen snow and the outside temperature hovering around 20 degrees below zero. Dressed for the winter's intense cold, Iron Hand and Joshua Dent stepped outside armed with their rifles, glanced upward at the clearing skies, struggled through the deep snows over to their horse corral, and let their stock out so they could feed. As the

302

animals moved out onto the open prairie and pawed their ways to the grasses underneath, the two men stood watch over their valuable horse herd. Two hours later, a shout caught the two men guarding the horse herd, happy to see two additional trappers coming their way in order to relieve them from their cold duties and enable them to return to the cabin to warm up.

Once back at the cabin warming up, Iron Hand and Joshua watched Gabe, Old Potts and Big Foot removing a small mountain of beaver *plus*, removing their willow hoops, folding the pelts skin side inward, and forming packs of the dried hides for the summer's transport to Fort Union. Before long, the packs of *plus* ready for transport and covered with tanned deer hides to prevent damage in transit, were stacked clear to the ceiling in the back of the cabin! Thus went the winter regimen among the seven trappers, only interrupted with almost daily forays onto the prairie to kill buffalo for meals, kill buffalo to surround with wolf traps for those pelts so generated, or trips to their winter woodpile to feed their fireplace for heat and cooking. Old Potts's idea of combining forces among the seven men for the protection, beaver trapping and company it offered, had indeed borne fruit of the sweetest of kinds!

With a long winter behind the men, spring beaver trapping became a welcome change. Initially, the seven men went as a beaver trapping group for the protection it offered and then when the swell of hides became too much to reasonably flesh and hoop during the evening hours, the trapping team was reduced once again to just the five best shooters of the group. That left Old Potts and Big Foot back at the cabin to watch over their valuable horse herd and flesh out and hoop all the men's daily catches. Finally, come late spring when the beaver went out of their prime, the men began preparing for their long trek back to Fort Union with their valuable furs and pack strings.

One day while currying the horses, Iron Hand noticed a slight depression in the ground near a giant fir tree. Stopping what he was doing, he remembered the story Harlan had told Old Potts about there being so many beaver along the

The Adventurous Life of Tom Warren

Porcupine that he had to leave some of his furs behind in a hidden cache next to his cabin...

Walking over to their cabin, Iron Hand took up a shovel, and enlisting the help of Big Foot and Gabe Dent, began digging in the spot next to the fir tree where the ground showed a slight indentation. After about 40 minutes of digging, Iron Hand and Gabe had broken through the log roof of an old trapper's cache and before you knew it, all seven trappers were pulling numerous items up out from the underground and heavily brush-lined cache! There were rusty items of cast-iron cookware that could be cleaned up and re-used, a rusty rifle, a keg full of rum that was still drinkable, several stoneware crocks full of much-appreciated honey, six packs of beaver *plus* that were still good, a keg of powder that had picked up moisture from being buried in the earth that had turned into a solid brick of gunpowder and was ruined, 20 rusty but still usable beaver traps, an ax, shovel, and a rusty saw! What Iron Hand had discovered was Harlan's reported cache from years earlier. Items that he had 'cached' because he had so many valuable beaver *plus* that he had to leave some of his goods behind because his horses had all they could carry with their 90-pound packs of furs! And since he was reported to be no longer living, the items in the cache now belonged to the seven men as they now saw it. As it turned out, the stoneware crock of rum was still good, welcome and gone in four days! The furs had been well-packed and were judged as still good and added to their already small mountain of packs as well. The crocks of honey were put to use right away since the men had long since run out of most of their original stores, and the rusty but usable beaver traps added to the trappers' inventory. The rest was stacked next to their cabin in case they had room to take it with them when they left the area.

With the mountain of packs the men had amassed over the trapping season, they discovered that they were going to use all of their packhorses, the two extra Indian horses procured from the dead Indians still lodged under the log pile in the

Terry Grosz

Porcupine River (the ones who had earlier shot the plate off Iron Hand's lap as he ate his breakfast one morning and had died shortly thereafter for their bad breakfast manners), and the livestock from the Bad Seed's group of Indians!

It was only then that the trappers realized the problems they had earlier with the Indians had been Heaven-sent. Heaven-sent because without the extra horses they had brought to the 'party', the men would have had to dig a cache and leave a number of their valuable *plus* behind, as had Harlan in his earlier days when trapping on the Porcupine River...

Wanting to get a 'jump' on the summer season and ahead of all the other trappers who would also soon be traveling back to Fort Union to sell their furs and re-provision, Old Potts began to get fidgety and eager to 'hit the trail'. One morning as Iron Hand prepared making breakfast at the outside firepit now that the spring weather had improved, he noticed that Old Potts had emerged from their cabin and was sitting alongside the fire warming up. Iron Hand also noticed, having lived with the older man for a number of years now and aware of his moods, that the old man had that faraway look in his eyes that morning. Iron Hand 'steeled' himself for what might be coming next from Old Potts and he did not have long to wait.

"Iron Hand, I think now is the time to pack up our gear and *plus*, make a run for the Missouri River in order to outsmart those who would take them from us, and see if we can make it back to Fort Union, sell our furs and see who of our fellow trapper friends is still above the ground."

Grinning, Iron Hand said to Old Potts, "Old Potts, can I finish making breakfast this morning first before you decide to up and skedaddle us out of here?"

Then Iron Hand ducked as Old Potts grabbed the first stick he could lay his hands upon and threw it at him in jest over being made fun of over his desire to be up and moving. But 'that' look in Old Potts's eyes said it all that morning. He had done what he wanted to do in seeing what the Porcupine River

The Adventurous Life of Tom Warren

had to offer, had 'kept his hair' and was now ready to move on and see what the other side of the frontier looked like...

Iron Hand just grinned over the morning's events, checked his Dutch oven biscuits for just the right amount of browning and crust they were showing, and then shouted, "Breakfast is ready. Come and get it before I throw it out." Nothing had to be thrown away...

Chapter 11: John Pierre And Sinopa's Revenge, "Wambleeska"

ON THE FIFTH DAY after Old Potts had gotten that faraway 'traveling look' in his eyes while sitting by the campfire one morning, the seven Porcupine River trappers were up, fully packed and on the trail back to Fort Union. Heading south from their secluded cabin towards the Missouri River while riding along the western side of the Porcupine River, rode Old Potts at the head of the trappers' long and heavily loaded pack string of horses, happily leading the way.

In front trailing five loaded horses rode Old Potts, followed by Otis Barnes and Big Foot trailing five horses each, followed by the Dent brothers riding side by side each trailing five horses, and bringing up the rear of the trappers' pack string rode Crooked Hand and Iron Hand, each also trailing five heavily loaded horses. Since Crooked Hand and Iron Hand were the best shooters in the group, they rode 'drag' in the pack string for a specific reason. That reason being the two best shooters of the group of trappers rode 'drag' because any attacking Gros Ventre Indians historically attacked the rear of any pack string encountered. This they did because attacking from the rear of any pack string usually created the greatest amount of consternation, enabled the least concentration of fire to be laid down upon the attackers and pretty much guaranteed the attacking Indians the least number of casualties in any such attack.

Old Potts's travel plans were to head due south until they reached the Missouri River, then riding along on the northern bank of that river, proceed easterly until they reached Fort Union. There they would trade in their hides, skins and *plus*, sell their extra horses and then decide what would be their next trapping season's destination.

Finally arriving at the confluence of the Porcupine and Missouri Rivers without incident, the group rested for a day in order to not stress out the heavily loaded horses and allow them to extensively graze in the river's rich bottomlands' grasses. Following that respite, the trappers headed easterly along the Missouri and after several days of slow travel because of the fully loaded pack strings, arrived at the confluence of the Wolf River and once again rested their livestock and let them graze under guard for a full day and evening.

All along their travels, Iron Hand had noticed several times the fresh tracks of numerous shod horses heading in their same direction of travel, and even sometimes paralleling their direction of travel. As he did, he also noticed that his 'sixth sense' was a little 'raggedy' but not outstandingly so. So Iron Hand ignored those normally trusted inner concerns because he had one hell of a pack string of rambunctious, Indian horses who had never been packed before, causing him a high level of concern. Besides, Iron Hand just figured his 'sixth sense' was also concerned with moving such a valuable herd of 'green pack animals' carrying packs full of *plus*, and that was what had caused them to be more than unruly at times.

However with all those 'somewhat concerning' fresh tracks, Iron Hand along with the other riders of his group, just figured they were also early arriving trappers returning to Fort Union to sell their furs and re-supply for the coming trapping season the same as them. Several more days of slow travel in order not to overly stress their heavily loaded pack animals and the trappers arrived at the confluence of the Poplar River. Upon their arrival, they crossed at one of their familiar shallow-water fords from previous times of use in that river. That evening, the trappers rested their horses and let them feed throughout the night in the river bottoms. But as they did, the men working in two-hour shifts managed to maintain a watchful eye over their horse herd, even though all of them had been hobbled to preclude anyone from easily running off with some of their stock. Two more days of easy riding found the trappers at another familiar ford, only this time they crossed over the Big Muddy River and then they once again rested in the Missouri River bottoms on the east side of the river just crossed.

That evening as the men rested their stock, they took turns bathing in the Missouri in preparation for their several days' out arrival at Fort Union. Then Crooked Hand managed to shoot and kill a fat barren doe deer coming down to the river to drink and come suppertime, their camp smelled Heavenly of roasting deer meat, Dutch oven biscuits, and a baking with the last batch of their dried and now soaking apples and raisins for another of Iron Hand's famous Dutch oven cobblers. Additionally, the men broke out their last stoneware jug of rum in celebration of their pending arrival at Fort Union with their load of some of the finest 'blanket-sized' beaver *plus* the men had ever taken since they had begun trapping west of Fort Union.

Taking longer than normal to completely bake the deeper than usual Dutch oven celebratory cobbler, the men dipped even deeper into their stock of rum and soon, no one even noticed the always present clouds of mosquitoes hovering over everyone anymore. Finally, the cobbler was done and Iron Hand began serving his fellow trappers sitting around their campfire from right out of the Dutch oven.

Just starting to dish out a portion to Old Potts since he was the oldest of the group, Iron Hand saw the strangest look

The Adventurous Life of Tom Warren

quickly cross Old Potts's face as he looked past and behind his favorite pie and cobbler maker. Not sure what the hell had caught Old Potts's usually sharp eyes, Iron Hand turned with the bail of his Dutch oven in hand and a serving spoon in the other, only to find himself staring directly into a large bore, leveled rifle barrel just inches away from his face!

Looking up in surprise past and down the leveled rifle barrel, Iron Hand saw the leering face of John Pierre holding that firearm, and all around him and standing behind with leveled rifles as well were the outlaw trapper's gaggle of seven evil friends with smiles of satisfaction splashed clear across all of their faces! Then Iron Hand 'flashed back in his memory banks' to the one and same John Pierre who he had stopped from beating little Sinopa, his sex slave, along the Missouri River some months back. The one and same John Pierre who had started a fight with Iron Hand back at the fort's warehouse and had badly lost. The same John Pierre who had been banned, along with his seven now-fired, once Company Trappers from Fort Union, forever. The one and same now holding him at bay with the end of his rifle barrel sticking in his face...

"Well, well, well, what do you suppose we have here? A bunch of McKenzie's ass-kissing pissant Free Trappers all caught with their pants down! That being the case, I say we kill the lot of them, take all of their furs and horses, and ride out of here rich men once the Hudson Bay people settle up with eight of us better trappers than all of them," snarled John Pierre through his tobacco-stained teeth behind lips curled up in a snarl.

Iron Hand, feeling the killing boil of bile building up inside him and pissed that he had not 'read' more into all the shod horse tracks he had been seeing along the Missouri for the last few days or had paid more attention to his 'sixth sense', slowly set down his Dutch oven with the still hot cobbler contained therein and slowly rose to his full height to face John Pierre.

"That's right, you giant piece of shit! Don't make any quick or false moves 'cause it will just get you killed before I am ready to do so. I think you and yours need to sweat for a few days before me and my boys finish off the lot of you and toss your miserable carcasses into the Missouri for the fish to enjoy," continued John Pierre, obviously enjoying for once having the upper hand over Iron Hand. But upper hand or not, John Pierre kept the end of his rifle barrel pointed at Iron Hand's chest, full well knowing that if he did not enjoy the current position of power, from the deadly looks he was getting from the giant-sized trapper, he, John Pierre, would already be on his way into the darkness that comes with eternity!

"Kill him! Kill him now, John! Don't let him get a-holt of you or he will crush you like a bug just like he almost did to me," said one of his men with a last name of La Rochelle.

"No, that is too good for this bastard. I want him to suffer a while before I blow him a new asshole with this here rifle. Then I want to be the one to toss his miserable carcass into the Missouri for what he did to me some time back at Fort Union. No one treats John Pierre like that and gets to live a longer life. Additionally, I get to give it to you big time for taking away my little Indian 'play-pretty'. Plus my boys still resent you for taking away their 'fun time' with her as well. Now you are going to pay for that as well, if I have my druthers," snarled John Pierre, now really enjoying himself because he finally held the upper hand over a much-hated individual in the being of Iron Hand...

"Tie this big bastard's hands and feet up and do a good job of it, or you will suffer more than my wrath," snarled John Pierre, as he instructed another trapper named La Duc to do the deed. "But be careful how you do it. Just make sure he does not get a hold of you, or you will rue the day because he is so damned strong and is reputed to enjoy killing those he does not favor with his bare hands," continued John Pierre. As he spoke, his eyes never left those of his prisoner and the end of

The Adventurous Life of Tom Warren

his rifle barrel never left being leveled at the center of the trapper's chest standing before him...

Moments later, Iron Hand felt his wrists being tied so tightly that it would be just moments before the fingers in his hands would be going numb. Then a rope was tied around Iron Hand's ankles, which was tied just as tightly as had been his wrists, after being forced to sit down. Then it was Old Potts's group of trappers' turn at being tied up as had been that of Iron Hand. Then the lot of the now tightly tied-up men were tossed into a squirming pile in the middle of a small clearing where they could be easily watched over by John Pierre's group of outlaw trappers.

Then John Pierre's men ate the cobbler meant for Old Potts's men and finished off the remains of the entire stone crock of rum in short order as well. Following that, John Pierre gave orders regarding who would be on guard duty over the now-bound trappers and assigned two men to watch over the prisoners at all times. Then the rest of the group started going through the packs of furs and provisions from Old Potts's group of trappers, helping themselves to what they wanted as they sacked the piles of packs and supplies.

Finished with their looting of the packs, the outlaw trappers returned to the group of tied-up men being led once again by John Pierre. "Alright, Boys, let's give it to them. We will teach all these pukes that it does not pay to bathe in our Missouri River." With that, all of John Pierre's men dropped their pants and as a group, urinated all over the helplessly tied-up men...

Iron Hand's internal seething was almost beyond his control! Yet he could do nothing and knew that if he tried, they would all be killed on the spot. So best to hold in check his killing temper and await his chance, if he ever got one... About then, a young Indian boy, obviously a slave-boy to the group of rough outlaw trappers, walked into the campsite struggling with a heavy stoneware crock of rum. About then, he stumbled on a tree root and dropped the jug of rum, which

312

Terry Grosz

did not break in the process. Instantly, John Pierre jumped up, grabbed a horse quirt from off a nearby saddle and commenced thrashing the young Indian lad unmercifully! Soon, the young Indian lad was on the ground crying out like little Sinopa had been doing the first time Iron Hand had observed John's cruel and mean-tempered behavior. Seeing that cruel act, Iron Hand could no longer contain himself.

"Quit beating that young boy!" yelled Iron Hand.

John Pierre, being surprised over being given an order, turned and saw who had just given that order and then aggressively stomped over to where Iron Hand lay tied up on the ground. Standing over Iron Hand, John's face twisted into a snarl, and it was then that John Pierre commenced savagely beating Iron Hand over and over again with the lead-handled horse quirt, until he passed out from the rain of smashing blows around his head and face!

The next morning, Iron Hand awoke from his vicious beating, only being able to see out of one eye clearly and barely able to move from all the dried and crusted blood all over his badly beaten and heavily bruised head and shoulders!

"Iron Hand, are you OK? Damn, I thought you were dead from all the vicious beating you took at that bastard's hands," quietly said a very concerned sounding Old Potts lying next to him.

Iron Hand tried to reply but found his mouth horribly busted up, several teeth missing and both lips split clear to his jaw bone! Trying to roll over and sit up, he found he was in so much body pain that it took several tries before he was successful. But he finally managed to sit up, only to get a hard kick in the center of his back from an unknown assailant until he heard a man's voice. It was La Duc and it was his turn to guard the prisoners. Because of resenting having to do so, La Duc took out his dissatisfaction and wrath on the unprotected back of Iron Hand. However, little did La Duc realize he had just signed his death warrant, especially if Iron Hand ever got

313

The Adventurous Life of Tom Warren

away from what was binding him and he got a hold of his scrawny neck with either of his powerful hands!

Finally, La Duc, tiring of kicking the hell out of Iron Hand's back, ambled off over to the fire where the rest of his cohorts were starting to have breakfast. Once again, Iron Hand tried sitting up because that seemed to help keep his badly damaged head from spinning off his shoulders like a penny top! It was then that Iron Hand heard a slight rustling in the dense brush behind him. Figuring he was in for another foot stomping, he braced himself! Then he felt a knife begin quietly cutting the rope that was binding his wrists and hands! Soon he was free, but his hands had been tied so tightly that he had hardly any feeling in them. Then he felt something like the thick handle of a knife being thrust into his constantly flexing hands as he tried desperately to get some feeling and movement back into them! Looking out his good eye, he could see all the men still eating breakfast a distance away and not paying any attention to all the men they had captured the day before, still all tied up and dumped into a large pile.

Then the rustling in the bushes was heard once again and this time Iron Hand got a glimpse of the young Indian lad fast disappearing through the brush like he feared for his life if ever caught doing what he had just done! Then Iron Hand could just see that the boy moving through the brush was the one and same young Indian who John Pierre was beating with a horse quirt when Iron Hand had told him to stop. In so doing, the beating of the Indian lad had stopped and then commenced on Iron Hand as he lay helplessly tied up from head to toe. Then Iron Hand realized the young Indian lad had more than likely braved another beating or even death for what he had just done and in so doing, was returning the favor to him for standing up for the young boy in his time of need! That act of utter bravery would not be lost on Iron Hand...

Looking over at the men around the fire having breakfast and seeing that no one was paying any attention to their prisoners, Iron Hand made his move. With his hands now

314

Terry Grosz

getting their feeling back, Iron Hand reached down and cut the ropes binding his ankles so now he could stand up and better defend himself if and when the opportunity arose. Then looking at Old Potts who was closely watching what Iron Hand was doing, he placed the butt of the knife into Old Potts's hands so he could cut himself free. Following that and in order not to bring any unwanted attention to the captured trappers, Iron Hand resumed his sitting up position like nothing out of the ordinary had just happened. Yet he full well knew that the rest of his group was now one by one making wise use of the knife just handed off to Old Potts earlier!

It was then that Iron Hand could feel the adrenalin racing throughout his body and rifle or not, someone was going to die for what they had done to him and his group of trappers! About 30 minutes later, here came La Duc to once again watch over the prisoners. When he did, he saw that Iron Hand was sitting up and looking right at him. Looking back at the 'prairie rattler'-like stare Iron Hand was giving La Duc, he failed to notice that the giant trapper was loose from his ropes and ready to strike the first thing that he got his hands upon...

"What you looking at, puke? Want some more of the same I gave you before breakfast, huh?" Without realizing that his prisoner was now free and flushed with adrenalin, La Duc kicked out at Iron Hand. In so doing, La Duc's last remembrance was seeing a blur of movement, feeling a roaring pain in his neck and then nothing but blackness, as Iron Hand grabbed his leg, jerked him suddenly down to the ground alongside him, and cleanly snapped his neck with one quick movement with his powerful hands!

Being that all the recent action had taken place on the ground below the level of the close at hand bushes, no one around the campfire took notice of La Duc's quick disappearance. Slowly removing La Duc's two belt-sash pistols, he handed one to Old Potts and the other to Crooked Hand. Both of those men, thanks to the young Indian boy and the use of his knife, were now also free. In fact, all of Old

315

The Adventurous Life of Tom Warren

Potts's group of trappers had passed the knife around and all were now free! Free but not moving because all of them realized they were, except for a knife and two pistols, basically unarmed.

To make any kind of obvious movement advertising that they were now free and to subsequently arouse suspicion from their captors still around the campfire, would court Old Potts's trappers nothing but death and destruction in short order! Then Iron Hand quietly slipped La Duc's rifle across his legs and handed it off to Gabriel Dent to use when the time came. Now three of the trappers were armed and Iron Hand heard the quiet 'clicks' as all three men holding a weapon, had just cocked them so they would be ready for instant deadly action when the time came and action on their parts was called for.

Still not having a plan for himself, Iron Hand was soon given an opening. John Pierre's closest ally from his bunch of cutthroats was a French-Canadian named La Rochelle. It was that man that Iron Hand could see coming their way to relieve La Duc from his guard duties, so he could go and get some coffee which was now ready for consumption.

La Rochelle walked over to where the prisoners were still making like they were all piled up, then stopped, turned around and took the time to begin to urinate nearby. When he turned around, he had his rifle jerked from his hands and tossed backwards by Iron Hand in the direction of the rest of Old Potts's crew! When his rifle was abruptly jerked from his hands and upon seeing Iron Hand materialize up out from the underbrush, La Rochelle's eyes showed an inordinate amount of white ringing his pupils, indicating his deep fear over what was coming next! He died without emitting one single sound, not only because of extreme fear upon seeing Iron Hand quickly reaching for him initially, but because it is hard to yell when one has his neck quickly choked closed and then snapped...

Then all hell broke loose in that Missouri River bottom in those next few seconds! John Pierre, looking in the direction

316

of his prisoners and seeing Iron Hand tossing his best friend La Rochelle off into the brush like a rag doll with a broken neck, roared out in surprise and then in extreme fury! Grabbing up a nearby musket, John came screaming and charging towards Iron Hand with 'blood in his eyes'! As he did and unarmed as he was, Iron Hand just stood there and hoped that the rest of his fellow trappers were as ready as they could be for the hell and fury that was now coming their way.

Since Iron Hand had already killed two of the total of eight trappers in John Pierre's group, the remaining five came charging in behind John Pierre with their rifles at the ready! John Pierre, being the bully and coward that he was, stopped a good ten feet shy from Iron Hand, raised his rifle and said, "Now you die, you piece of shit. I should have killed you earlier but any time to die is a good time for you, you bastard!"

"ZZZIIPPPP—THUNK!" went an arrow straight into John Pierre's left ear, dropping him to the ground in less than a heartbeat! Then before Iron Hand and his fellow trappers could get involved in the fight with the onrushing five remaining armed trappers, about 20 Blackfoot warriors materialized up from their places of hiding in the dense brush surrounding the campsite, and showered the air with arrows, all aimed at the oncoming rush of John Pierre's group of trappers!

Seconds later after the screams had died away, out from the brush not more than 20 feet away from the still-wiggling in his death spasms John Pierre, rose Spotted Eagle with a stern and killing look on his face, as he now held an empty bow! As he did, his warriors swarmed over the dead and dying trappers from John Pierre's group of outlaws, scalping and knifing as they went! There were several muffled screams as several still-living trappers were scalped alive, but soon all was stilled, hearts and the like, on the evil trappers' side of the camp!

Old Potts's trappers, realizing they were outnumbered and figuring they were all going to die at the hands of the attacking Indians, figured on defending themselves to their last breath.

The Adventurous Life of Tom Warren

Then Iron Hand, after recognizing Spotted Eagle, yelled out to his fellow trappers, "Don't move or shoot! They are our friends. They are Spotted Eagle's braves, so don't shoot! Put down your guns!"

Then Spotted Eagle, still in his killing emotion of the moment, walked over to John Pierre, dropped his bow, then reaching down, slit his throat and then scalped him where he lay still wiggling in his spasms from suffering a violent death! Finally standing up and holding John Pierre's bloody and very recognizable long-haired scalp tied back with a bright yellow ribbon, Spotted Eagle said clearly for all to hear, "This is Sinopa's Revenge!"

When Spotted Eagle spoke those words, Iron Hand remembered back to when he had taken little Sinopa away from John Pierre and the rest of his raping fur trappers. The indignities she had suffered at the hands of John Pierre and his kind had now been laid to rest, that was if Spotted Eagle holding up a bloody scalp meant anything to the knowing and understanding observer...

Then Spotted Eagle walked over to his Brother, Iron Hand, saying, "My Brother, it seems once again you needed the help from your Indian Brother. What happened to your face and head? You look like you have been using your head to 'chop down trees'," he said with a big, all-knowing grin. Then the two 'Brothers' hugged in recognition and relief that the past moment in time was now lost to the ever-blowing prairie winds...

After all John Pierre's trappers' bodies had been tossed into the Missouri River, Spotted Eagle and a couple of his warriors treated John Pierre differently. Shortly afterwards, John Pierre's stripped naked body swung silently from a long rope tied to a stout cottonwood limb overlooking the Missouri River. There he swung quietly out in the open, 'waiting' for the ever-meat hungry black-billed magpies to find him and 'rejoice' over finding such a wonderfully sized, 300-pound dinner waiting for them and all of their buddies... A closer look at the naked man slowly swinging in the air 'waiting' for his soon to be aerial friends, the magpies, was noticeably missing his manhood. 'Manhood' which had been removed with the aid of a very sharp knife and tossed into the Missouri River for the crawfish to enjoy…

That afternoon celebrated over cups of rum, the trappers and Spotted Eagle's 20 warriors feasted on roast venison, Iron Hand's style of Dutch oven biscuits slathered in honey or sugar and finally, two Dutch oven baked apple pies, courtesy of John Pierre's provisions from his own pack string. That evening around a blazing campfire and more cups of rum supplied from the now dead Pierre's provisions, Spotted Eagle told his tale as to how and why he had managed to track down and save his Brother, Iron Hand, and avenge his wife, Sinopa, in the process.

Enjoying the attention as everyone waited for him to begin speaking, Spotted Eagle finally began telling his story with the following words. "Long ago, my cousin, Chief Mingan, asked me and 20 of his best trackers and warriors to hunt down two young and wayward members from our own tribe who abused the very young and old of my band. Because of their abuse of the young and very old from my band and then stealing a number of our best horses, Chief Mingan from the Medicine Lake Blackfoot Tribe asked that I lead a party of warriors, hunt down these two brothers who became known as the "Bad Seeds", and kill them for what they had done to our people."

"For many moons, my fellow warriors and I hunted the Bad Seeds unsuccessfully. They were very clever and we could not find them and do as Chief Mingan had ordered. Then along came my Brother and his fellow trappers and in a fight along the Missouri River, they hunted down and killed a number of the bad Indians who were preying on the white man trappers, killing them and taking their furs. In that fight, one of the Bad Seed brothers was killed by my Brother, Iron Hand. But the other older Bad Seed brother got away and we have been hunting him ever since."

The Adventurous Life of Tom Warren

"A moon ago, we finally discovered where the remaining Bad Seed and his Gros Ventre brothers were hunting down and killing the white men trappers. But when we finally tracked them down in the Porcupine River country, we discovered all of them had been killed by what appeared to be white man trappers. But it was really hard to tell who had killed the Bad Seed and his fellow Gros Ventre brothers, because our Great Bear Brother, the grizzly, had found their bodies first and had eaten many of them. Fortunately, he did not eat the scarred head of the Bad Seed and upon finding that lying around their old campsite, we figured our hunt for him was over but another one had just begun. So, we followed the trail of the white man trappers who had killed these bad people and found their cabin where they had been living. But they had already left the area and were heading down towards the Missouri River. So we followed their trail until we got to the great river, and then followed it even further as it led toward Fort Union," continued Spotted Eagle.

Then pausing in his storytelling for the effect it would have on his listeners, Spotted Eagle finally continued saying, "But then as we did, we ran across other Bad Seed white men trappers who appeared to be hunting those trappers who had killed the Bad Seed from our tribe. I was curious as to why the Bad Seed trappers were following the other white man trappers, so we followed them to see why they were following, and because we needed to get to Fort Union as well for more supplies, powder and lead ourselves."

"A day later, we came upon the Bad Seed trappers once again and saw that they had taken the other white men trappers that we had been trailing, prisoner, and one of those prisoners was my Brother, Iron Hand! I then knew we needed to stop the white men Bad Seeds, and began moving into their camp to ambush them in order to save all of you and my Brother. For a while we saw that my Brother was taking matters into his own hands and making good his escape. But he was discovered and when he was, we decided we had best help my

Terry Grosz

Brother at that moment and we did. As we moved in closer, I discovered that one of the Bad Seed white men was none other than the one who had abused my wife, Sinopa, in her younger years, and I vowed to personally kill him and have all my warriors kill all the other Bad Seeds as well. That we did and now here all of us sit around the campfire as brothers in happiness, now that all the Bad Seed white men trappers are the bait for the fish and birds," finally ended Spotted Eagle's story. When he had finished speaking, there was hardly a word among the trappers and warriors out of respect for the young sub-chief who had brought them into battle and had saved his Brother, Iron Hand and his friends... Plus, Spotted Eagle was able to settle a long festering sore in being able to avenge the many wrongs against his wife, Sinopa, caused by a bad white man trapper now covered with many black and white birds enjoying a meal....

Then all of a sudden, there was a lot of loud yelling and a great commotion up by the corral holding all of the trappers' horses! One of Spotted Eagle's braves had discovered the young Indian boy who had been all but forgotten in the killing that had previously occurred, digging through a pack full of provisions looking for something to eat! Moments later, Spotted Eagle's brave brought into the area of the campfire a struggling young Indian boy by the scruff of his neck. The brave walked over to Spotted Eagle and roughly deposited the young Indian onto the ground in front of Spotted Eagle, as if the Indians' leader was to make a decision on whether he lived or died being that he had been part of the Bad Seed trappers' camp.

Iron Hand, who had been quietly sitting on a log by the fire, all of a sudden realized his 'sixth sense' was roiling around inside him like he had not felt it doing so in many years! Rising to his feet to intercede, since the young Indian boy had been the one who had slipped Iron Hand the knife allowing him and his fellow trappers to escape, in case Spotted Eagle wanted the boy killed as well, he suddenly froze in mid-step.

321

The Adventurous Life of Tom Warren

Spotted Eagle had observed something about the young Indian boy's dress and instead of speaking to him in the Blackfoot tongue, began speaking to him in the language of the Blackfoot's nearby Sioux Indian neighbors. Upon his hearing his native tongue being spoken, the boy instantly began intently listening to Spotted Eagle. As the rest of the trappers and Blackfoot warriors intently looked on, Spotted Eagle kept conversing with the boy. After a few more moments of conversation with the boy, Spotted Eagle turned to the on-listening trappers saying, "The boy's name is "Wambleeska" in the language of the Sioux. His name translates into "White Eagle". White Eagle tells me that his father, two uncles and his older brothers took a number of the younger men from their village many moons ago and were out hunting buffalo. While butchering out their kills, they were approached by what they thought were friendly white trappers. The Sioux began trading fresh buffalo meat with what they thought were friendly traders and then all of a sudden, the adults were all killed by the trappers and the six younger boys were all taken as prisoners. White Eagle says the bad trappers then later sold off all the young boys including himself, to other trappers and traders and the people who work for what they called "The Queen" in the country to the north. He was kept by the bad trappers so he could help around the camp and tend to the horses. He says he has been with the trappers for over 11 moons since his capture. But he also says that he was treated badly by the leader of the bad trappers, who beat him many times in the past and did not feed him very well. He also says that when the bad trapper leader beat him after capturing the 'good trappers' and the big trapper with the heavy beard told the trapper leader to quit beating him, that the large trapper was then beaten badly trying to save him. White Eagle says that is why he gave the large trapper with the heavy beard and long hair a knife so he could let himself go and not get any more beatings."

Then Spotted Eagle did something that surprised everyone around the campfire. "Iron Hand, since you saved this boy

Terry Grosz

from being beaten by John Pierre like you did my Sinopa so long ago and then he saved you by giving you a knife so you could cut your ropes so you could escape, I think he is now your responsibility! He has no living parents and you should have a 'son'. I think that is the way it should be and that is what The Great Spirit is telling me to do with him"...

Still standing as if to intercede if Spotted Eagle was going to kill the boy because he had been with John Pierre's trappers, Iron Hand, upon hearing those words regarding the young Sioux boy being given to him as a son by Spotted Eagle, stood there dumbfounded for a few moments. Then Iron Hand noticed that everyone around the campfire was looking at him expecting him to say or do something regarding what had just happened. It was then that a feeling of 'family' like he had not had for years, slowly enveloped him like a warm bear robe sleeping fur! Following that warmth of having a 'family' moment flooding over his body, Iron Hand noticed that his 'sixth sense' was also warmly rolling around in him like he had not experienced since the day he had married the love of his life back in Missouri years earlier!

Then without really knowing why, Iron Hand all of a sudden found his right arm and hand being slowly extended towards the young Sioux Indian boy in a welcoming sort of way. Seeing that gesture coming from the trapper that had saved him from a beating from John Pierre earlier, the boy slowly walked away from Spotted Eagle and over to Iron Hand. Taking the boy's hand, Iron Hand, without a word being spoken between the two of them, walked White Eagle over to the campfire and saw to it that the young man had something to eat. With that gesture on the part of Iron Hand, everyone else around the campfire began talking, eating and drinking like nothing out of the ordinary had just occurred. But something out of the ordinary had occurred, as Iron Hand found years of sorrow over the loss of his first family almost being lifted and in a strange and good way, melted away from

The Adventurous Life of Tom Warren

his inner person and memories with the addition of the young man now into his life!

For the next two days, Spotted Eagle's band of warriors and Old Potts's trappers camped together, resting, eating life-sustaining buffalo meat at every meal and letting their horses 'put on the feed bag' with rich Missouri bottom lush grasses as well. That was also the time when Iron Hand discovered that in his 11 months of capture under John Pierre, his new 'son' White Eagle, had learned to speak passable English, having learned and been taught that language by all of his captors.

Because of White Eagle's ability to understand English, Iron Hand found himself quietly teaching the young man many things from the English language, as well as the proper use of firearms and how to care for saddling and packing horses. The rest of Old Potts's trappers also enjoyed those two days of rest, especially when it came to watching Iron Hand and White Eagle 'taking to each other' like a beaver does to water and fast developing a normal 'father-son' frontier relationship.

Come day three after the killing of the evil John Pierre and his like in kind trappers, Spotted Eagle, his mission over in discovering the death of his village's Bad Seed and now the killing of John Pierre while saving his white man Brother, made ready to leave and go back to his village near Medicine Lake, his wife and young family. Sitting on their horses with his warriors and making ready to leave, Iron Hand bid his Brother, Spotted Eagle, safe travels back to his village near Medicine Lake. As he did, Spotted Eagle leaned down from his horse and quietly told Iron Hand, "Care for White Eagle as if he were your own, because he now is as a gift to you from The Great Spirit." Iron Hand grinned over hearing those words from his Brother and had already pretty much realized the same…

Standing there with White Eagle watching Spotted Eagle and his warriors riding off to the northeast, White Eagle said, "Father, he is a brave and good man." Hearing the word 'Father' used for the first time coming from White Eagle, and remembering the loss of his biological son to smallpox years earlier, brought tears to Iron Hand's eyes and a newfound joy to his heart as well...

By noon the following day, Old Potts's trappers were saddled, had their pack strings loaded and assembled for the continuation of their trip to Fort Union. Once again, Old Potts took the lead leading a long pack string, followed by Otis and Big Foot doing the same, accompanied by White Eagle now leading a pack string as well. Next in line came the Brothers Dent leading fully loaded pack strings, followed by Iron Hand and Crooked Hand riding 'drag', trailing two longer pack strings of packed horses and John Pierre's trappers' riding horses in tow as well. As the heavily loaded pack strings left their Missouri River camp of 'death and renewal', they rode by a large, now bloating in death, naked man, quietly hanging and gently swinging from a stout cottonwood tree limb overhanging the swiftly flowing river! In passing such a gruesome sight, all the riders noticed that the man's naked body was heavily covered by 'wing-flapping' and happily feasting black-billed magpies and ironically, a "murder" of crows...

Two days of travel later and the learning experiences that came from that, gave Iron Hand pause and a proud smile over his new son's performance. White Eagle had taken to the love and teaching moments that Iron Hand had shared with him like a river otter did to chasing a fat rainbow trout... Then finally seeing the fort coming into view, Iron Hand could see his new son standing up in his shortened stirrups, eagerly looking at a sight like something that he had never seen before. Looking back once again, Iron Hand could see the look of wonder in White Eagle's eyes seeing the fort for the first time and had to smile. However, behind that smile lay a 'world' of questions without answers whirling around in the man named Iron Hand. Questions that would soon require answers, if Iron Hand was to understand the meaning of the restlessness of the 'sixth

The Adventurous Life of Tom Warren

sense' in his body and now racing around in the annals of his mind.

Once again, their friend McKenzie from Fort Union, as was his tradition, met the oncoming strings of trappers approaching the walls of his fort in anticipation of the business in furs the arriving trappers would be bringing. Then all of a sudden, realizing who the newly arriving trappers were, McKenzie got an even-wider grin on his face, especially seeing the large herd of valuable horses being trailed into the fort and once again the story to be they represented. The story the long string of horses represented, since this same band of trappers had left his fort a year earlier with less horseflesh than they were now bringing back into the area.

"Potts, you old scudder, how the hell do you and your trappers manage to do this every year you come back to the fort? For the last three trapping seasons, you four trappers have left the fort with less horses than you come back with. What is the story this time?" asked an eager for the tale to be told, McKenzie.

"It is a long story, Mr. McKenzie. However, if me and mine are invited for one of those fine suppers you and your Chinese cooks manage to put on in this God-forsaken place, then I could be encouraged and nudged to tell the tale and what a tale it is," replied Old Potts with a huge grin.

"Bring your caravan into the walls of my fort so your furs can be counted and graded by my Clerks and after that bit of business is transacted, you and yours will be my honored guests for supper this evening. Then damn your old hide, with enough of my rum, I hope that will loosen up your tongue so I can hear what kind of a tale you have to tell me this time," said McKenzie, obviously glad to see that his old friends still had their hair and were bringing him some more rich furs, riding and packhorses as well.

About then McKenzie noticed White Eagle leading a pack string and not remembering him as part of Old Potts's crew of

326

trappers, pointed to him out of surprise saying, "Say, Old Potts, who is your new member riding with your gang of trappers?"

"That is a story that surely needs telling. However, Iron Hand will have to advise you about this young lad and how we came about him," said Old Potts, as he rode his string of animals up and into the fort's central courtyard where several Company Clerks were waiting.

As Iron Hand rode into the fort's interior with his pack string, he rode up alongside Crooked Hand and handed him the reins to all his horses. Leaning over towards Crooked Hand, Iron Hand said, "White Eagle and I need to make a little trip over to one of the warehouses housing all the candy and such." Then with a wink and a smile from Crooked Hand, Iron Hand dismounted and had White Eagle take his pack string over to another waiting Company Clerk so he could count and grade out his furs. Then taking White Eagle in hand, the two walked over to one of the warehouses holding much of the sugar, honey, candy, canned jams, and the like. Walking in, Iron Hand saw White Eagle's eyes almost 'explode' open over all that he was seeing spread out before him on all the shelves and counters.

Taking his new son over to a section of the store holding the sweets, Iron Hand told White Eagle to try some. Hesitant to do anything over something he had never seen before, Iron Hand kept up the encouragement until White Eagle took a piece of hard candy and tried it. Moments after popping the hard candy into his mouth, he got a huge smile on his face and like many other young kids, was hooked for life!

That evening seated around McKenzie's supper table, White Eagle's eyes never were less than fully wide open over seeing all the riches of food and things like he had never seen before, like tablecloths, napkins, silverware, expensive crystal drinking goblets, and the like. But he was a quick learner and one could see him watching others and then following suit in what they were doing. That night, White Eagle slept in a bed for the very first time in his life. Then not liking its softness,

327

The Adventurous Life of Tom Warren

he removed all his covers and slept on the wooden floor next to the bed!

Chapter 12: A "Family" No More -- Old Potts, Big Foot And Crooked Hand Choose To Stay

HAVING MOVED OUT FROM THE CLERK'S CABIN in the fort, Old Potts's crew and the Brothers Dent chose to make camp in the Missouri River bottoms, along with the rest of the American Fur Company Trappers and Free Trappers that were arriving on a daily basis. One morning early, Iron Hand and White Eagle, now being taught how to cook the white man's way, were found together as 'father and son' making biscuit dough for their Dutch ovens and the breakfast to be. By now, the strong trapper's coffee was boiling away and a number of buffalo steaks were deliciously 'fat-spattering' away over a bed of glowing red coals, as they loosely hung from their iron cooking stakes.

About then, Old Potts emerged from their lean-to and took a seat on a sitting log next to the fire in order to warm up. As he did and being taught daily regarding proper white man's manners when around their elders, White Eagle poured a steaming cup of coffee, walked over to where he was sitting and handed it to Old Potts.

The Adventurous Life of Tom Warren

"Thank you, White Eagle," said Old Potts appreciating the young man's manners, but that was not why the old man was up so early that morning. He had been watching Iron Hand as he went about his duties and could see that he was 'mouthing' a piece of buffalo steak that was almost too big for him to swallow and when he did, he was not too sure if he liked it or not.

"Iron Hand, can we talk?" asked Old Potts.

"We always have been able to, ever since I have known you. What is on your mind?" asked Iron Hand, as he began 'hand forming' another biscuit for his waiting and ready Dutch oven.

"You seem to be chewing on something that is too tough or big to swaller. What is it that is ailing you and sticking in your craw?" asked Old Potts.

Iron Hand stopped making his biscuit, turned and said, "It is that obvious, is it?"

"Yep, it is that obvious to anyone who has at least one good eye and has known you for a while," said Old Potts, as he quietly eyed Iron Hand for his reaction to his probing personal question.

Laying the biscuit into the bottom of the Dutch oven alongside all the other biscuits, setting the cast-iron kettle over a bed of coals, and placing the lid just removed from the fire over the top of the pot so they could bake, he then shoveled some more coals onto the lid, turned and looked at Old Potts, who was looking intently right back at him.

"I have been doing some deep thinking. Now that I have a young man that I am responsible for, I have been wondering if living out here in this place where danger is at most every corner on a daily basis, if this is what I want for him. He has already lost his parents and older brothers, along with being mistreated by John Pierre's bunch, and I am a-wondering what is next in line for him in this frontier killing and dying department," said Iron Hand slowly, as if a little hesitant to share such deep inner feelings with anyone until he had made

330

up his mind and 'had ridden off the rough' over all of his concerns.

Continuing after a few more moments of quiet introspection, Iron Hand said, "Old Potts, I just feel a strong responsibility to raise up White Eagle correct and proper like and give him something other than the possibility of an arrow in the guts, a bullet in the head, getting him busted up in a horse wreck, being in the clutches of a mean-assed grizzly bear, drowning in a fast-flowing river, or freezing to death in the mud and cold water of a beaver pond. There has just got to be more for him than just an unmarked grave way out here on the frontier or ending up as grizzly bear scat out on the prairie somewhere. I already have lost one family by bringing them from the civilized part of the country out into the wilds of early St. Louis and having them contracting smallpox and dying from it several days later from its effects. I really don't relish the idea of doing the same to this young man, only out here on the frontier in the way it often happens, especially to one whom I now consider my son. I just feel I have been given a second chance at having some semblance of a family and I don't want to risk losing what I now have. Can you understand why I am having these misgivings and why I am feeling what I am feeling?"

After a few moments of blowing across the lip of his coffee cup before he took a sip, Old Potts said nothing in return to Iron Hand's question. Then he said, "I ain't never had a family other than what I now have here in Big Foot, Crooked Hand and you. So I don't know how to factor in a kid any too well. However, I think I know what you are feeling and I think I understand. That being said, what do you plan to do about what you are now facing?"

"I think I would like to take my share of the furs to St. Louis and sell them. Then I would like to go to Astor's fur house in St. Louis, take my share of the fur money I have coming to me from our previous years' trappings, build myself several keelboats, and go into the freighting of provisions

The Adventurous Life of Tom Warren

business up and down the Missouri and Mississippi Rivers. I say do that because I think with this frontier opening up, there would be a good business opportunity to be had in shipping provisions upriver to the likes of you and the boys, not to mention any other pioneers who want to come up here and settle in this vast, resource-rich wilderness. Plus, before I left St. Louis several years ago and hooked up with you and the boys, I heard tell of a steam-powered boat being invented called a 'steamboat' that could soon drive itself upriver with a set of sternwheel or sidewheel paddles. When that happens, that will open up the west and I would like to be a part of that history," said Iron Hand, with just a tinge of excitement over what he figured could be happening in the future.

"I knew it! That be that military training you got back at that Army academy years back. I just figured you having that faraway look these last few days might be because of that," said Old Potts with a grin of realization. "Well, if that be the case, I say you and White Eagle go for it!" Then after another pause to blow on his coffee to cool it down some more, he said, "If you two go off and do that, we could pick up Otis Barnes as your fill in. He is a damn good and frontier-steady man, and would fit in with our bunch like some of that homemade jam I am going to have on one of your biscuits this morning, if you would just get them out from that 'Dutch' and serve them," said Old Potts with a big grin.

Then he got a frown on his face and said, "How you be going to do all of that? I mean, going all that way back to St. Louis through some mean-assed Arikara Indian country and damn hostile 'war-hoops' at that?"

"The way I figure it, I can join up with some other trappers who want to go back to St. Louis in order to sell their furs for a better price and in a larger group for the protection that offers. With a group like that, we could fight off or bluff our way home to St. Louis," said Iron Hand.

About then Iron Hand became aware the Dent brothers, Big Foot, Otis and Crooked Hand had been quietly listening to

332

the entire conversation with Old Potts from the front of their nearby lean-tos.

"Well, the cat is now out of the bag," said Old Potts with a shake of his head. "So, I guess it is time for some Iron Hand special biscuits with some of that homemade jam slathered all over them that we picked up the other day at the fort's warehouse. What do you say? Then all of us can talk it over since we are all 'family' here."

During breakfast and afterwards, Iron Hand discussed his feelings and hoped-for plans with his fellow group of trappers. They were understandably sorry to contemplate the loss of one of their own but also understood fully why Iron Hand felt the way he did. They also included Otis in their conversations because he had indicated an interest in further trapping and seeing some more country, versus running with the Dent brothers trying to locate and kill Black Bill Jenkins and the rest of his killing clan. Finally, it was decided to bring Otis into their group of trappers, provision up for the four of them for another trapping season, and then the original group of Old Potts's trappers right out of the blue came up with a surprise deal for Iron Hand to consider.

Old Potts and his group of trappers realized they had more than enough credit with McKenzie from past years of trapping successes to carry them on for as long as they wished to remain on the frontier trapping. As such, Old Potts, Big Foot and Crooked Hand suggested that Iron Hand and White Eagle take all of the year's trappings as a stake in any business deal he wished to get involved in back at St. Louis. However, if he got into supplying the Missouri River trading posts with goods from St. Louis with the boats he wished to buy, he was to bring up the essential supplies they would need for each trapping season for his old crew at the St. Louis prices. Additionally if they wished to get out of the trapping business sometime later, he would give all of them a free ride back to St. Louis. Lastly, the three original trappers from Old Potts's group wanted Iron Hand to take ALL of the money they had coming being held

The Adventurous Life of Tom Warren

for them in the St. Louis fur house, and use it in his new business. That way if and when his former trapper friends returned to St. Louis, they would have some steady income derived from Iron Hand's hopefully thriving supply business to comfortably retire on. Iron Hand could hardly believe what he was hearing from his friends, but seeing the wisdom and potential in their proposal, agreed to it.

Taking a big bite from one of Iron Hand's still too-hot-to-eat biscuits and then chewing rapidly to avoid getting mouth-burned, Old Potts said, "Besides, it makes no difference in the long run how much money we all have back in St. Louis if we all get our topknots removed and left out on the prairie for the birds and wolves to pick at. If that be the case and God's will, then you and that there boy can have it all." With that statement out into the cool morning's air, the rest of the trappers mouthed their approval as they too were now mouthing a piping hot biscuit slathered with rich strawberry jam recently purchased from the fort's stores for a princely sum...

The next day, Old Potts and his group of trappers met with McKenzie and had him draw up the papers of credit from their previous year's trappings that the American Fur Company owed them and signed all of the funding owed over to Iron Hand in proper document format. At McKenzie's suggestion, $5,000 was left in his fort's account in case the trapping market bellied up, leaving Old Potts, Big Foot and Crooked Hand up high and dry when it came to having any funds to fall back upon so they could continue making a living.

That settled, the entire crew plus their new partner, Otis Barnes, went shopping for those provisions the four new partners would need for the coming trapping season. Additionally, it was decided by the new group that they would return to their previous trapping grounds up on the Porcupine since they already had a well-built cabin for all the men, their provisions and fur storage. Following the end of their shopping for those needed provisions, several crocks of rum were brought back to their camp, and while Iron Hand put together another of his buffalo stew and biscuit dinners all to be topped off with an apple cobbler, the rum flowed freely among the men.

However, the trappers had no more than sat down to their supper, when the Dent brothers returned to camp in an excited mood. They had word from several Free Trappers who had been trapping in the Medicine Lake area that they had run across four individuals fitting Black Bill Jenkins and his three red-headed brothers' descriptions! As such, they requested help in getting together all their needed provisions the next day and the day following that endeavor, they would leave the fort for the Medicine Lake country in their quest to run Black Bill and his kin down, and kill the lot for killing their parents, uncle and aunt several years earlier back in Missouri.

Two days later, the Dent brothers left Fort Union en route to the Medicine Lake area on the hunt for Black Bill and his kin. But not before all of the men pitched in and cast up a small mountain of pistol and rifle balls in case the Dents ran into some kind of killing grounds that needed a lot of flying lead in order to make their way through it... Little did Iron Hand realize that he would, almost a year later, run across the Dent brothers' six beautifully matched buckskins in an unlikely place and under unlikely circumstances. Additionally, little did Iron Hand and White Eagle realize what would be the ramifications of that unlikely meeting taking place along the shores of the Missouri River with a sizeable force of Arikara Indians, several kegs of rum and a handful of swivel guns.

Two days later, word had gotten out around the fort that Iron Hand and White Eagle would be taking a year's worth of fur trappings via pack string down to St. Louis for sale. Come nightfall that same day, four Free Trappers rode into Old Potts's camp, dismounted and asked if this was the campsite of one Free Trapper named Iron Hand. Iron Hand, cooking around the firepit, allowed that he was Iron Hand and what could he do to help them. The four men exchanged handshakes

The Adventurous Life of Tom Warren

and then one man named "Adam York" said, "It's not what you can do for us but what we four can do for you. We had planned on going to St. Louis so we could get a better price for our furs as well and then getting out of the trapping business altogether, take our proceeds, buy a farm and go back to farming for a living. If you would care to join us, from what we hear you are a good man to have around, especially with that Hawken you are carrying. We really could use the extra firepower going through all that Indian country we must travel through to get to St. Louis, if you would care to oblige us. Care to join us, my friend?" he asked.

That evening, Old Potts's group and the Brothers York had supper together and feasted on buffalo steaks, Dutch oven biscuits and apple cobbler. During that meal, the men planning on making the long trip to St. Louis with their *plus*, got to know one another fairly well. In just that short period of time over supper, it became apparent that the York brothers were good men and had been very successful trappers. With the men travelling down the Missouri River and through deadly Arikara, or Sahnish, Indian country with extra shooters, made a lot of sense because now there would now be five shooters and a boy to defend the fur train heading south. That evening over several cups of rum, it was decided that the Brothers York and Iron Hand would join forces and see if they could safely force their ways south through Indian country all the way to St. Louis, and keep their furs and 'topknots' in place while doing so.

That bit of good luck was followed with even more before the five men could finalize their plans. Two more Free Trappers wanting to head south to St. Louis to sell their furs had heard that the Yorks were going to do so as well. The next day, all of the men met and found that they were agreeable in timing and temperament to make a trip to St. Louis, thereby allowing the formation of an even larger group of now seven shooters and a young Indian boy learning to shoot as well. With that kind of firepower, most wandering groups of hostile

336

Terry Grosz

Indians encountered would hopefully think twice before making any kind of a frontal assault on such a well-armed band of determined fur trappers. However, any kind of a night attack instituted against the seven trappers by a superior force would understandably be a 'horse of another color'…

The next day, Iron Hand and White Eagle provisioned up for the several months' long and dangerous trip to St. Louis. Then as a precaution, Iron Hand had all their horses re-shod by the fort's blacksmiths and purchased extra shoes, files, nails and picks as well. Then Iron Hand taught White Eagle how to cast a small mountain of bullets for their rifles and pistols around their campfire after supper had been served and eaten. Following that, a trip the next day was made to the fort's warehouse storing their gunpowder, and several kegs of fresh powder were procured for the trip as well. Wanting to eat well on their trip to St. Louis, Iron Hand made a trip to the fort's cast-iron portion of their warehouse and purchased several Dutch ovens and three three-legged frying pans. Now they were finally ready to begin a rather long and dangerous trip south to St. Louis, the frontier capital of the known west.

Typical of Old Potts, he once again got antsy to be on his way back to their cabin on the Porcupine once he saw Iron Hand getting ready to leave for St. Louis. Well, that and the fact of the loneliness he would soon be feeling, being without a very dear friend and excellent trapper, namely Iron Hand, as a member of his party, fired up his 'traveling-moccasins'. Two days later, Old Potts, Big Foot, Crooked Hand and Otis 'Otie' Barnes said good bye and wished each other well when it came to 'keeping one's hair'.

Iron Hand stood on the fort's walls watching his dear friends and 'frontier family' as they finally disappeared off into the wilderness. He really felt a twinge of guilt for letting his friends leave without him, then caught himself with the realization he was about to make a new life for himself and a young Indian boy who had nobody but him going for him.

The Adventurous Life of Tom Warren

Besides, The Great Spirit had willed such a venture, it was now about to happen, and it was not nice to piss off The Great Spirit.

Two days later, the York brothers--Adam, Arnold, Andrew and Alexander, Jim Tweedle and Robert Caster, Iron Hand and White Eagle assisted the fort's personnel as they ferried all the men's furs, provisions and horses across the Missouri River to its eastern side. However, during that operation, Iron Hand took out some time to discuss his future keelboat plans that had been spinning around in his head with his good friend McKenzie. Upon hearing Iron Hand's keelboat upriver shipping plans, McKenzie told Iron Hand that any and all goods he could bring upriver to Fort Union, he would purchase the same at fair prices. Upon that the two good friends agreed, and then Iron Hand got back to his duties of transferring his goods across the river for the trip to St. Louis. Those endeavors took the entire day, so the men camped on the far side of the river that night, exhausted from all the work getting across the river safely but typical of the men of the day, happy to be off on another dangerous adventure.

Chapter 13: The Brothers York, The Arikara, White Eagle Comes of Age, Iron Hand, And St. Louis

THE NEXT MORNIN, WAY BEFORE DAYLIGHT, Iron Hand, now the designated camp cook, was into making biscuit dough and staking venison steaks over the campfire. As he continued making breakfast, the Brothers York, White Eagle, Robert Caster and Jim Tweedle saddled all their riding horses and packed their pack animals for the day's long travels to the south.

"Breakfast is ready," yelled Iron Hand, as he set his biscuit making Dutch ovens off the coals, so the biscuits would not burn and would be ready to serve once the men arrived to eat. With the rush of hungry men, Iron Hand quickly realized that when he cooked in the future, he had better have lots of chow handy because from the actions of his fellow trappers that morning, he had a mess of 'chow hounds' in camp with 'no bottoms in their bellies'!

Soon, the only sounds coming from that trappers' camp were the sounds of happy men making 'merry' over a frontier meal served piping hot and with lots of it to go around. In fact, the deer killed the night before by Adam York had its entire hindquarters and back straps consumed by those seven hungry men and one Indian boy, along with two Dutch ovens full of biscuits! Following breakfast, the camp clean-up and finishing with the stock, the men found themselves ready to leave the Missouri River headwaters area and proceed southerly along the river through Indian Territory towards St. Louis.

For the following week, the trappers followed the Missouri River east-southeasterly as they and their animals settled into their daily traveling routine. As they did, they passed vast herds of elk, buffalo and antelope, usually killing one of those animals in the afternoon for their supper meal that evening. Then as Iron Hand and White Eagle processed the most recently killed animal and set up their camp, the rest of the men unpacked and unsaddled their livestock, curried them down, hobbled and let them out to water and graze. After supper and right at dusk, the men brought their livestock back into camp for the evening and taking turns, one man kept the campfire going all night and an eye on their valuable animals since they were now deep in hostile Indian country.

Then come the following morning, Iron Hand and White Eagle tended to the making of breakfast and the rest of the camp's cooking duties, while the rest of the men checked their pack animals over to make sure they weren't getting 'sored' up over being improperly packed, and let them out to graze and water once again. Once breakfast was served, the men quickly ate their meal and gulped down their trapper's coffee, then began packing their animals once again for the day's daylight-to-dusk journey. As they did, Iron Hand and White Eagle cleaned up and packed their camp's cooking and sleeping items, and then began saddling all the men's horses for the day's travels. Thus began the daily regimen for the next several months as the men finally turned almost due south as they followed the mighty Missouri across the vastness of the living prairie.

After several weeks of travel, the trappers found themselves in the vicinity of the Cannonball River where it entered the Missouri River from the west. Skirting a large herd of buffalo, Arnold York shot and killed a cow buffalo for their supper that evening. While the rest of the men stopped their travels, let their stock graze and waited, Jim Tweedle and Robert Caster quickly removed large portions of meat from the animal's hindquarters and its back straps. When the men were finished with the butchering process, they left the remains of the animal for the wolves, magpies and ravens. With that meat loaded up on top of several sturdier pack animals, the men made for a stand of nearby cottonwood trees in order to make their evening camp. That evening as the men made camp, Adam and Andrew York took all their pack and riding animals hobbled as they were, and watched over them as they grazed throughout the nearby adjacent river bottoms. As they did, Iron Hand and White Eagle prepared their campsite, started a fire and made ready their evening meal. That was when the two York brothers brought the horse herd into camp earlier than they normally did.

Looking up, Iron Hand could see the two men as they brought their livestock back into camp, as they continued looking long and hard over their shoulders. Sensing danger of some sort, Iron Hand walked over to his rifle, cradled it in his arms and then walked out to meet the two men. Once there, Iron Hand waited for the two men to dismount in order to see what had been obviously bothering them back out on the prairie.

"Indians! We just saw two Indians on horseback from afar watching us as we tended to our horses," said Adam York, as he continued looking out across the prairie from the semi-seclusion of their evening camp hidden away in the cottonwoods. For the longest time the heavily armed men watched over the York brothers' back trail looking for what had alarmed them. After a while and not seeing any sign of the Indians following them, the men relaxed their guard a bit but not too much.

The Adventurous Life of Tom Warren

Come nightfall and aware of the Plains Indians' hesitation and beliefs in not attacking once darkness descended upon the land, the men relaxed even further. Iron Hand, with his Hawken lying close at hand, continued with his preparations in making the evening meal, as the rest of the trappers hobbled all of their horses and tied them to a long lead rope, so they would not wander off during the night or be easily spooked off if they were disturbed by Indians or a meat-eating critter with an affinity for horseflesh.

After supper and when the men were sitting around their campfire drinking coffee or smoking their pipes, Iron Hand took the moment to share with them some of his historical knowledge from his military days regarding the local Indians in whose country they currently were camping. Pulling up a saddle to be used as a chair by the fire, Iron Hand began with what he knew about the local Indians whose territory they were now traveling through and needed to be aware of their hostile tendencies.

"Men, as all of you know, we are now traveling through the territory of the dreaded Arikara Indians or as they are also known, the Sahnish. A number of years back, the U.S. Government began a program of bringing a number of tribal leaders from many of the Plains Indian Tribes back to the eastern United States. This was done in an effort to familiarize those Indians, many from warring nations with the white man, with just how powerful the United States was, as well as how powerful the Great White Father in Washington truly is," started Iron Hand.

"One of the chiefs brought back to the eastern United States was Chief Ankedoucharo of the Arikara Nation. If I remember my military history correctly, he was a very popular and much-loved chief by all of his people. Well, as hell would have it, that chief died shortly thereafter of unknown causes in Washington, D.C. My guess is the chief got sick and died because of some white man's disease that he picked up and had no immunity against, or for eating some of the white man's

Terry Grosz

contaminated food. Well as usual, the powers to be in Washington, fearing what the rest of the Arikara Indians would say about the untimely death of their much-beloved chief, didn't bother telling that nation of Indians about the death of their great chief until about a year after he had died. That stupidity really upset the entire Arikara Nation and they have been at war with the white race since 1823, and I would imagine that state of war will continue until they are all killed off or are so decimated from the white man's numerous diseases, that they no longer are capable of being a threat to the travelers in this area of the country," quietly advised Iron Hand.

(Author's Note: The Arikara Nation, as a result of the untimely death of their great chief Ankedoucharo, remained in a state of war with all white men from 1823 until as late as 1863. However, from 1836-37, that tribe was stricken and decimated by the third epidemic of smallpox at their village below the Knife River in current day North Dakota near old Fort Clark. Today the remnants of that nation of Indians share a reservation with two other smaller tribes of Indians, also previously decimated by the white man's numerous diseases. Early on those three tribes, because of their reduced numbers, affiliated with each other in order to hold off the numerous depredations from the more powerful and adjacent bands of Indians from the Sioux Nation.)

"So, as a matter of course, as long as we are traveling through the lands of the Arikara, I suggest we double up in everything we do and that means riding 'drag' on our pack strings, to those who remain awake all night guarding our camp and the horse herd," once again quietly advised Iron Hand.

A thoughtful murmured agreement followed from all the men sitting around their campfire that evening and Adam York and Jim Tweedle drew the first all-night camp and horse herd-guarding assignments, especially after observing several Indians earlier that afternoon watching on. With that, the tired and saddle-sore men retired to their sleeping furs and braved

343

The Adventurous Life of Tom Warren

the clouds of ever-present mosquitoes throughout the night. However, before retiring for the night, Iron Hand, alerted by the stirring of his 'sixth sense' took White Eagle off to one side and provided him with his very own pistol, which had been previously loaded with buck and ball since he was still learning how to shoot and shoot accurately.

"Son, you and I have been practicing loading and shooting ever since we arrived at Fort Union some time back. I want you to start carrying a pistol from now on, especially when we are in the land of the Arikara. You being from the great Sioux Nation automatically makes you an enemy of the Arikara Nation, and they will kill you just as surely as they will kill any of us if they get the chance. However, as you well know, this is a single shot pistol. You will only have one shot and if you need it to defend yourself, you must make your one shot really count. I have taught you over these last weeks on how to shoot and load and you have done very well. But just remember, you only have one shot and then you will be at the mercy of who you just shot at if you don't kill him. So aim to kill if you have to use this pistol by shooting at your adversary's largest body part. Also remember what I have said when it comes to shooting, 'aim small, miss small'," advised a very serious Iron Hand.

The look in White Eagle's eyes said it all when Iron Hand was instructing the young man in the art of survival, and being allowed and required to rather quickly grow up. White Eagle had come to love and trust Iron Hand and when his 'father' spoke, he listened. However, that evening as his father was speaking to him, the survival genetics from his people's 10,000 years of living as a primitive in the wilderness came roaring back to him, and he realized that what he had just been told was for a very serious reason and as such, White Eagle had committed such words from his father to 'stone' in his heart and memory banks…

The following morning before daylight, like in other mornings past, Iron Hand and White Eagle rose, started a

Terry Grosz

campfire, made coffee, staked out buffalo steaks on the cooking sticks, and began the process of heating up the Dutch ovens so they could make up a mess of biscuits in short order. The horses were allowed to graze close to camp and when it became time for breakfast, the men needed no further calling to come and partake of the meal at hand. After breakfast, Iron Hand and White Eagle cleaned up their cooking items and then carefully packed their cast-iron cooking gear, as the other six men saddled their riding stock and packed their pack animals.

Finally ready to go, the men formed up in their respective pack strings, including White Eagle and they streamed out from their cottonwood grove and out onto the prairie in the face of an ominous looking red-orange rising sun to the east, portending a serious weather change. Within minutes, the trappers were strung out and heading south, as they picked their ways through the numerous herds of buffalo, elk, prairie wolves and several small herds of bighorn sheep. (Author's Note: Bighorn sheep were originally found on the plains of North America. That was until they were quickly shot out for their fine tanned skins and great eating flesh. They have since, as a species, continued to live in the rugged mountains of North America, Mexico and Canada.) That morning was cloudless, a warming and soft breeze was blowing from out of the northwest, the air was still summer warm, and it was good to be in the saddle once again and heading towards their final destination of St. Louis.

Come around three in the afternoon, dark black and bluish-colored clouds had been forming in the northwestern skies and as they did, the men being frontier weather-wise, grew uneasy over what they were observing. Summer thunderstorms while out on the prairie could be extremely dangerous for anyone afoot or on horseback. As the men watched the dark forming clouds billowing upwards, they continued their southward migration hoping for the best come the rest of the afternoon travels. However, soon the summer heat began dissipating and now a cooler wind began blowing out from the northwest, as

The Adventurous Life of Tom Warren

the low-hanging and now obvious storm clouds were ominously gathering and heading their way.

Iron Hand, riding drag with Alexander York, began having his 'sixth sense' warnings once again as he continued along and finally after another long look at what was forming in the northwestern skies, rode forward with his string of pack animals until he met up with Robert Caster who was in the lead that day with his pack strings.

"Robert, what say we turn our pack strings towards that large grove of trees along the Missouri River and seek shelter there before this storm is upon us? I have heard a few way-far-off rumblings coming from that direction and I think we are in for a darn good and dangerous summer thunderstorm if we aren't careful," said Iron Hand.

"I think that is a damn good idea. I too have been watching those storm clouds building up for the last hour or so and I also think we best seek shelter while we can," replied Caster, as he continued looking to the northwest and the oncoming gathering storm.

Standing up in his stirrups, Caster waved back at the trailing riders getting their attention and then pointed towards the grove of trees along the river to their west. Then sitting back down in his saddle, Caster and Iron Hand turned their pack strings and headed for the cover the cottonwood trees along the river would offer them in the face of the oncoming storm and the usual heavy rains that accompanied such a potential violent event.

Minutes later, the entire string of trappers, their fur packs and provisions were under the covering cottonwoods. There the men hurriedly unloaded all their packs and stacked them in a central pile. Once stacked, several buffalo hides were placed over the packs for their protection in the face of the soon to be expected heavy summer rains. Then while four of the men double hobbled all of their horses, the remaining men strung several ropes between numerous close at hand cottonwoods and constructed their sheltering lean-tos. As they did, Iron

Terry Grosz

Hand and White Eagle gathered up numerous dry limbs from beneath the trees and stacked them near a soon-to-be firepit. As they did, the men now noticed the cottonwoods beginning to lean away from the winds of the fast bearing down upon them storm as it began moving over their hastily erected sheltering encampment. Dragging in their saddles and sleeping furs, the men hurriedly placed them under the covering lean-tos as well. Then all the men made sure all their horses were lead-rope tied to the trees so they would not be frightened off from the overhead noises generated by the now hurtling down upon them, violent summer thunderstorm.

As the winds coming out from the northwest began increasing, dust, falling leaves and cottonwood 'drift' began flying through the air as did numerous large and cold drops of rain. Soon the loud rumbling of thunder and ominous cracks of lightning could be heard all around the trappers' encampment. About then across the prairie to the west, the men could see an onrushing darkness that ran from the clouds clear down to the ground, announcing that a heavy wall of rain was soon to be upon them in all of its fury! With that, the men ducked under their lean-tos and not a minute too soon, as the rains now came down with a battering vengeance!

As the rains came down, the thunder rumbled and the bolts of lightning sizzled through the sky, as the men huddled under the cover of their lean-tos and watched as Mother Nature did what she did best with one of her violent summer thunderstorms spawned out on the prairie. Then 'thunder' of another type was heard by the men as herd after herd of frightened buffalo, spooked by the close at hand strikes of lightning, made themselves known in the close presence of the heretofore adjacent peacefully resting and feeding animals. Moments later, 'brown thundering carpets' of heaving up and down in their familiar running-gait, stampeding buffalo added to the din, until one could hardly hear himself think or even talk in a normal tone of voice.

The Adventurous Life of Tom Warren

Like in many Northern Plains weather events of such magnitude, they soon lessened in their intensity and moved on to 'other battlefields' and once again did what 'they' did best. As for the group of trappers, they soon crawled out from under their lean-tos and looked all around them in amazement and surprise. Tree limbs and cottonwood leaves by the inches deep were strewn about their encampment. Their horses all had their ears up and were snorting and moving around at the end of their ropes nervously, as the 'buffalo-thunder' had now also moved off into the vastness of the plains, soon to be lost to sight and sound.

Removing an armload of dry wood from under his lean-to, Iron Hand made for an area previously cleared off for the location of his firepit and soon had a warming fire crackling away in the now cool and damp evening air. As Iron Hand and White Eagle tended to their camp duties, the rest of the men gathered up their rifles, ever mindful of the Indians in the area, released their hobbled horses and escorted them out onto the plains so they could graze until suppertime.

About an hour later, Iron Hand, with his supper ready to be eaten, sent White Eagle out onto the prairie on foot to gather in the men and horses. Once the men and the horses had returned, Iron Hand had the coffee in a rolling boil, the last of their buffalo meat almost cooked over stakes and two Dutch ovens ready and full of biscuits ready to eat. Within minutes after their arrival, every man around the campfire was busy making sure his 'little guts weren't being eaten by the big ones' because of the lack of plentiful 'grits' of which they were partaking.

After supper and while Iron Hand and White Eagle cleaned up their cookware, several of the men gathered up a huge supply of limb wood so they could keep a fire going all night in light of the possible Indian menace in the area. Since the men had not worked a full day and were not overly tired, they now had more time that evening to sit around the campfire, smoke their pipes and talk over the day's events and the many

Terry Grosz

adventure memories dredged up from their storied pasts as trappers on the frontier. Finally, the men drifted off to their lean-tos in ones and twos to get some sleep because tomorrow would be another long day on the trail, as the two designated night guards hunkered down for the long night at hand of guarding the horse herd and the rest of the encampment.

Way before dawn the next morning, Iron Hand and White Eagle were up and now had a roaring fire going to ward off the night's chill and damp air. The two men who had been up all night were now in their sleeping furs trying to grab a few minutes of sleep since Iron Hand and White Eagle were now up and could watch the camp and their horse herd. About an hour later, the false dawn could be seen in the east and the smell of Iron Hand's baking biscuits and coffee pervaded the air around the trappers' campsite. However, there was no great smell of cooking buffalo meat because the men had not had the chance to go out and kill another cow the evening before because of the oncoming fierce summer storm. The men would just have to make do with a load of strong coffee and a huge mess of Dutch oven biscuits to tide them over for the day.

"HI–YI–YI–YI–YI!" yelled an Arikara Indian, as he rode his horse right into the trappers' camp at full gallop and bailed off his horse right on top of Iron Hand from behind, slamming him violently onto the ground! "ZZZIIPPPP—THUNK" went an arrow into the side of a tree just as White Eagle stepped past it carrying another load of firewood, missing him by just mere inches! More yelling Indians followed as they galloped their horses right into the trappers' camp, all the while trying to stampede off the trappers' saddle and packhorses as well! The next Indian leading the charge of Indians galloping at full speed into the camp on his horse tried riding over the top of White Eagle!

"BOOOM!" went White Eagle's pistol, blowing that Indian clean off his horse and causing him to fall right into the blazing campfire! That was when Iron Hand roared up from off the ground after killing his assailant who had jumped onto

The Adventurous Life of Tom Warren

his back from off his racing horse and in the process, had died for his efforts. That Indian attacker's awkwardly twisted head off to one side of his neck, told the story of the strength of Iron Hand when his full fury was aroused in a life or death battle! Iron Hand then grabbed his always handy rifle by the fire, swung it towards the group of oncoming racing Indians on their horses and blew the lead Indian clean off his horse. When he shot that lead Indian, the bullet from his Hawken had gone clear through his chest at such close range, that it exited his body and had slammed into the second hard-charging Indian in line, killing him outright as well!

By now, the trappers' camp was in a frenzy of hard-charging horseflesh, yelling humanity, flying bullets, trappers' horses going crazy over all the fury of battle and breaking their tie lines and stampeding off, curls of black powder smoke rolling across the campsite from all angles, violently falling Indians smashing onto the ground, flying tomahawks and pieces of body tissue being blown off its owners as the huge .50 and .54 caliber slugs from the trappers' rifles did what they were meant to do!

The first of the trappers to fall was Robert Caster with a lance from an Indian on horseback, who surprised him just as he emerged from his lean-to, driving it clear through his body! That Indian killing Robert Caster was almost the next one to die. Caster was lanced clear through his body by an Indian who had a huge knife or tomahawk scar running from his left forehead, diagonally across his nose and across his right cheek, who ducked just in time as Iron Hand, swinging the ax used to chop firewood, missed him completely. Dropping the ax and grabbing a pistol from his belt-sash, Iron Hand swung the barrel in the direction of the speeding Indian trying to make his escape. Just as Iron Hand pulled the trigger on his pistol, the Indian laid his torso down across his horse's neck to present a reduced target and Iron Hand's killing shot missed the Indian lying over his horse's neck and struck the racing horse in the back of its head! Down went the horse, throwing the Indian onto the ground in a flying cloud of dirt and debris! However in an instant, that Indian jumped up running and disappeared into the pre-dawn darkness surrounding the trees, never to be seen in the remainder of the fight! However, months later, Iron Hand would cross trails with that same heavily scarred Indian and that time, he would not just kill the man's horse...

Adam York, from lying in his sleeping furs having been up all night as one of the camp guards, shot the next Indian in line on horseback from his sleeping position and out through the end of his lean-to, blowing him clean off his horse and onto the ground in a crumpled, lifeless pile!

Jim Tweedle, getting late into the fracas, shot one Indian off his charging horse and was so close to him, that he was felled and knocked out cold by the Indian's falling body right on top of him!

Andrew York, running out from his lean-to with his rifle in hand, was run over by another hard-charging Indian's horse and knocked asunder. But not knocked out, as he drew his pistol since his rifle had been knocked from his hands, and shot another Indian riding by who was trying to ride over Iron Hand from behind!

Another Indian running into the trappers' camp on foot with an upraised tomahawk heading for White Eagle to kill him was shot in the head by Arnold York from just two feet away! When the musket ball struck that Indian's head, his brains were splattered all over a surprised White Eagle, who was still trying to reload his pistol!

A large-in-size Indian racing his horse into camp, was slowed by having another Indian's riderless horse run in front of him in panic and in so doing, was swept off his horse by Iron Hand violently swinging his now empty Hawken's heavy barrel into that Indian's mid-section! That hefty blow swept that hard-charging Indian clear off his horse and threw him into the still-roaring campfire. When he fell, that Indian impaled himself on one of the vertical cooking stakes! That Indian screaming in pain over the hot metal cooking post now piercing through his mid-section and stoutly holding him in place with his flesh steaming and burning in the campfire, had his screams cut short when Iron Hand stepped forward, reached in and broke his neck with a quick and violent twist of the dying man's head!

Two more Indians trying to cut the tie ropes on a string of horses so they could be run off and captured, were both cut down simultaneously by two quick rifle shots being fired by Arnold and Adam York with their pistols at such close range that both Indians had parts of their bodies flung over the rest of that string of horses! When that happened and the black powder smoke clouds rolled over the still tied and frantic horses because of the intense fighting going on all around them, along with flying blood and snot from the two dead Indians, they spooked, broke their tie ropes and bolted off onto the adjacent prairie and into the fast-breaking dawn!

Then a lone wolf could be heard howling from out on the prairie somewhere over all the commotion going on at the trappers' campsite, as a deathly quiet silence settled over that camp as the last Indian attacking the men or trying to get at their horses, joined the Cloud People!

It was then that Adam York yelled, "Is everyone alright?" "I am alright," said Arnold York, "and I can see Alex trying to reload one of his pistols so he is OK," he continued. "Caster is dead," said his friend Jim Tweedle, "but I am alright." "I still have my hair," said Andrew York, "and I can see White Eagle still trying to reload his pistol." "I am OK," said Iron Hand, as he was in the process of tossing the two dead and partially burned Indians out from his campfire, and pulling his cooking iron out from the still-steaming body of the Indian that had impaled himself but later died from an Iron Hand-administered broken neck...

"You Yorks need to saddle up what horses we have left and go after the ones who broke their tie down ropes and fled during the heat of the battle. We can't afford to lose a single horse," said Iron Hand as he bodily picked up another dead Indian and tossed him away from his campfire seating area.

"Come on, Guys. Iron Hand is right. We have to get all of our horses back or we can't get all of our furs to St. Louis," said Adam. With that, the York brothers dug out their saddles from their lean-tos, saddled up on the remaining and still-tied horses and disappeared out into the dawn looking for their still hopefully hobbled and nearby now-feeding horses. As they did, Jim Tweedle and Iron Hand picked up Robert Caster, removed the spear still thrust through his body and laid him off to one side. In the meantime, White Eagle, now with his pistol reloaded, was gathering up and tying off all the Indians' horses standing around that he could catch. About an hour later, Iron Hand could hear a number of horses approaching their campsite. Grabbing his since-reloaded Hawken and now wearing a brace of fully loaded pistols, he waited to see who was coming into camp. Soon he could see a number of familiar pack and riding horses coming back into camp followed by the four York brothers herding them along.

"We got every damn one of our horses back, plus three belonging to those damn killing Indians," said Alexander York. "However, that tall Indian with the badly scarred face got into the river bottom and before we could kill him, disappeared like a gust of wind." With that, the Yorks lit down and commenced re-hobbling all of their stock and tying them up to a makeshift rope corral.

Then all of a sudden the Yorks stopped what they were doing and looked over into their camp at Iron Hand. Since the fight, he had come down from his killing emotions and was now hard at work. Not quite believing what they were seeing, the Yorks walked back into their camp to find hot coffee boiling away, fresh horse meat now roasting away on cooking stakes, two Dutch ovens baking biscuits, and all 12 of the dead Indians carted off and stacked into a pile outside their main camp area...

"What the hell?" said an amazed Adam York.

The Adventurous Life of Tom Warren

Without missing a beat, Iron Hand looked up from his cooking and said, "What is the problem, Adam? You men have to eat and since one Indian's horse was killed outright in the battle, White Eagle and I butchered him up and now we have fresh meat to eat since none of you brought back anything except a mess of horses and I always make biscuits because everyone likes them. So we had a little action this morning and except for Robert, we all survived. The way I see it, we still have a long ways to go to get to St. Louis and a man cannot travel on an empty gut. So, breakfast is ready if any of you care to join White Eagle and me, but you best get to it before it is all gone. White Eagle and I have developed a monster-sized hunger after this morning's excitement, so the four of you had best hurry afore it is all gone..."

It was amazing how hungry all those men were, especially in light of losing one of their own in the fight and almost all of the rest of them being killed as well. But that was the frontier and 'here today and gone tomorrow' always seemed to be the mantra of the day, and that was now the case for their friend Robert Caster...

(Author's Note: Robert Caster's unmarked grave, based on the written history of the area, except for a large pile of stones placed there to prevent the wolves and bears from digging him up, is still visible in that grove of cottonwood trees to this very day over 100 years later. I know this because while working in that state as a Special Agent for the United States Fish and Wildlife Service's Division of Law Enforcement in 1975 and aware of that area's history, I chanced upon it after an hour of searching in the described river bottom and spent a few 'quiet' moments with Mountain Man Robert Caster...)

Once the men were finished with their breakfast and knowing that if the 12 dead Indians did not return soon, someone would be out looking for them, so they quickly went about their business at hand. As Iron Hand and White Eagle dug Robert Caster's grave next to their campsite, the other men gathered up rocks for placement over the dead trapper's grave so the wolves and bears would not dig him up. Knowing the Indians would do the same when it came to digging up the dead man if they figured the stones were a gravesite, several of the men gutted out the Indian's dead horse and placed its guts in a covering fashion over that pile of stones. Then they cut up the dead Indian's horse into pieces and placed those over the pile of stones as well. When they were finished, the gravesite just looked like a pile of dead horse parts and hopefully would garner no more attention from searching Indians other than a passing glance. It worked...

Although now late in the morning, the men loaded all of their gear onto their horses and left the area quickly. Once again, Tweedle took the lead, followed by White Eagle who was now responsible for leading and trailing the Arikara Indians' horses, and the rest of the men fell in behind, now trailing longer pack strings because Caster was no longer among them to trail his share.

For the next two weeks, the trappers hustled their pack strings in a slightly southeasterly direction as they followed the Missouri River in an attempt to throw off any Arikara Indian followers looking for them after their earlier recent battle further to the north, in which a large number of their 'kin' had been killed and left in a pile for the critters to partake...

Late one afternoon, the trappers came upon a small herd of buffalo resting along the Missouri River breaks and since it had been a long day, Adam York killed a cow buffalo so the men could have some fresh meat for their supper. As the rest of the trappers held up their travels, Iron Hand and Arnold York cut out steaks and back straps from the buffalo and loaded the slabs of meat onto two of their stouter packhorses. Just as the men started to head for the Missouri River breaks and set up their camp for the evening, they found themselves being quickly surrounded by 23 whooping, yelling insults and making obscene hand gestures, Sioux Indians. Apparently they had heard Adam York shoot and kill the cow buffalo and had come over to investigate. When they did and discovered a string of

The Adventurous Life of Tom Warren

valuable horses loaded with packs of furs and only guarded by six trappers and a young Indian boy, they quickly surrounded their 'lucky find'.

Suspecting a fight being close at hand, each trapper bailed off his horse and standing behind their individual horses, prepared to fight to the death if necessary. When they did, Iron Hand, having used Old Potts's rule of making sure the first horse in a trapper's pack string was carrying an extra rifle and at least two pistols in holsters, grabbed a pistol from his pack animal and handed it to White Eagle, who was now standing by his 'father's' side.

"Here, 'son', take this extra pistol and now you have two. I don't know what tribe these Indians are from but if they are more Arikara, they will kill you in a heartbeat since you are a Sioux and they are a nation of Indians that are hated by the Arikara. Save the last shot for youself if we lose this fight because what they will do to you will not be pleasant," said a worried Iron Hand, because the trappers were so outnumbered and had now been caught out in the open in a defenseless position.

For the next few minutes, the Indians slowly circled the trappers who were now bunched up in the middle of the circle in a defensive fighting position expecting the worst. Then the band of Indians stopped circling the trappers, seeing they were prepared to fight for what was theirs and just sat there in a big circle looking on at their trapped 'prey'. Then one of the Indians rode his horse forward a few feet towards the trappers, stopped and then began speaking in a loud tone of voice directing his words at the trapped men. When he did, Iron Hand thought he must be the leader of the group of Indians confronting the trapped trappers and figured when the shooting started, he would single out that man and kill him on the spot. Iron Hand so figured that if he did, and knowing how the Indians hated getting into a battle and losing their leader, would many times retire from the field after suffering such a loss. Retire from the field of battle because the Indians

356

considered the loss of a leader a very bad sign. Laying his Hawken over the saddle of his horse, Iron Hand took a clear bead on the Indian talking to the circle of trapped trappers and held it there waiting for the Indians to make their move. As Iron Hand did, he was all of a sudden surprised to see White Eagle riding his horse into his 'line of fire' that he was holding on the Indian doing all the talking!

With that, Iron Hand froze over seeing his new son riding out to meet with the Indian leader. Then he really hunkered down and took a deadly bead with his rifle on the Indian to whom White Eagle was riding towards in case he made any kind of a deadly move in the direction of his new son! Reining up alongside the Indian leader, Iron Hand could see White Eagle was speaking with the man. For the longest time, the two conversed with each other and then White Eagle turned his horse around and slowly rode back towards the ring of his trapper friends.

Dismounting, White Eagle proudly walked over to his father and said, "Father, they are from the mighty Sioux Nation. Because I could understand what their chief was saying, I figured they were a band of my people and since none of you speak Sioux, I took it upon myself to talk to him. I hope that was alright with you."

"They say we are trespassing on their lands and killing their game without being allowed to do so. They also say they will fight the white man trappers unless they pay for walking on the ancestral lands of the mighty Sioux and killing his buffalo. Chief "Ten Bears" says we must pay in horses if we want to live and continue on our way. Plus, he says he sees that we are not the friends of the hated Arikaras because he sees that we have taken some of their horses. I told him that we are not friends with the hated Arikaras and that we took their horses after we killed many of them in battle. When I said those words, he seemed pleased but he said we must still pay to be on the lands of the Sioux. He also said we will not be allowed to pass unless we pay in horses. He says we all will

The Adventurous Life of Tom Warren

be killed if we do not listen to his words and wants to know why the mighty white man sends a boy to speak to the mighty Chief Ten Bears instead of showing a man to talk. So I told him that I came of my own will because I am the only one who speaks their language and that is what a Sioux warrior would do. That seemed to please him. Then I told him that we would give them horses in order to peacefully pass through their lands," said White Eagle in a calm tone of voice, like what he had just done was nothing out of the ordinary.

"White Eagle, we need all of our horses if we are to take all of our furs to St. Louis," said Iron Hand, now perplexed over what his son had just promised. By now the other trappers from their group had gathered around to hear White Eagle's words and just shook their heads in disbelief over his offer of horses, which would leave them desperately short of the needed pack animals.

Then White Eagle said, "Father, I am leading a string of 11 Arikara horses which are not carrying a single pack of furs. I figured in order to pass, we could give all of those horses we captured in battle and that would make Chief Ten Bears very happy and then he would allow us to safely pass. Besides, that way he and his warriors could go back to his village with enemy horses and count many coups in the eyes of his villagers. That is why I agreed to give them horses that we do not need. Did I do wrong, Father?"

For the longest time, Iron Hand just stood there dumbfounded over what his 11-year-old son had just done. Then he realized if Ten Bears was a man of his word, the trappers could safely proceed along their way and only lose 11 horses for which they had little use anyway other than selling them once they arrived in St. Louis... Looking over at the rest of his party, all Iron Hand could see was relief and big smiles over what White Eagle had just, in his 11 young years of thinking, negotiated with a fierce warrior such as the renowned Sioux Chief Ten Bears.

The next thing Chief Ten Bears saw was a young Sioux 'warrior' leading 11 Arikara horses towards him. However, as he did, Iron Hand kept his Hawken 'tuned in' on Chief Ten Bears in case he was not a man of his word. If his word was not good, Iron Hand figured his aim with his Hawken would be and in that case, would make sure the chief never ever again negotiated with any trappers under such trying circumstances.

Chief Ten Bears took the lead rope from the hands of White Eagle and then reached across his saddle and gave the young Sioux 'warrior' facing him a typical 'hand and arm' shake out of respect for the boy's courage in the face of superior numbers. Then with a yell of success, Chief Ten Bears and his 22 warriors thundered off towards the east, leading a string of 11 Arikara horses and a story to tell the elders back in the village. Little did Chief Ten Bears realize that he had just missed joining the Cloud People only because of his honorable and good word spoken to a young man on a warm day out on the boundless prairie.

That evening around their campfire, the trappers celebrated their good luck in having a young White Eagle in their midst. Little did the trappers realize that in White Eagle's mind, since he had already slain a hated enemy in battle, he did nothing more than what was expected of a Sioux warrior. Sitting around the campfire that evening after supper, Iron Hand also realized that his son of just 11 'summers' was already in his eyes, coming of age.

Another month of hard travel without any more trials and tribulations other than a few horse wrecks and another dousing rain from a late summer thunderstorm, found the men about to enter the famed western 'jumping off' city of St. Louis!

Since the men had arrived so late in the evening at their destination, they chose to camp their last day of travel along the Missouri River. There the men bathed in its waters and then retired to their camp, as Iron Hand prepared their last meal on the trail after a long and dangerous journey. That evening, the trappers feasted on staked venison, hot Dutch oven biscuits,

The Adventurous Life of Tom Warren

coffee, and later on, drank the last of their rum from Fort Union as they celebrated the end of a long, dangerous and hard trail of travel. However, because of the excitement the next day would bring, the men got little sleep as they sat around their campfire, smoked their pipes and visited over their many experiences and 'pratfalls' as Mountain Men trappers in the wilds of the frontier.

As the rest of the party saddled and packed their horses, Iron Hand and White Eagle prepared their last breakfast on the trail. Since they were low on their supplies, that breakfast consisted of just Dutch oven biscuits and hot coffee, then the men saddled up and headed for St. Louis.

Riding down a muddy street, Iron Hand stopped the fur caravan and asked a stranger where they could find the fur house of Astor's American Fur Company.

"Keep a-heading that-away, Stranger, and you will see her in about 300 yards on the right side of the street. You kain't miss her, she is a big yeller building with a black sign hung out front announcing 'Astor's Fur Company'."

Iron Hand thanked the man and then amid a lot of stares from a number of folks walking or riding by, the fur trappers finally pulled up in front of the yellow warehouse. When they did, a Company Clerk, upon seeing the arrival of the long string of heavily packed horses, had the men ride their mounts right up into the center of the warehouse where they could be offloaded and the furs easily counted and graded.

Four hours later, Iron Hand, Adam York and Jim Tweedle were holding bank drafts against the First Mercantile Bank of St. Louis for their furs and hides just sold. Additionally, Iron Hand was holding an additional bank draft for a little over $72,000 from a two-years' previous trip in which Old Potts and company had waylaid a nest of trapper-killing Indians along the Missouri, and ended up with over 60 horses and their packs of furs that the renegade Indians had stolen from a number of other now dead trappers!

Terry Grosz

After visiting the bank and collecting their monies, the men agreed to meet at a nearby boarding house and eatery come dark for supper. With that, the men separated their ways except for Iron Hand. There in the bank meeting with the bank president, Iron Hand banked in his name $72,400 and except for $100 in Spanish silver dollars, walked out the door with White Eagle to see the town.

For the next three hours, Iron Hand, now asking to be called "Tom Warren", visited shop after shop so White Eagle could see how the white man lived and what in life he favored. In fact White Eagle got his first taste of ice cream and couldn't seem to get enough of that kind of white man's food.

However, it was starting to get dark and close to suppertime. Plus, a few biscuits and coffee for breakfast so long ago was now 'wearing thin on their slats', so Tom and White Eagle headed for a place in town to eat, trumpeted as "Mrs. Davis's Fine Food –– All You Can Eat –– 50 cents".

Walking into the noisy and filled with other white folks eatery, Tom and White Eagle took a table with five extra empty chairs so when the Yorks and Tweedle arrived, they would have a place to sit and eat as well. Tom ordered a draft beer, his first in over three years and for White Eagle, the first glass of cow's milk that he had ever tasted. The two of them then sat around for about an hour, getting hungrier by the minute as they waited for the Yorks and Tweedle to arrive.

Little did they realize, their friends were in jail! They had gone to a saloon, gotten rip-roaring drunk and into a fight with a number of patrons, who had objected to how they smelled! Soon the local constable was on the scene and moments later, his friends were on their way to the local lockup for being drunk and disorderly in a public place!

Finally tiring of waiting and getting hungrier by the moment, Tom and White Eagle ordered supper, another draft beer for Tom and now a sarsaparilla for White Eagle. The whole time, Tom noticed that a lot of people in the eating establishment kept casting sideways glances at the two of

361

The Adventurous Life of Tom Warren

them. Tom figured it must have been the raggedy clothing he was wearing and swore to himself that tomorrow he would buy and commence wearing clothing like everyone else was wearing. After all, he was no longer a Mountain Man but a soon to be businessman in the shipping business, if he had his druthers.

That was about when Tom's 'wheels came off his wagon'! About then, the front door of the eating establishment flew open with a loud 'BANG' and in walked seven burly men all talking, laughing and swearing loudly. From the cut of their clothing and coarse manner of talk, Tom figured they were men who made their living on the nearby boat docks of St. Louis or were some form of Missouri River boatmen who were rough in their cut and ready to howl.

Looking all around the now crowded eating establishment and seeing no place for them to sit, the suspected river men began talking even more loudly and grousing about the lack of space for 'real men' to sit and eat. That was when the largest of the men spotted Tom and White Eagle sitting at a table with five empty chairs.

"How about the two of you getting your asses out of them chairs and let a bunch of real men sit down so they can eat?" bellowed out the "Man Mountain Dean"-sized man, who was the notorious and much-feared locally, famous brawler and river man, "Mike Fink"!

About then one of the female cooks from the kitchen arrived carrying two plates of food for Tom and White Eagle. "Hello, my name is Betsy Davis and my Mom and Dad own this place. If you two haven't been here before, just ring the little bell on your table and we will bring you more food if you want more," she sweetly said.

"Hey! I just told these two to get their lard asses out from those chairs so all of us could sit down there and eat. Besides, what the hell is a damn good-for-nothing dirty 'Injun' doing here in a white man's eating place anyways? Little Lady, you best get this damn 'war-hoop' out of this place established for

362

serving white people, and fast. His kind does not belong here among us decent white folk. Now if you don't move him, I will," bellowed for everyone's benefit, Mike Fink.

With those words and Miss Betsy telling him that her mom served everyone who wanted to eat just as long as they paid for their meal, Mike Fink, famed Missouri River boatman, just pushed her small frame aside and lurched his massive grabbing hands out for White Eagle in which to forceably remove him from the premises. All the time the big bully was running his mouth, White Eagle just kept his eyes on his father to see how to react to the harsh words being hurled their ways at the supper table. That was when Mike Fink, famed and feared Missouri River boatman, was jerked upward from behind as he reached for White Eagle, was spun around, kneed in the groin and with a right cross from Tom that would have felled an Army mule, Mike Fink hit the dirty wooden floor and was out for the count!

Mike Fink's compatriots, surprised at just how quickly Tom had erupted up from out of his chair and had felled their boss and much-feared river brawler with one punch, recoiled backwards in concert and fell against a nearby wall. Then almost in unison, the remaining six men drew wicked-looking, long-bladed knives and made threatening moves towards a still standing over Mike Fink, ex-Mountain Man named Tom Warren! When they did, they found themselves in a crowded eating establishment that had now gone silent as a tomb over the action at the front door. That was when the remaining river men found themselves instantly staring at two leveled pistols being held by Tom and another one being leveled at the men by White Eagle!

"Here! Here! You men lay down those knives. You know Miss Sylvia does not cotton to any kind of violence in this here eating establishment," said a large and muscular man named Thomas, wearing a much-used apron, who had upon hearing the start of a fight, come out from his kitchen at a dead run with an upraised meat cleaver in hand! Behind him was also another rather stout-looking individual named Clifton, also

363

The Adventurous Life of Tom Warren

coming at a dead run and he was carrying a double barreled ten gage shotgun with both hammers pulled back and ready to 'dance if someone played the right tune'!

Then little Betsy Davis stepped in between the warring men once again saying, "You river rats can take this outside and I mean right now! But if you persist, my brothers Clifton and Thomas will see to it that the seven of you never eat here again or anywhere else for that matter! And if my dad Marshall gets wind of this, there will be hell to pay just as sure as my mother is the best cook going in the whole darn city of St. Louis. Now, unless someone wants a good dose of buckshot from my brother's shotgun in their hide, you men leave and I do mean now! And you two, put away those pistols, sit down and enjoy the supper I brought you. Because if you don't, I will be offended and think that the two of you don't like my style of cooking," said the tiny women named Betsy...

With those words hanging heavy in the air coming from such a tiny little gal, Tom put his two pistols back into his belt-sash and so did White Eagle. As they did, the river men picked up their leader who was still out cold from Tom's angry blow and removed him and themselves from the eating establishment. As they hurriedly exited the place carrying their boss, they fired back with the words, "Mountain Man, this ain't over for you and your Injun. We will see you later when you don't have this little bit of a women watching over you and those two brothers of hers standing guard over the three of you. We will have our due and pound of flesh from your miserable carcasses, you can rest assured of that." With those words, they disappeared into the night and the folks who had grown so quiet in the eatery, broke out into cheers and loud clapping over what they had just witnessed, namely a single large man and a boy facing down seven burly, ill-tempered and ill-mannered river men...

Keeping a 'cocked' eye on the front door the whole time in case the mean-spirited river men came back, Tom and White Eagle finally enjoyed their supper. In fact, all five platefuls between the two of them! Just as the two of them were on their second piece of apple pie, over to their table walked Mrs. Sylvia Davis, owner and head cook for the eatery.

"I would like to apologize to the two of you for the poor manners of some of my patrons. That darn Mike Fink and his crew of ruffians always bring trouble wherever they go. As a gesture of my thanks for not shooting up my place, your suppers are on me," said Sylvia with a friendly smile.

"That won't be necessary, Ma'am," said Tom with his typical, heavily bearded smile. "They weren't any bother, just a little loud and lacking in proper manners," he continued.

Then Sylvia took another long and examining look at Tom and then said, "Say, I do believe I know you. Aren't you Tom Warren, Miss Jeannie Warren's husband?"

"Yes, Ma'am," said Tom quietly, surprised over being recognized after so many years of his absence from the St. Louis area.

"Whatever happened to that little gal? I ain't seen her in several years now that I think about it," she continued.

"Yes, Ma'am. She and my young son died several years ago from the 'pox'," said Tom quietly.

"Oh, I am so sorry. She was a real God-fearing women and well-respected around hereabouts. No wonder I haven't seen her in church for such a long time," continued Sylvia, without realizing how much hurt she was 'digging up' from Tom's past.

Quickly changing the subject, Tom said, "Say, me and the boy here will need to rent a room for a few days. Do you happen to have any that are available?" he asked.

"Sure do. That will be 50 cents per night for the two of you," replied Sylvia.

"When we finish dinner, where do we go to get our rooms?" asked Tom.

"Just come to the front desk and either my son Clifton or Thomas will fix the two of you right up. Say, may I also suggest that the two of you also purchase a couple of baths as

The Adventurous Life of Tom Warren

well?" said Sylvia with a knowing smile, realizing the two folks in front of her had just either come off a boat from New Orleans or had just returned from the frontier...

That evening, after a shopping trip to the local mercantile for some new clothing for the two of them, Tom and White Eagle enjoyed their first night's sleep without a cloud of mosquitoes hovering overhead. Oh by the way, the hot baths, use of lots of soap, a luxury they never had on the frontier, and a heavy dose of lilac water sure felt and made everyone smell one hell of a lot better and more in keeping with the locals...

Terry Grosz

Chapter 14: Keelboats, Reunited, "Buckskins" And Gabe's Rifle

THE NEXT MORNING AFTER A FINE BREAKFAST at "Mrs. Sylvia's", Tom and White Eagle went to the livery stable and retrieved their horses. Then with a few directions as to how to find one's way to the keelboat builder's docks and ways, Tom and White Eagle set out for the boat docks below the bluffs of St. Louis. After a little trial and error, the two of them finally found themselves in the office of one "J. Dawson and Sons, Esq. – Boat Builders".

Walking into a sparse-looking two-room office, Tom introduced himself and White Eagle and then asked for a "Mr. Dawson".

"You are looking at him, Stranger. What can I do you in for?" asked an all-business-like James Dawson.

"I hear from those at the Astor Fur Company, the man at the livery and Marshall Davis at the local boarding house and eatery where we are staying, that you build the best keelboats on the Missouri. Is that true?" asked Tom, "and if that be so, we maybe can do some business."

"You heard right, Stranger," replied Dawson quietly, as he began sizing up the giant of a man standing in front of him.

367

The Adventurous Life of Tom Warren

"My name is Tom Warren and this here young man is my adopted son, White Eagle. We are 'late' of the fur trapping trade from the frontier, and now I am interested in going into the business of shipping provisions up the Missouri to those making a go of it on the frontier, be they trappers or settlers," said Tom. "But in order for the two of us to get into such a trade, we need to have a vessel or two under our feet and the crews to power them," he continued.

"Well as I said, you heard right, Stranger. My sons and I build the best keelboats going along the Missouri. They are made only from the finest virgin oak, hickory and chestnut trees that we can find, and the lumber used therein on my boats is kiln-dried before planking to reduce warping. I only build them to order, though. However, I would suggest if you really want a good one and one for use on this here river for all seasons, that it be from 60 to 75 feet in length, have the beams run from 15 to 18 foot depending on the length of boat needed, and three to four feet when it comes to the depth of the hold. My keels also run clear from the bow to the stern for the strength in design it brings. My cargo boxes run from four to five feet a-top the deck, and run about 48 to 60 feet in length depending on the length of boat you would want. All of my boats have a sturdy mast far forward to carry a full sail for use when the winds are right or for holding a rope or cable when cordelling is called for. Bear in mind, the forward mast on all of my boats is built stout enough to be pulled by anywhere from 20-40 men while travelling upstream when the vessel is fully loaded. You won't find any keelboat better made along these here docks when it comes to a well-built vessel," said James, with just a lilt of 'stud horse proud' in the tone of his voice.

"What kind of time frame are we looking at? I ask, because I will need two of them built in time so we can run the Missouri all the way up to Fort Union during the annual high water flows, because I will be in the business of carrying a full

Terry Grosz

load of provisions for Kenneth McKenzie, Factor for the American Fur Company at Fort Union," said Tom.

For the longest time James just looked at Tom as if sizing him up for being truthful and a serious customer. After all to his way of thinking, purchasing and then operating a keelboat was not one for the faint-hearted or one without the proper monetary backing, much less when it came to two boats.

Then James Dawson asked, "What length you be looking at, Stranger, were I to build them for you here at my ways?"

"I would be looking at two keelboats that were each 75 feet in length and everything else standard in structure," replied Tom firmly.

"Are you a military man?" asked James Dawson, as he gave Tom a sideways look as if trying to figure out the huge man standing before him before his conversation went any further.

"Why do you ask?" asked Tom.

"The manner in which you speak tells me you be either military or ex-military," replied Dawson with a serious look back at Tom.

"You are correct. I am ex-military, U.S. Army, Topographical Engineers," replied Tom.

"I thought so. You military guys are all alike, all business and no pleasure. If you want two keelboats, I will need $2,000 per boat on the 'barrel head' in order for me and my boys to get started. Then I will need an additional $2,000 per boat upon finishing and in coin of the realm, not paper, on credit or trade," replied Dawson with a stern look back at Tom.

"That sounds fair to me," said Tom. "I will have the money to you just as soon as the bank can give me the needed funds and then I will be back shortly," replied Tom. With that, a handshake followed closing the deal and off Tom and White Eagle went back to their new bank to withdraw the needed funds in order to put a down payment on the boats so Dawson and his sons could get started. After all, come spring thaw and high water on the Missouri, Tom needed to have his boats built,

369

The Adventurous Life of Tom Warren

have them fully loaded with provisions, and his cordelling crews lined up so they could be on their way to Fort Union at the right time of the year and without delay.

At the bank, Tom withdrew $4,000 Spanish silver dollars (the accepted coin of the realm in those days on the frontier), or about 250 pounds of silver! With a rented buckboard from the livery, Tom and White Eagle returned to Dawson's boatyard, made their down payment against two keelboats and were informed that Dawson would make a request to the local lumber mill for the needed kiln-dried oak and hickory lumber to be shipped over the next couple of days by wagon to his shipyard.

"See why I thought you be military? 'Johnny on the spot' when it came to making any business deal," said Dawson with a smile over the business just generated.

With that chore out of the way, Tom then made a special surprise trip to the local cemetery. It took him a while to locate Jeannie's gravesite but he finally did. Alongside her grave also rested another smaller tombstone for one Christopher Warren, age two...

Tom and White Eagle spent several hours that afternoon pulling weeds around the two gravesites, laying flowers and in general cleaning up and rebuilding the small white picket fence surrounding the gravesite. As they did, Tom noticed several different bunches of dried up and old wilted flowers which had been placed by the two gravesites. He had no idea as to who could have put flowers around his family's gravesites, but it made him 'smile' inside that someone other than him had remembered his two passed and dearly loved souls as well.

Then it was off for more riverboat business at hand. That afternoon, Tom and White Eagle visited several supply houses and established lines of credit at each facility so when the day came for loading his new keelboats with provisions, there would be no unnecessary delays. By then it was suppertime and White Eagle and Tom returned to "Mrs. Davis's Boarding House and Eatery" for their evening meal. When they did, one

370

guess as to who was waiting there for them but the four York brothers and one Jim Tweedle, all with sheepish looks spread clear across their faces and a mess of downcast and embarrassed-looking eyes…

During supper as it turned out, the five men had a tall but sad tale to tell. They were all dead broke after trapping for a year, amassing a huge pile of valuable fur and pelts and after only spending one day in the 'sin bins' of St. Louis, had lost it all! They had gone to a local saloon and whorehouse and had enjoyed themselves apparently immensely. Then fully into their 'whiskey cups', they got to gambling at a gaming house and losing like there was no tomorrow. Then foolishly, they kept betting more of their hard-won fur monies, doubling down and trying to get back what they had just lost but in the end, all five of the men had lost everything! That led to accusations and recriminations of the house being crooked and having cheated them out of their fur monies. Soon several punches had been thrown over the cheating accusations and shortly thereafter the five fur trappers found themselves not only dead broke but in the local 'hoosegow' for fighting and being drunken, disorderly and for smelling so badly!

If their tale of woe hadn't been so sad, it would have been almost laughable. That evening, Tom purchased their evening meal since his fellow trappers were now dead broke and in the process, he got one hell of a good idea. After mulling over his idea for a few moments while eating two pieces of Mrs. Sylvia's wonderful homemade apple pie served by the petite Miss Betsy, Tom verbally approached the five men with his 'off the cuff' idea. He did so as they were collectively in the process of demolishing several pies because they had not been fed anything while spending their evening in jail the evening before…

"I have an idea and would like to discuss it with you five 'jail birds'," said Tom with a smile. With those words uttered by Tom, he got another five sets of sheepish-looking, apple pie-smeared grins from his fur trapper friends. "Here is what

The Adventurous Life of Tom Warren

I am going to be doing in the future. Back at Fort Union, I took at their recommendations, the money earned by Old Potts, Big Foot, Crooked Hand and myself through several years of trapping successes kept here in St. Louis by the fur house, and am investing a portion of it into purchasing two keelboats. With those boats I planned on going into the shipping business supplying provisions to those upriver on the frontier and at Fort Union."

Letting that announcement sink in, Tom then said, "Today White Eagle and I made a down payment on two keelboats and spent the rest of the day lining up lines of credit at two local supply houses so when it is time, I will load those boats up with the type of provisions that are needed on the frontier, head upstream and sell those goods to McKenzie at Fort Union. I made that deal with McKenzie because he was not happy with either the service or the costs associated with his current keelboat operators, namely one Mike Fink, who operates here from out of the St. Louis area. The one and same Mike Fink that White Eagle and I almost had a deadly run-in with our first night here in this very same eatery. In fact, if it hadn't been for the help from Mrs. Davis's two built like bull buffalo-sized sons, Clifton and Thomas, I am not sure White Eagle and I would have been here today because there were seven of them and only the two of us."

"Lastly, McKenzie, in light of our supply agreement, sent down with me a packet of letters of credit concerning our previous earnings with Old Potts and company as trappers, and a request that the fur house drop Mike Fink as their annual Fort Union supplier with his keelboats and employ us in their stead," advised Tom. By now with all that had been said, the five friends had stopped wolfing down the best pies they had eaten in years and were raptly listening to what Tom had to say.

"With all of that information in mind, here is my proposal to you guys. In order to successfully operate both boats and a ground crew leading pack strings up through Indian Territory

Terry Grosz

while I am sailing alongside upriver, I need a number of good and experienced men that I can trust. If you guys are interested, I could use all five of you as my partners in this venture. I would need another boat 'captain' along with me and a damn good crew to take the valuable horse and mule strings upriver with us. Then each night we would come ashore from the boats and the ground crew herding the pack animals along the river will have previously set up camp and have meals ready for us."

"Then once we get to Fort Union, we can sell our horses and mules to McKenzie for his Company Trappers and the Indians to buy, offload and sell our goods from the boats and then take all of their accumulated furs back downstream to Astor's fur house in St. Louis. But I can't do all of that without a good and trusted crew. That is why I suggested taking in all of you 'pie-faces' as my partners. What do you have to say to that proposal, especially in light of the fact that you guys don't like staying in the local jail?" said Tom, with his characteristic big ole grin the men had come to know so well.

The five men, all with mouths still full of pie and not chewing at that point, finished what they had in their mouths and then began excitedly talking among themselves. It was apparent from all the talk flying back and forth that they felt they were kind of backed into a corner concerning their now-bleak futures there in St. Louis. They had already spent or lost all of their money from a full year's trapping up on the frontier. Now they were dead broke and did not have any 'stake' on where to go and try anything else, so they kind of had trapped themselves. However, from all the energetic talk flying back and forth, Tom could see that the men trusted him and figured teaming up for another venture out on the frontier appealed to all of them, especially in light of the fact they already had a gutful of city life, St. Louis style...

Soon Tom's idea had borne fruit among his ex-trapper friends. All five men decided to 're-up' with Tom and 'throw their hats wholeheartedly into the ring he was proposing'!

The Adventurous Life of Tom Warren

Soon, all the men were excitedly talking all at the same time about their next set of adventures and how to best 'get it done and get on their ways' with life. That was all except for White Eagle. He was helping himself to his fifth piece of Mrs. Sylvia's great homemade pie and keeping Miss Betsy busy delivering the same to the men's table.

For the next two months the men worked in concert on their newest venture, and found the tasks lying before them rather daunting when it came to starting up a new business basically unknown to all of them, being trappers and all. Splitting up into smaller crews, one crew began purchasing quality horse and mule herds for the trip northward. That same crew saw to making sure all the horses and mules were properly shod for the trip, broken to ride, lead or pack, and that all the appropriate pack and riding saddles had been purchased and fitted to their respective animals. Another crew began purchasing and setting aside supplies to be carried by the land and horse crews, so they could provide sleeping and feeding resources to the boat and cordelling crews at the end of each day's travels. Another crew saw to the procurement of provisions to be loaded and carried by the huge 75-foot keelboats for sale up on the frontier at Fort Union once the time came to shove off upriver. Finally, all the men together, after advertising in the local paper of the forthcoming keelboat spring trip to Fort Union, began selecting the men needed for manning the boats and comprising the needed 40 men for the two keelboat cordelling crews.

As the boats neared completion, Tom and White Eagle attended to last minute details and supervision, and to personal structure changes or improvements. For one, Tom had installed swivel guns mounted on each boat's bow and stern and one atop the roof. At first, most everyone working along the docks and wharves, upon seeing the large bore, brass swivel guns mounted on the boats, laughed at the newcomers to the keelboat business. Then word began trickling in from various returning trappers of the death and destruction being

374

caused by the aggressive and warlike Arikara Indians up and down the Missouri River on the returning fur trapper fur trains. When that word of death and destruction to traveling bands of trappers returning to St. Louis with their furs by the Arikara Indians along the river became common knowledge, the laughing at the placement of those three swivel guns on each keelboat slated for upriver travel, slacked off and then stopped entirely.

One day in early spring when White Eagle and Tom were making improvements on the boat's sleeping quarters, they heard a great ruckus starting up on their particular boat above decks. Emerging from his place below decks, White Eagle appeared on deck first and was immediately grabbed by several men swearing about damned Injuns and was immediately tossed overboard into the icy and swiftly flowing spring runoff roaring down the Missouri River! As he was being grabbed and in the process of being tossed overboard, White Eagle had enough sense to yell out a warning to Tom, who was still below decks working and unawares of what was ongoing!

Racing up onto the deck, Tom was grabbed by the same two men who had just tossed White Eagle overboard, one on each side as he emerged from the boat's cabin and then 'bum-rushed' for the rail and a trip overboard into the icy spring runoff as well! When that happened, Tom went from being surprised to instant fury befitting a much-surprised grizzly bear in its day bed!

Bringing both of his muscular and powerful arms together in a 'thunderclap of energy', each assailant hanging onto Tom's arms was brought together like a hammer on the face of an anvil! Both men who had been hanging onto Tom's arms were immediately stunned by not only the strength Tom showed in an instant, but were physically stunned when they were flung together into each other with such force! That was when the two burly men who were knocked together with such force dropped to the deck stunned. The next thing they felt was their bodies' extreme shock and reaction when they in turn

The Adventurous Life of Tom Warren

were flung overboard into the icy waters of the Missouri during the spring flood-stage runoff!

Then Tom felt someone grabbing him from behind and pinning his arms to his sides so he could not defend himself! Tom was then violently whirled around, struck in the face and stunned by another man standing there waiting for his chance to strike, once Tom had been immobilized!

"That will teach you, you son-of-a-bitch, for taking away our annual keelboat supply business up the Missouri to Fort Union!" bellowed a mangy-looking river man.

It was then that the full fury inside Tom rose up and presented itself for all to see. Bending over and quickly reaching down between his legs, Tom grabbed the legs of the man who had been holding him from behind, jerked them upward and then quickly sat down, HARD on the assailant when he landed flat-backed on the deck of the keelboat! When he did, he heard the man behind him who was still holding his arms, have his ribs snap as Tom's weight smashed downward onto the man's mid-section! With that man screaming out in pain and now out of the fight, Tom quickly rolled away, ducking a vicious kick aimed at his head from the man in front who had just struck him in the face! As Tom rolled off to one side, he grabbed that man's booted foot and jerked him down to Tom's level. Following that move, a single powerful right cross from a fist of a now totally maddened Tom, sent that man into 'la-la land' as well! Lunging to his feet and concerned over what had happened to White Eagle, Tom could see him swimming quickly under the next set of docks towards the shore underneath. It was then that Tom was struck from behind and knocked forward against one of his swivel guns! Quickly turning and seeing his assailant lunging towards him, Tom swung the swivel gun barrel in his direction and with a loud "OOFFFPH," jammed the end of the gun barrel into the man's mid-section, knocking out his wind! If that didn't do it, when that man hit the icy river water after being flung overboard by a now truly furious Tom, nothing would!

Terry Grosz

Quickly turning once again to face any kind of oncoming danger, Tom finally realized who his numerous assailants were, now observing who was coming right at him in a dead run! Charging his way with an ugly look spread across his face, came the locally feared giant of a river man, Mike Fink! When he did and got into Tom's range, Fink ran into a set of hammer fists, one of which broke his jaw and dropped him to the deck like a 'head-shot deer' hit with a .50 caliber, speeding lead ball fired from a Hawken rifle!

It was then that Tom saw his guys, the Yorks and Tweedle, fighting with a number of other Mike Fink-supported river men and their friends at the bottom of the gangplanks leading up onto the two new keelboats! Tom, seeing nothing but red over the unprovoked attacks laid against him and his crew, was such that when he hit the docks, the first two men he met saw nothing but stars by the time they hit the dock's decking! By then, Tom had evened the odds and shortly thereafter, the rest of Fink's drunken crew who had unwisely attacked Tom and his crew out of pure damn meanness over losing their contract for keelboat deliveries up to Fort Union, were wishing they were elsewhere. Then out of their own pure damn meanness, Tom and his crew of mean-assed ex-fur trappers who had outlived the many dangers on the frontier as a matter of course, threw the rest of the entire Fink crew into the swiftly flowing Missouri River and watched them desperately trying to get out from its icy waters before they got hypothermia and died or floated off downstream...

Then once again, Tom remembered little White Eagle swimming in the river and ran over to the edge of the dock to see if he could see him. White Eagle was nowhere to be seen! Then Tom heard his voice saying, "Down here, Father!"

Looking over the edge of the dock, Tom could see White Eagle clinging to one of the dock's pilings underneath his feet. Lying down on the dock, Tom leaned over, reached down over the edge of the planking, grabbed White Eagle's outstretched hand and lifted him up onto the dock to safety.

The Adventurous Life of Tom Warren

It was about then that four constables rode up on their horses after hearing there was a riot on the docks next to Tom's two new keelboats. That time 11 of Fink's friends and Mike Fink himself groaning all the way, were marched off to jail for inciting a riot. When confronted by the head constable, Tom declined to file any charges saying, "They had learned a lesson and with the swim most of them had taken in the icy river, I figured that was punishment enough." Both Tom and the constable had a good laugh over the end results of the fight on the boats and docks with Mike Fink's motley crew of river men. Then the chief lawman took the Fink rabble to the lockup for the night, since he himself had witnessed in part the fighting, and did not need Tom to file a complaint with the city for disturbing the peace.

With that, Tom and his crew, along with a wet as a hen White Eagle, adjourned to Mrs. Sylvia's for some hot soup and fresh homemade bread made daily from her ovens. When they did, White Eagle went up to their room in the boarding house, changed his clothes into something drier and then joined the men for lunch. There Tom complimented White Eagle for giving him a warning even though he was en route to an icy swim in the Missouri River at the time. White Eagle smiled outside and inside over his Father's compliments. Tom did as well, as he was watching White Eagle truly growing into that of a young man and certainly that of a warrior.

Come the day of departure for Fort Union several weeks later, Tom watched his first boat swing away from the dock and his crew of 20 cordelling men began pulling the boat upriver, as Jim Tweedle surveyed from the top deck his vessel's progress. Soon Tom's boat swung away from the docks as well, and he too was en route to Fort Union after many months of boat building, planning and the like that comes with initiating such an endeavor.

As for the Brothers York, they had left with their pack and riding animals at first light and were expected to meet the keelboats and make camp some 12 to 15 miles upriver at the end of that first day. In so doing, they would have their evening camp set up, supper cooking away and ready for the arrival of the tired cordelling crews walking along the eastern bank of the Missouri pulling the two fully loaded keelboats along.

For the next two months, the two keelboats made progress up the Missouri River en route to Fort Union. Depending on the weather, the two crews made anywhere from 2 to 18 miles per day of cordelling or poling, and aside from a few horse wrecks and getting hung up on a sandbar or two, the trip upriver was uneventful and without Indian problems.

Finally Fort Union hove into view and being across the river from the cordelling crews, the keelboats were then paddled and poled their ways across the river and docked. As was the tradition anytime a keelboat arrived carrying supplies, a band of Indians showed up to trade, or a string of trappers arrived at Fort Union carrying their packs of furs, McKenzie made it a point to be on hand and personally greet each and every one of them. That day was no different as Jim Tweedle and Tom Warren arrived with their keelboats and docked alongside the docks of Fort Union.

"Iron Hand, you old scudder, you made it! Who would have thought? Once you were one of my best Free Trappers and now a river boatman on the Missouri. Also, I see you converted Jim Tweedle to becoming one of yours as well. Well, you could have done worse than Tweedle but only if you had grabbed off a mangy old muskrat and put him in charge instead of him," laughed McKenzie easily.

"It is good to see you once again as well, Mr. McKenzie. Say, I have the York brothers across the river with a herd of horses and mules along with the rest of our cordelling crew. I am sure they are itching to get across this river and get into some of your rum stores because we ran out last week. Would it be possible to have your men send your company flatboat across the river and ferry my men and their animals across?" asked Tom.

The Adventurous Life of Tom Warren

"I can have that done. Say, are any of those horses and mules for sale? Because if they are, we sure can use them," asked McKenzie.

"Every one of those animals were brought along on this trip for sale to you here at the fort. We kind of figured you could use them, so when my men purchased them from around the St. Louis area, I had them only purchase those animals that we considered 'hell-for-stout', knowing how you and your Company Trappers would be using them under the toughest of conditions. Additionally, about half of my cordelling crews want to stay on here at the fort or continue working as Company Trappers, so you best be prepared to have lots of answers for those men," advised Tom, as his deck hands tied up the keelboat to the dock's pilings.

"Well, I sure can use the extra men around here at the fort as well as employing some as my Company Trappers. As you probably remember, I was losing about a quarter of my trappers annually to Indians, horse wrecks and those damn mean-assed grizzlies, so those new men are more than welcome as replacements," replied McKenzie. "Say, how about you and your crew of men having supper with me and some of my Company Clerks this evening? If that would work, I will get my cooks moving, kill a buffalo and start preparing for a grand time. Besides, I can't think of a better way to welcome the first arriving keelboats of the season with our much-needed supplies than a damn good feed and several cups of rum. By the way, I hope you brought one hell of a load of good grade rum in the holds of those boats. It seems the Indians and our trappers can't get enough of it," continued McKenzie. "Oh by the way, now that you are here, I have a special surprise for you for this evening. Just you wait and see what I have in store for you," said McKenzie, with an all-knowing smile spread clear across his heavily whiskered face.

That evening, the entire force of men who had brought the keelboats all the way from St. Louis to Fort Union were honored at McKenzie's grand celebration and supper in the fort's courtyard. As McKenzie had predicted, the two kegs providing the cups of rum were the centerpiece of attraction for everyone with a cup in hand that evening.

Then through the front gate came the surprise McKenzie had in store for Iron Hand. In marched Old Potts, Crooked Hand, Otis Barnes and Big Foot! Upon seeing his old comrades, Tom was almost speechless, then there was a lot of handshaking and backslapping once the men from the original 'family' were all together once again! For the rest of that evening, the old 'family' of trappers were talking so much, that anyone else wanting to join in could hardly get in a word edgewise...

The next day as the two keelboats were being unloaded, Tom sat on the riverbank with his old buddies and filled them in on how he was doing and the status of the original three trappers' investments given to him by the men a year earlier. For the next hour, Tom filled the three men in on how he had invested their joint monies accrued from their trapping successes and what shipping plans were in the future. All three men seemed surprised over what Tom had managed to achieve in such a short period of time and were well pleased in how he had invested the group's monies. They were even more surprised when he apprised the three men that he still had over $30,000 of their original funds back in the bank in St. Louis!

That was when Old Potts dropped a surprise on Tom. "Iron Hand," said Old Potts, "we three need to take you up on your offer of a free boat ride back to St. Louis, so we can avoid traveling through all that Arikara Indian country and getting our hair lifted. We three, like you, have had enough of this country and have seen what we wanted to see. It seems that the winters up here are making our bones ache more and more every time we venture forth in the howling prairie winds and blowing snows. We three have decided to go back with you to St. Louis if you will have us. Then we plan on buying us some land and the three of us settling down together since none of us have any living kin to our knowledge. We have made some

381

The Adventurous Life of Tom Warren

good money this last trapping season and coupled with what can be provided to us from our share of the keelboat business, that should about do it for us in retirement. We don't need much other than some time to ourselves and not worrying about our next meal, how cold the beaver trapping waters are getting, or being eaten by a mean-assed bear and being left behind as just 'bear scat' out on the prairie. So what do you say, Iron Hand? Do we get that promised boat ride back down the Missouri to St. Louis, or do we have to chance it by a-going through that damn Indian country and risking our hair?"

For the longest time, upon hearing those 'going home words' from Old Potts, Tom just sat there on the riverbank and looked from one to the other of his dear friends to see if they really meant it. Finally seeing that the men were serious, Tom rose from his place on the riverbank saying, "Old Potts, Crooked Hand and Big Foot, the three of you are more than welcome to travel downstream when I leave and enjoy your ride clear to St. Louis! That also includes space for you, Otis, if that is what you desire."

"No, I think I will be staying. My home is out here on the frontier and I even have a group of trappers all lined up who still need another trapper for their foursome," replied Otis. "In fact, my two brothers Oliver and Sterling are here and with the addition of my dad, Rufous, we four will have a go at being a family once again and this here beaver trapping thing," he continued.

Then there was more backslapping and eventually back on Tom's boat, a number of cups of rum drunk by the now celebrating, together once again, foursome of 'family'...

Finally, after the two keelboats had been unloaded except for the supplies the boat crews needed for the long trip back to St. Louis, the reloading of furs from the fort's accumulated stores began. For the next three days, the fort's Company Clerks and men from the old cordelling crews who had been hired by McKenzie and were staying on at the fort, loaded the tons of packs of furs, hides and pelts to be shipped back down to St. Louis to the American Fur Company warehouses.

Once they had loaded the keelboats to the 'gunnels' with all the peltries, Tom's 'sixth sense' began kicking in for some unexplained reason. That time, that 'sixth sense reason' included the need for more rum... Knowing his 'sixth sense' was seldom wrong, Tom had four extra kegs of rum loaded back onto his keelboat, and that extra amount of rum now loaded, for some reason seemed to settle down his 'sixth sense' demons. Then because of the need to leave before the Missouri's waters got any lower in order for the keelboats to get over a number of sandbars after the spring runoff had been reduced, Tom made ready to leave and head back to St. Louis just as soon as possible.

The day before his departure, McKenzie approached Tom with a packet of letters and his fur house draft to be drawn from the fur house's bank upon his return to St. Louis for his transport payment of all the peltries from the fort. Additionally, McKenzie handed Tom a long list of his needs come the following spring to be delivered to the fort as well upon his return. Then with a big grin, McKenzie made a request. McKenzie wished to have another celebration and supper for all of Tom's remaining crew before they left for St. Louis.

That evening, a grand party was held in celebration of another successful trapping season for the fort's trappers and the fact that their warehouses were now filled to the brim with new provisions for the year ahead. That evening, the two casks of rum were once again the center of attraction followed by the entire roasted carcass of a cow buffalo and all the homemade pies the men could eat, prepared by McKenzie's now three Chinese cooks on his staff.

After the grand celebration, Tom took McKenzie off to one side and arranged for him to supply free of charge all of his Indian Brother Spotted Eagle's needs for the year, once he came into Fort Union on his next annual trading venture.

The Adventurous Life of Tom Warren

Additionally, Tom arranged through McKenzie, for him to provide a shopping trip for little Sinopa for anything that caught her eye and she wanted, free of charge of course...

The next morning right at daylight, Tom's two keelboats shoved off from Fort Union's boat docks and were quickly swept out into the river's currents. With Tom were the remains of his cordelling crew who wanted to return to their homes in St. Louis, his boat crews including the Brothers York and Jim Tweedle manning the second keelboat, Old Potts, Big Foot and Crooked Hand. With Tom at the rudder, he found time in teaching White Eagle, now his 12-year-old son, on how to 'read' the water and learn to steer the boat downstream. In so doing, Tom allowed White Eagle to steer his keelboat while learning how to avoid the dangerous sandbars and the river's floating and dangerous logs, snags and the numerous buffalo carcasses. Come nightfall on that first night of travel, the two keelboats tied up on a sandy island out in the middle of the Missouri River as a precaution against any surprise attack from Indians. As supper was being prepared on the two boats for their crews, Tom yelled across to Tweedle and had him load all three of his swivel guns with buck and ball. When asked why, Tom just replied, "Because my frontier 'sixth sense' is telling me to." With that, Tom then moved to his three swivel guns and taught White Eagle how to safely load the weapons in case the need for their use arose downriver. Then picking a floating bloated buffalo carcass drifting by mid-river, Tom had White Eagle practice shooting at the carcass as it drifted by and scattering off the feeding ravens with the cabin roof swivel gun.

(Author's Note: Just for history's sake, during the winter months, many buffalo used the iced-over river to cross from side to side in order to get to better feeding grounds. Come spring and thinner ice, many crossing buffalo broke through the ice and drowned. Come ice out, many of those frozen in the ice buffalo carcasses broke free and floated down the

Terry Grosz

Missouri River, becoming commonplace hazards for any boats plying the river.)

For the next two days, the keelboats made good time drifting downriver. The weather was nice allowing the use of the sail on both boats and good and fast progress was made. Then on day three, Tweedle's keelboat ran aground on a small sandbar. Seeing his other boat ahead of him running aground, Tom quickly dropped his anchors stopping the downriver progress of his boat alongside Tweedle's grounded boat. As the two boat crews worked with their poles to free the grounded boat, they saw about 20 unidentified Indians riding along the far riverbank trailing about 30 horses and mules heavily loaded with packs of furs!

The Indians, upon seeing the two stalled keelboats midriver, stopped and watched the white men in the process of freeing the grounded boat. In so doing, the Indians made many hand gestures that they wanted the boats to come ashore and trade. Tom, watching the Indians carefully making gestures indicating they wanted the two boats to come ashore and trade for the packs of furs they were trailing on their horses, detected his 'sixth sense' at work once again...

However 'sixth sense' aside, the Indians were trailing a huge number of obvious packs of furs and if he could trade for them, that would enrich his trip back to St. Louis just that much more. Holding a hurried conference with the other boat crew, Indian treachery aside, they all indicated a willingness to court the possible danger of trading with the band of Indians because there were a large number of armed boatmen to keep the order and safety of all concerned. Besides, "they had six swivel guns, didn't they?" were the common utterances among the crew members. They did so, because to be able to trade what little they had, mostly rum of low value for furs of high value, would enrich each man's share of money on the trip, so the boat crews wanted to take the risk and 'test the waters'.

Against Tom's cautious judgment but knowing he could trade off their 'sixth sense'-ordered kegs of rum for packs of

385

The Adventurous Life of Tom Warren

high value furs made good sense if he could safely pull it off, he relented and opted for the trade. However in the thought processes, Tom figured the Indians had probably run across a number of trappers bringing their furs to St. Louis for sale and had killed and robbed them. This 'darkness of source point acquisition' Tom had deduced just by looking at the motley band of Indians.

The Indians were traveling light and none sported any kind of beaver trapping gear other than the numerous packs of furs on their horses and mules. The more he looked at them, the more he figured he was looking at numerous packs of stolen furs that represented the hard work from many dead trappers! Having been there and almost robbed by renegade Indians earlier in his life, that alone made him angry. Then the previous 'military training' in him took over as a crazy plan spun through his mind.

Casually walking over to the edge of his boat, Tom beckoned Jim Tweedle over to his boat as the two of them floated on the Missouri River. "Jim, I feel that those Indians have attacked and killed a number of trappers who were bringing their furs down to St. Louis for the better prices they could get for them. Somehow along the way, they must have been attacked by this band of Indians, robbed and killed. I think all of those packs of furs we are seeing were stolen and their original owners are now dead as the last buffalo you shot. I would like to get those furs if I could somehow prove they were stolen and those Indians are nothing more than the killers of a bunch of innocent trappers. Here is what I propose we do. We have 24 men still on board these boats who are experienced shooters and frontiersmen. I want you, and I will do the same, to alert all of the men on our boats to quietly arm themselves without making it obvious and prepare for battle with these Indians if necessary. Then have three of your men casually act like they are working on your boat near the swivel guns. That way if the crap hits the fan once they get nearer, or I am able to determine these Indians are the killers of innocent trappers, we can be near and ready for battle, starting with the collective destructive power from those six swivel guns. The bottom line is that if these Indians are sniping off the trappers as they head for St. Louis, they will continue doing so and killing off more of the same as long as they are rewarded with those valuable packs of furs. That needs to stop and that means right here and now, so that the killing of innocent trappers stops as well," said Tom, with a new quiet determination in the tone and tenor of his voice.

"Here is what I propose to do. We need to find a river channel that is deep enough for our boats to navigate that runs right along the bank. That way we can bring our boats right along the shore and still be protected with our swivel guns and the rest of the armed men on board. Then I propose to take ashore several kegs of our extra rum, set it up and make like we are their friends. In so doing, I will make it look like we wish to trade for their furs. As I do so, I will try to discover how they came by those furs by walking among them looking around to see what I can see. If I am able to discover that the furs have been stolen from now dead trappers, we will kill the lot of these Indians so their depredations on other human beings will stop here and now. If I can't make that determination, we can trade our rum for whatever furs it will bring us and then just move on, letting the Indians trade their furs as they see fit," said Tom.

"I will take Adam York with me when we anchor next to some sort of a secure landing place and the two of us can take ashore several kegs of rum for the Indians to drink. Maybe as they are drinking and getting all liquored up, Adam and I can discover if they are 'good or bad' Indians once we find evidence of wrongdoing on their horses or in the gear they are packing. If we feel they are nothing more than killers, I will give you a signal if I am not back on the boat or able to do so, and you can have our men kill the lot of them," concluded Tom.

The Adventurous Life of Tom Warren

With those plans being set into motion, Tom went back onto the bow of his keelboat and through hand gestures, now that the other keelboat was off the sandbar and floating free, for the Indians to meet them further down the river. When Tom had finished gesturing to the Indians, they seemed to understand and began herding their pack strings downriver and watching the boats all at the same time. About half-a-mile downriver, Tom spotted a channel that ran alongside the riverbank by a wide sandy stretch of the river. Slowing the vessel down with the use of their long poles, Tom finally maneuvered and anchored his boat next to the shoreline, as Tweedle nosed his keelboat right in behind Tom's so they could communicate among themselves and yet provide a covering fire for each other if the Indians attacked their anchored boats or they decided to 'go to war'.

As they waited for the Indians still coming downriver to arrive where the boats had anchored, Tom and Adam York lugged a keg of rum off Tom's boat and set it up tactically between the two boats so if necessary, the six swivel guns could concentrate their fire on the area of the rum keg's close at hand location. Then Tom returned and retrieved a number of cups so the Indians would have the means to consume the rum once they had arrived.

About then, the Indians began streaming off the riverbank onto the sandy plain adjacent the river and the two anchored keelboats. Lounging around on board appeared the boatmen but unknown to the arriving Indians, the men had laid their rifles and pistols down where they would be unobserved but in easy access if the need arose for their immediate use.

Spotting the keg of rum sitting out in the open, the Indians bailed off their horses and while their pack strings milled about on the sandy bench of land adjacent the two anchored boats, Tom greeted the arrivers by holding out a handful of drinking cups. Immediately the Indians made use of the cups with a lot of shoving and gesturing of their appreciation for the free drinks as they swarmed around the rum keg.

388

Terry Grosz

However, as they milled about, Tom had determined that they were the dreaded Arikara Indians by the manner of their dress! As Tom moved about through the ranks of the Indians' riding and pack stock, he began looking for any signs leading to the intent or previous behavior of those now gathered around and pushing and shoving each other near the keg of rum.

Then Tom's eyes fastened upon two packs being carried by a bay. Both packs had the burned-in brand stamped into the deer hide covering the beaver *plus* with the name "DENT"! Moving closer, Tom recognized the brand of his friends, the 'Brothers Dent', burned into the pack covers! Then his eyes discovered the very evidence he had been looking for regarding the Indians' guilt or innocence when it came to their acquisition of the many packs of furs. Standing together in a clump, Tom spotted six beautifully matching buckskin horses! The one and same horses proudly owned by the Brothers Dent when Tom had first met them back at Fort Union the year before! The same six buckskin horses Gabe Dent had told Tom they would never part with unless they were dead, because that was all the two brothers had left of their family's heritage! Walking back to his boat in grim determination, Tom was stopped dead in his tracks by what he saw hanging from the mane of another Indian's horse. There were three fresh scalps from white men, and red-headed white men at that! Then Tom remembered the Dent brothers seeking a man named Black Bill Jenkins and his three red-headed kin, who had not only killed the parents of the Dent brothers but their aunt and uncle as well! Having seen enough, Tom started to head back to his boat once again when he passed another Indian's horse, and in the rifle scabbard he spotted an 1803 military rifle with the words "Gabe's Rifle" artfully carved in its stock! Tom then realized that these Indians had probably killed off the Dent brothers and had not only taken their furs, but their coveted six buckskin horses that meant so much to the two brothers as well!

The Adventurous Life of Tom Warren

Picking up his pace, Tom walked by the much-used keg of rum smiling at its users like all was well when all of a sudden he slowed his pace. There looking right at him from just a few feet away was an Indian with a large scar running from the left side of his forehead, clear across his nose and down the man's right cheek! It was the same Indian that had lanced Robert Caster during their first battle with the Arikara on their trip down the banks of the Missouri en route to St. Louis! It was all Tom could do to keep from drawing his pistol and killing the leering, now in the process of getting drunk, killer of his friend and fellow fur trapper, Robert Caster!

About then, Adam York walked up alongside Tom saying, "Tom, there are at least a dozen fresh scalps from white men tied to a number of these Indians' horses' manes so they can dry out! Let's get the hell out of here before they rear up and kill us where we stand and hang our hair on those horses' manes as well. You were right, these Indians are catching trappers coming into St. Louis with their furs and killing them!"

Without another word, the two rivermen casually walked away from the group of Indians still gathered all around the keg of rum, up the gangplanks onto their respective keelboats and then pulled up the gangway so no one could run up them and attack the boatmen. Then Tom gave the pre-arranged signal to all the men manning the six swivel guns to be ready, turned and then looked back on the raucous scene around the rum keg on the beach.

When the right moment came when all the Indians were knotted around the keg in one jumble of pushing and shoving humanity, Tom gave the final signal and instantly six swivel guns boomed out their loud reports and lethal messengers of death in the form of buck and ball was hurled on its way. At that range, the group of Indians literally exploded into hundreds of body parts as pounds of lead balls tore into their bodies from just 30 yards away! When the buck and ball arrived, in addition to the Indians, the keg of rum exploded into sprays of alcohol and thousands of wooden splinters! In fact there was so many lead balls concentrated into such a small area, that five nearby horses fell to the deadly blasts as well. Then the rest of the boatmen grabbed their hidden rifles and poured another deadly volley of lead into anyone who even remotely looked like there was any life left in their carcasses. Then there was nothing left but black powder smoke drifting over the area, the acrid smell of burned gunpowder, and the smell of burned blood and flesh, along with the silence from the world of birds in the immediate surroundings over the tremendous blasts that had just rent the normal prairie quiet...

As the boatmen quickly reloaded their rifles, the swivel gunners reloaded their pieces as well just in case they had missed someone. However, the only heartbeats left on the sandy shore were those of a number of pack and riding horses, nervously moving about because of all the loud explosions and now, the fresh smells of blood.

Then the work began as the men quietly descended from their keelboats and surveyed the scene of destruction. As for the large group of Arikara Indians guzzling the rum before they were blasted into...well, let's just say there was a lot of 'fodder' for the wolves, coyotes, magpies and ravens if they weren't bothered with the fact that there weren't very many large pieces left 'standing'. As for the five horses inadvertently killed by the blasts, they were unpacked and left to also lie as the next set of meals for those predators of the plains as well.

It wasn't long before the men had stripped all of the horses of their packs of furs and had loaded them on top of the decks on the keelboats because all their storage space below decks was already jammed full of furs, hides and peltries from Fort Union. Since there was no way to bring all the Indians' packhorses home with them on the boats, they were let loose to roam and live on the prairie. That is except for the six matched buckskins that meant so much to the Dent brothers. They were unpacked and with the use and aid of a planking

The Adventurous Life of Tom Warren

system, led up onto Tom's boat. There they were tied off, because Tom had decided he would keep those beautifully matched horses on his new farm that he had purchased just before leaving St. Louis on his first trip to Fort Union with his two keelboats carrying their loads of provisions. He would do so out of respect for his friendship that he held for the Dents and for the memories they cherished so much when it came to the remains of their family's heritage. As for Gabe's special rifle, Tom also kept that in memory of the two men who were hellbent in destroying the ruffians who had so completely destroyed the Dent brothers' families and had apparently died trying to fulfill their vision quest.

The rest of the trip to St. Louis was made without mishap, with the boats' arrivals being made several weeks later. Once there, the furs were unloaded and taken to Astor's American Fur Company warehouses, and Tom was reimbursed for his transport of Fort Union's furs downstream to St. Louis. As for the 72 packs of furs recovered from the destruction of the predatory Arikara Indians who had been ambushing and killing returning trappers, they were sold to the American Fur Company and the profits shared equally among the men on the two keelboats who had taken a chance and had avenged those fur trappers who had died such an untimely death.

The scalps of the white men taken from the Indians' horses were buried in a common grave on the bluff overlooking the scene of death and destruction. All the personal property taken from the Indians was piled up on the beach and burned. As was usually the case out on the vast prairie, when so much fresh meat was on the scene, it was mostly gone as a result of the many visits by the land and aerial predators within a week. All that remained were a number of pelvises, long bones and a number of grinning skulls, all of which had been much picked over by the 'little people'. Come the following spring flood on the Missouri River months later, all signs of such a violent struggle had disappeared under the sands of time because of the high water effects that came with the annual flood waters.

Terry Grosz

(Author's Note: The six matched buckskin horses were later released on Tom's Missouri farm located several miles to the west of St. Louis. There they lived out their lives peacefully and upon their individual deaths, were buried under the warm Missouri soil by Tom and his family together as they had lived. Gabe's special rifle resides as family property to this very day in the hands of Tom's descendants.)

As came to light several months later from four other trappers taking their furs to St. Louis, the Arikara Indians killed by Tom and his keelboat men had not killed the Dent brothers. The four trappers traveling to St. Louis discovered where the Dent brothers and their two Indian wives had been killed in a small camp along the Missouri River. Two of those trappers had recognized the Dent brothers' remains and from all signs around their campsite, they had been killed by at least four white men from ambush and their furs taken, as well as had been their horses. The trappers, hellbent on avenging the Dent brothers' deaths, followed the four white men killers to their camp located several miles further down along the Missouri River, and discovered that they in turn had been ambushed and killed by the Arikara Indians! The trappers viewing the freshly killed white men who had killed the Dents, reported there appeared to be one very large individual with a heavy black beard and three other white men who had traces of flaming red hair still on their skulls even though they had been scalped! One of the trappers who knew of the Dents' search for a large black-bearded individual with his three red-headed kin, surmised these were the ones the Dents had been pursuing all those many years. It then was postulated from the evidence left in the camp scenes, the Dents and their Indian wives had been surprised and ambushed by the very four men they had been tracking since their kin had been killed previously in Missouri. Then Black Bill Jenkins and his red-headed kin had in turn been attacked and killed by the renegade Arikara Indians. The one and the same bunch of Indians led by the scar-faced one who had lanced Robert Caster back in Tom

393

The Adventurous Life of Tom Warren

Warren's camp earlier when he and his fellow trappers were traveling to St. Louis with their trappings...

Then in turn, Tom Warren discovering the Dents' personal property and matched buckskin horses in the hands of the scar-faced one, he and his boatmen in retribution had sent that trapper-killing bunch of Arikara Indians off to join the Cloud People in their land of eternity...

As for the Brothers York and Jim Tweedle, they joined forces, and upon Tom's following spring return to Fort Union with the year's provisions, worked their ways north via Tom's keelboats as well. They then returned to trapping on the frontier once again and several years later, none of them ever returned to Fort Union, but disappeared into the vastness known as the Great American West.

Old Potts, Crooked Hand and Big Foot bought a homestead next to Tom Warren's 1,200-acre farm and settled down. There they lived comfortably on their retirement returns from Tom's keelboat business. When they all passed together years later due to a bout of food poisoning, no relatives showed up to arrange for their burial. Tom and his family took control of the bodies and instead of having them buried in St. Louis's Potter's Field, interred them in his family's personal cemetery. There those three fur trappers reside to this very day on the Warren property.

Terry Grosz

Chapter 15: Flowers In The Graveyard And The 'Sixth Sense'

TOM'S NEXT TRIP in one of his keelboats took him to Hermann, Missouri, where he dropped off barrels of whiskey, sacks of milled flour and barrels of salt that had been shipped upriver all the way from New Orleans. Upon his return trip to St. Louis, Tom found his 'sixth sense' acting up once again and for the life of him, he couldn't figure out why. There didn't appear to be any danger nearby, the boating weather was excellent and they hadn't had any Indian problems near about for years. Finally writing off his intense inner feelings, he continued sailing downriver back to St. Louis and docked his keelboat near his other boat. Leaving the boat and his captain in charge of loading the boat for their next trip, Tom rode over to Mrs. Davis's boarding house and had lunch.

Still concerned over his 'sixth sense' feelings, Tom rode back to his farm, checked in with his farm manager and found nothing amiss regarding the management of his properties. Tom then checked in on his son's school teacher and discovered that White Eagle was doing very well in school; in fact, he was at the top of his class in mathematics, reading, language studies and deportment. Still not able to figure out

395

The Adventurous Life of Tom Warren

why he had a 'bout' of his 'sixth sense' acting up on his trip back from Hermann, he offhandedly headed for the cemetery where his wife Jeannie, and his son Christopher, lay buried. He had not been to their gravesites for some time and felt it necessary to clean up around their graves and maybe plant some more flowers, because when Jeannie was alive, she had loved her flowers.

Tying off his horse to the stone cross on another man's nearby tombstone, Tom walked over to Jeannie's grave and just stood there for a few moments looking down at her headstone and that of his son's, thinking of what his life might have been had they both lived. Then catching himself and remembering why he was there, he knelt down and began pulling the weeds from around the two graves' headstones.

As Tom removed a number of weedy plants from his son's gravesite, all of a sudden his 'sixth sense' began kicking up inside him and he thought, *What the heck is going on?* Pausing and looking all around him, he saw nothing in the way of any danger in the quiet cemetery, so he commenced pulling more weeds from around the tombstone of his son.

Moments later Tom heard, "Hello, Tom. Gosh, it has been a long time since I saw you last."

Whirling around on his knees, there just a few feet away quietly stood a beautiful-looking young woman in a blue gingham dress holding a handful of flowers! She had shoulder-length reddish-blond hair, eyes that were so pretty and deep blue that they would catch anyone's attention just looking at her, and a smile that was nothing but purely captivating! It was then that Tom recognized his sister-in-law, Linda Johnson, quietly looking down at him. Tom had not seen her since his wife's and son's funerals years earlier and not until this day, ever since his trip into the frontier as a Free Trapper.

Standing up and dusting off his pant legs from all the weeding he had been doing, he said, "Good afternoon, Linda. Damn, woman, I haven't seen you in years. How have you been?"

"I have been doing just fine, Tom. Like you, I too have also lost the love of my life when he was kicked in the head by a mule he was shoeing, but other than that, my son David and I have somehow managed. I hear tell that you are now in the keelboat business, now that you are back from being a trapper up near Fort Union. How is that going?"

"It is going great. White Eagle and I have all the business that we can stand and it is increasing," he said, as an unfamiliar warmth kept flooding throughout his body.

"Who is White Eagle?" Linda asked demurely.

"White Eagle, why, I guess I could say, is my adopted 12-year-old Sioux boy whose family is all dead. I adopted him when I was a fur trapper and he has been with me ever since," responded Tom, still having trouble getting over just how pretty Linda looked.

"Oh, I would like to meet him. He is the same age as my David, and I bet the two of them would have a lot to talk about," said Linda with a beautiful smile.

When she smiled over hearing about White Eagle, Tom's 'sixth sense' strangely made 'its presence' more than strongly felt inside of him once again.

"I will tell you what. I have a chicken supper and all the trimmings planned for me and my son this evening. Why don't White Eagle and you come over and have supper with us, and that way the two boys can get acquainted as well?" asked Linda with another of her captivating smiles.

For a moment, Tom just stood there like a 'goose out of water' not knowing quite what to say in response to her very genuine-sounding 'invite' to supper. It had been such a long time since he had spoken more than just a few words to or with a woman and he found it surprisingly hard to respond. Then he realized it was his turn to speak and without knowing why, he mumbled out the words that he and White Eagle would be glad to come…so the boys could get acquainted…

"Great! I live just a quarter-mile down Nelson Road from your new place. It is the white two-story home with the red

The Adventurous Life of Tom Warren

trim. How about you and White Eagle coming around six and just bring your appetites and I will supply the down home cooking," said Linda, accompanied by another one of her great smiles.

Tom stammered out his words of acceptance and then for the next 20 minutes, the two of them cleaned up around his wife's and son's gravesites in quiet silence, after Linda had placed the flowers she had brought onto her sister's grave...

Then just as quickly as she had appeared, she returned to her buckboard and was gone!

Untying his horse, for the longest moment Tom stood there trying to figure out what the hell had just happened. He had just met his sister-in-law after a number of years of being apart, she was now single, she was beautiful, White Eagle and he had just been invited to supper, and did I say she was beautiful...

Tom's horse had never been ridden as hard as he was that day back to his new farm. Then Tom took a bath and trimmed his massive beard so he didn't look like a mad grizzly bear on the prod. About then, White Eagle arrived home from attending school and he too was all spruced up so he would be presentable and make a good impression come suppertime... It was then that Tom realized that his 'sixth sense' had not 'died down' ever since his meeting with Linda at the cemetery...

That evening, Tom and White Eagle had supper with Linda and David. As luck would have it, both boys, being boys, hit it off like clockwork! After supper the two boys took off to explore the woods around Linda's home, leaving Tom and Linda alone.

Dreading that moment, however, Tom found Linda very pleasant to be around and she made him feel relaxed and comfortable around her. Soon the two of them were talking about old times and their lives since they had been apart for so long after Tom's wife's death, and when he had retreated into the frontier as a trapper in order to try and forget his loss. Linda in turn, shared with Tom her ordeal in the loss of her husband and trying to make ends meet after his untimely death.

Then the two of them adjourned to the front porch and sat on the swing together talking about their two boys. About then, here came the two boys both covered with mud from 'stem to stern'. The tale soon came out the boys had gotten into the nearby creek and were catching crawdads, hence being muddy from head to toe. However, the great thing in Tom's eyes was that the two boys had hit it off so wonderfully and seemed so comfortable around each other.

It was pretty late before Tom and his mud-covered White Eagle left Linda's. But when the two of them left, it was with another supper 'invite' in their back pockets!

For the next fast-paced month, Tom and White Eagle found themselves almost nightly supper guests of Linda and David. Tom was finding Linda very interesting to say the least and the two boys found themselves almost like 'blood-brothers'!

Several months later as Tom helped Linda do the evening's dishes and the boys were out and about throwing rocks at the bats in the barn, Tom finally found himself and the time to 'talk' to Linda. Sitting out on the porch swing in the warmth of an early fall evening, Tom finally found the courage to reach over and hold Linda's hand! When he did, she did not pull her hand back...

Looking over at him, Linda quietly said, "I am not my sister. But I can love you every bit as much as she did if not more!"

Two months later, Tom and Linda were married and the void left in each of their lives with the earlier passing of their spouses, was truly filled. In so doing, the boys found that each now had a 'brother' and each now had a mom and dad...

In the years following, Tom's shipping business flourished and he graduated from keelboats into his first steamboat, named "The Linda". Shortly thereafter, business on the Missouri and Mississippi Rivers grew so much that Tom

The Adventurous Life of Tom Warren

purchased his second steamboat or paddle wheeler named "The Fur Trapper".

Years later when Tom finally retired as a riverboat captain, his two sons White Eagle and David replaced their dad as riverboat captains on "The Linda" and "The Fur Trapper", and became successful businessmen in and of their own right on the Missouri River. In fact in later years, both White Eagle and David took their respective paddle wheelers up the Missouri River and supplied provisions to Fort Union and brought passengers, furs and buffalo hides back to the markets in St. Louis until the beaver fur market crashed around 1843, when the beaver top hats were replaced with those made from silk and became the 'rage'.

In his later years, Tom went on to serve as a distinguished Congressman and later as a Senator in and for the Great State of Missouri.

When Tom "Iron Hand" Warren, Mountain Man, passed years later, he was buried on his family farm next to Old Potts, Crooked Hand and Big Foot... Once again, the four trappers and 'brothers in kind' were together in death as they had been in life and were now eternally pursuing the "Celestial Beaver and Chasing the Wind"...

THE END

A Look at Josiah Pike by Terry Grosz

Josiah Pike, captured by the Sioux! A novel, but also an in depth look into the western wilds and the life of the tribes. Terry Grosz has an extensive background in the wilds of the West, as a California Fish and Wildlife warden and later a U. S. Fish and Wildlife agent, serving from California to North Dakota. His contact and interaction with the tribes of the West, gives him in-depth knowledge of Indian life, and he passes that along in Josiah Pike. The book is a novel, but written in a style that not only entertains but educates. It's dense with tribal life, yet vibrant and compelling as Pike is captured then taken to the bosom of the Sioux.

Available from Wolfpack Publishing and Terry Grosz.

The Adventurous Life of Tom Warren

About the Author

Whether as a professional in the field of wildlife law enforcement or as a prolific writer, Terry Grosz has distinguished himself with a kind of passion, dedication, integrity and professionalism that often exemplify Humboldt State alumni. The beginning of his 32-year career in wildlife law enforcement came in 1966 with the California Department of Fish and Game in Eureka. After several years and a transfer to Colusa, he was hired by the U.S. Fish and Wildlife Service (FWS), moving into increasing responsibility for conservation and wildlife law enforcement in successively larger geographic regions, from jurisdiction over the central half of Northern California to finally Assistant Regional Director for Law Enforcement where he supervised FWS's wildlife law programs covering 750,000 square miles.

When Grosz became the FWS Senior Special Agent, he wrote regulations, policy and procedures, responded to congressional inquiries, provided advice, guidance and expertise. But it wasn't just a desk job. He also traveled throughout Asia assisting foreign governments in curtailing the smuggling of wildlife and establishing cooperative international law enforcements programs. In all the various positions held by Terry, he supervised agents who protected wildlife from being smuggled or imported illegally into the US, protected eagles from being poisoned or trapped, and more.

In 1998, Grosz retired from the FWS and began a second career as a prolific writer, and has since authored and published

402

fourteen wildlife law enforcement memoirs and seven historical novels. Clearly, he's got a lot of material to work with. Many of his stories have hilarious moments and hair-raising adventures, some others are sad and tragic, they are all about the men and women who work as wildlife conservation officers trying to preserve our natural heritage for future generations.

Find more great titles by Terry Grosz and Wolfpack Publishing at http://wolfpackpublishing.com/terry-grosz/

Made in the USA
Lexington, KY
16 August 2019